Hegels	Hegel's
Philosophie des	Philosophy of
subjektiven	Subjective
Geistes	Spirit
BAND I	VOLUME I

Hegels
Philosophie des
subjektiven Geistes

HERAUSGEGEBEN UND ÜBERSETZT

MIT EINER EINLEITUNG

UND ERLÄUTERUNGEN

von

M. J. Petry

*Professor der Geschichte der Philosophie an der
Erasmus Universität in Rotterdam*

BAND I

EINLEITUNGEN

D. Reidel Publishing Company

DORDRECHT : HOLLAND / BOSTON : U.S.A.

Hegel's Philosophy of Subjective Spirit

EDITED AND TRANSLATED

WITH AN INTRODUCTION

AND EXPLANATORY NOTES

by

M. J. Petry

Professor of the History of Philosophy,
Erasmus University, Rotterdam

VOLUME I

INTRODUCTIONS

D. Reidel Publishing Company

DORDRECHT : HOLLAND / BOSTON : U.S.A.

Library of Congress Cataloging in Publication Data

Hegel, Georg Wilhelm Friedrich, 1770–1831.
    Hegel's Philosophie des subjektiven Geistes.

    Added t.p.: Hegel's Philosophy of Subjective Spirit.
    English and German.
    Bibliography, v. 3; p.
    Includes indexes.
    Contents: Bd. 1. Einleitungen. Bd. 2. Anthropologie. Bd. 3. Phenomenologie
und Psychologie.
    1. Mind and Body. i. Petry, Michael John. ii. Title. iii. Title: Philosophie
des subjektiven Geistes. iv. Title: Philosophy of Subjective Spirit.
B2918.E5P4 1977   128'.2   77-26298
ISBN 90-277-0718-9 (set)
ISBN 90-277-0715-4 (Vol. 1)
ISBN 90-277-0716-2 (Vol. 2)
ISBN 90-277-0717-0 (Vol. 3)

PUBLISHED BY D. REIDEL PUBLISHING COMPANY,
P.O. BOX 17, DORDRECHT, HOLLAND.
SOLD AND DISTRIBUTED IN THE U.S.A., CANADA AND MEXICO
BY D. REIDEL PUBLISHING COMPANY, INC., LINCOLN BUILDING,
160 OLD DERBY STREET, HINGHAM, MASS. 02043, U.S.A.

TYPE SET IN ENGLAND BY WILLIAM CLOWES AND SONS LTD., BECCLES
PRINTED IN THE NETHERLANDS

PREFACE

The foundations of this edition were laid at the University of Bochum. The readiness with which Professor Pöggeler and his staff put the full resources of the Hegel Archive at my disposal, and went out of their way in helping me to survey the field and get to grips with the editing of the manuscript material, has put me very greatly in their debt. I could never have cleared the ground so effectively anywhere else, and I should like to express my very deep gratitude for all the help and encouragement they have given me.

It has been completed in the Netherlands, — in a University which is justly proud of both the liberal and humanistic traditions of its country and its close links with the enterprise and accomplishments of a great commercial city, and in a faculty engaged primarily in establishing itself as a centre of inter-disciplinary research. I have found these surroundings thoroughly congenial, and can only hope that the finished work will prove worthy of its setting.

It is in fact the first critical parallel text edition of any part of the *Encyclopaedia of the Philosophical Sciences in outline*. Its publication in this form would have been impossible without the financial assistance of the *Nederlandse organisatie voor zuiver-wetenschappelijk onderzoek*, and I should like to say how very grateful I am for the generous way in which they have backed the project. I have been most impressed by the manner in which the publishers have handled the planning of the volumes, and the care and expertise with which they have seen them through the press. In this respect I should like to make particular mention of Mr. Patrick Wharton, who by thinking well ahead and anticipating problems with remarkable efficiency, has made the checking of the proofs almost pleasurable.

The work owes a very great deal to my family. My sons, Martin and Andrew, have done most of the really heavy work involved in preparing the indexes, and without the help of my wife the whole thing would have taken very much longer to complete. She has not only helped me to solve many knotty problems in the translation, but has also helped me to decipher the manuscripts and check the transcriptions and proofs, as well as typing out the greater part of the final text. It is to her that I dedicate these volumes.

M.J.P.

Rotterdam
September 1977

Volume One

INTRODUCTIONS

Volume Two

ANTHROPOLOGY

Volume Three

PHENOMENOLOGY AND PSYCHOLOGY

INHALTS-ANZEIGE (BAND EINS)

CONTENTS (VOLUME ONE)

To Helga

INTRODUCTION

a. EMPIRICISM

The empiricism of the Hegelians is not, of course, the ordinary naïve empiricism of the collectors, observers and experimenters, nor is it the bewildering empiricism poured forth in the pretentious oratory of Schelling and his school. It is an empiricism which is *aware of its faults*, and which openly and candidly *admits* its inner contradictions! Hence its instructiveness, hence its constituting the proper and indispensable introduction to metaphysics. — J. F. HERBART.

Herbart threw out this remark at the end of a lengthy, fascinating and extremely hostile review of the second edition of Hegel's Encyclopaedia. To us, accustomed as we are to hearing about the Hegel who was content to interpret the world rather than changing it, or the problems involved in sorting out the Phenomenology, the Logic and the academic ditherings of the Jena period, it has a paradoxical ring. How could anyone in his right mind accuse Hegel and his followers of being empiricists? To Herbart's contemporaries however, to those who were watching the way in which Hegel's teaching was affecting his pupils, to those who were actually feeling the effect his philosophy was having upon the climate of opinion prevailing in their particular disciplines, the review in question gave voice to a general impression, and, indeed, to a general distrust, of an enigmatical and disconcertingly fashionable manner of thinking. Very few had any real grasp of the broad principles of Hegelianism, but everyone could see that it could not possibly be regarded as dealing in mere abstractions, or as indifferent to empirical facts.

In the introduction to the Logic, the first part of the Encyclopaedia reviewed by Herbart, Hegel dwells at some length upon what he considers to be the three aspects of all thinking. He distinguishes here between its simply subjective or psychological foundation, its concern with the truth of general objects, and its being itself true or objective in that it is no longer at odds with its subject matter. He then points out that in the Phenomenology, published twenty years before, he had been concerned primarily with developing the dialectic of spirit from its initial or most simple appearances to the standpoint of philosophy. He admits that this essentially formal procedure had been complicated by his having had to presuppose such concrete forms of consciousness as morality, social ethics, art and religion, and, — perhaps more significantly if we remember that he is introducing his 'Encyclopaedia of the Philosophical Sciences' in outline, — that its implications for, "the development of the content, the general objects of the various philosophic sciences", need re-stating. This leads him into a lengthy consideration of the ways in which we think about objectivity. He apologizes for presenting them in such a loosely reasoned and merely historical manner, but suggests that since the initial clarification of the simplest aspects of the thinking fundamental to cognition can be most conveniently carried out in the Logic, it is only fitting that this science should be so introduced.[1] Once again, there are three basic determinations. In our *ordinary* daily activity as *conscious beings*, as in all the sciences and the early stages of philosophy, we are unaware of there being any disparity between what we are thinking and what we are thinking about. We remain convinced that when we reproduce the content of our sensations and intuitions in thought, we are conceiving of things as they really are. As *empiricists* however, we are no longer content with regarding what we think as being true. While still accepting the content of our sensations and intuitions as a valid foundation, we consciously exploit experience by fixing upon facts in order to get a firmer grasp of objectivity, and distinguishing and co-ordinating them in order to elicit general propositions and laws. As *philosophers*, our primary concern is to reconcile ordinary consciousness with empiricism, to show that it lies within the very nature of thought and subjectivity to be inseparable from being and objectivity. Since philosophy presupposes empiricism in the same way that empiricism presupposes realism, these three ways of thinking about objectivity are complementary.[2] Taken as a whole, a syllogistic unity of complementary attitudes, processes or procedures, they also constitute the basic principle or broad foundation of the whole systematic exposition of the Encyclopaedia. At the end of the Logic, before he goes on to demonstrate their effectiveness in working out a philosophy of the natural sciences, Hegel summarizes their significance under the general heading of the Idea, pointing out that in their cognitive aspect they involve the receptive analytical procedure of breaking material down and making it intelligible, and the synthetic procedure of interrelating what has been grasped, while in their dialectical or philosophical aspect they are involved in combining and co-ordinating this cognition.[3]

[1] Enc. §§ 19–25.
[2] Enc. §§ 26, 37, 64.
[3] Enc. §§ 213–244. Cf. the more exhaustive account in L. Logic, pp. 755–844. There is a lucid and perceptive critique of Hegel's conception of these analytical, synthetic and dialectical procedures in K. Düsing's 'Das Problem der Subjektivität in Hegels Logik' (Bonn, 1976), pp. 295–335.

Empiricism therefore constitutes an essential and integral part of the basic methodology of Hegelianism. Analyzed and assessed as it is, both at the beginning and at the end of the Logic, there can be no doubt that Hegel did indeed conceive of it as "the indispensable introduction to metaphysics". Unlike most of his immediate predecessors, – Descartes, Locke, Berkeley, Hume, Kant and Fichte, unlike the vast majority of his followers and interpreters, and unlike many modern philosophers, he did not regard the subject-object antithesis of a theory of knowledge and the concomitant questioning of the status of commonsense realism and empiricism, as the unavoidable starting point of any well-founded philosophy. In these lectures on the Philosophy of Subjective Spirit, though not of course endorsing all the claims made by the materialists and by Condillac, he has a good word to say for their approach to the body-mind problem (II.19; III.91), and he makes it quite clear right at the outset that he is dissociating completely from the whole post-Cartesian enterprise: "Aristotle's books on the soul, as well as his dissertations on its special aspects and conditions, are still by far the best or even sole work of speculative interest on this general topic. The essential purpose of a philosophy of spirit can be none other than re-introducing the Notion into the cognition of spirit, and so re-interpreting the meaning of these Aristotelian books" (I.11). It is the sciences based upon experience which stimulate the mind to master the wealth of empirical detail by interrelating the various disciplines and so eliciting their form. Philosophy does not simply question the status of their empirical content, it recognizes and uses it, and brings it into a more comprehensive categorial relationship.[4]

At the end of August 1827, Eduard Gans, one of Hegel's pupils, while visiting Goethe, was drawn into conversation by him on the teaching of philosophy at the University of Berlin. Gans reported the drift of the discussion as follows: "He was of the opinion that if philosophy regarded it as its duty to devote itself to the actual subject-matter of the topics it is concerned with, it would be so much the more effective if it also had dealings with the empiricists. This would, however, always give rise to the question of its being possible to combine, at one and the same time, the work of a distinguished investigator and observer with that of a worthwhile universaliser and co-ordinator... He had no doubt about the extent of Hegel's knowledge of natural science and history, but he could not help asking whether his philosophical ideas would not lose their categorical nature on account of their having to be constantly modified in accordance with fresh discoveries. I replied that a philosophy can certainly not pretend to tie up thought for ever, that it merely gives expression to its time, and is, moreover, ready to transform what is typical of itself into a fluid development, in accordance with fresh historical advances and the discoveries that they entail. Goethe appeared to be pleased with this modest claim for

[4] Enc. §§ 9, 12, 16.

xii · *Hegel's Philosophy of Subjective Spirit*

philosophy."[5] One might very well question whether Gans's conception of total historical relativity is completely satisfactory. Even the empirical sciences themselves, though they are certainly historically determined, would appear to give evidence of a not entirely negligible factor of non-historical stability, and although we can hardly accept Hegel's claim that the categorial relationships of his Logic are entirely non-historical, it is difficult not to admit that they are essentially so. Goethe had, however, raised a crucial issue, and our concern with the timeless abstractions of Hegel's Logic or the non-historical element in his philosophical system as a whole, ought not to lead us into any denial of the historical contingency of the empirical knowledge which constitutes such an essential part of their overall significance. It would certainly be impossible to make any sense at all of these lectures on Subjective Spirit were we to do so. Hegelianism could only make itself ridiculous by trumpeting around Blumenbach's conception of racial variety, Pinel's approach to insanity or Humboldt's views on comparative philology, simply because Hegel happened to develop his conception of these empirical disciplines on the basis of their insights and researches. This is not to say that he had misjudged the merits of his contemporaries, or that modern anthropologists, psychiatrists and philologists might not consult the ideas of their early nineteenth-century counterparts with profit, but only that Hegel's appreciative philosophical assessment of their empirical work does not eliminate its historical contingency. When dealing with any particular discipline, he simply accepts the empirical knowledge it involves in its current state. Hermann Drüe, in a recent attempt to relate Hegel's Psychology to the present climate of research in this discipline, has actually reached the refreshingly direct conclusion that, "an empirical psychology based upon Hegel's manner of thinking would be incapable of distinguishing itself from any other".[6]

In so far as Hegel's interpreters have been concerned exclusively with the third aspect of thinking, — with its being true or objective in that it is no longer at odds with its subject-matter, with his claim that it lies within the very nature of thought and subjectivity to be inseparable from being and objectivity, they have quite rightly concentrated upon the texts of the Phenomenology and the Logic. This is certainly the central problem of Hegelianism, and the relationship between these two works has proved to

---

[5] E. Gans 'Rückblicke auf Personen und Zustände' (Berlin, 1836), p. 310. Cf. G. Nicolin 'Hegel in Berichten seiner Zeitgenossen' (Hamburg, 1970), no. 518.
[6] H. Drüe, 'Psychologie aus dem Begriff' (Berlin, 1976), p. 75. In Germany, there has been some increase in realistic conceptions of Hegel's empiricism during the past few years, see L. B. Puntel, 'Darstellung, Methode und Struktur' (Bonn, 1973); pp. 248–58; D. von Engelhardt, 'Hegel und die Chemie' (Wiesbaden, 1976). Cf. M. B. Foster, 'The opposition between Hegel and the philosophy of empiricism' ('Verhandlungen des Dritten Hegelkongresses', ed. B. Wigersma, Tübingen and Haarlem, 1934), pp. 79–97.

be a fertile field for research. — Is the Phenomenology to be regarded as the *introduction* to the Logic? If so, is it to be interpreted existentially, historically or systematically? And what are we to make of Hegel's claim that the beginning of the Logic is indeed the beginning of all systematic philosophy? Is it, on the other hand to be regarded as the *presupposition* of the Logic? If so, how are we to interpret Hegel's claim that the mature system is truly encyclopaedic and self-sustaining? If it is justified, are we in fact dealing with two complementary self-sustaining cycles? Does the Phenomenology itself presuppose a Logic, and is everything therefore the presupposition of everything else? If so, how can there possibly be a beginning anywhere? Since Hegel never completely repudiated the Phenomenology he published in 1807, why should he have introduced a truncated version of it into the Encyclopaedia? Would it not be better to treat the whole issue as simply a matter of Hegel's own *historical* development?[7]

It seems to have been the search for an answer to these questions that led so many of Hegel's immediate followers to concentrate upon re-expounding the Philosophy of Subjective Spirit. A contemporary critic characterized their whole corporate enterprise as, "an undisciplined fooling about with empty concepts, which not infrequently lapses into being completely scatter-brained",[8] — and the judgement was not entirely unjustified, since none of them seems to have grasped the significance of Hegel's insisting that any worthwhile philosophical exposition must be based upon the commonsense realism of the analytical and synthetic procedures employed as a matter of course by any effective and responsible empiricism.[9]

By concentrating upon the Phenomenology, the Logic and what they thought were the general principles of the mature system, to the exclusion of

---

[7] H. F. Fulda has recently revived this whole nexus of problems: 'Das Problem einer Einleitung in Hegels Wissenschaft der Logik' (Frankfurt/M., 1965); 'Zur Logik der Phänomenologie von 1807' in 'Hegel in der Sicht der neueren Forschung' (Darmstadt, 1973), pp. 3–34.

[8] F. Exner (1802–1853), 'Die Psychologie der Hegelschen Schule' (Leipzig, 1842), pp. 108–109. Cf. K. Rosenkranz, 'Widerlegung der von Herrn Dr. Exner gegebenen vermeintlichen Widerlegung der Hegel'schen Psychologie' (Königsberg, 1843), and the reductionist criticism of C. H. Weisse (1801–1866), 'Die Hegel'sche Psychologie und die Exner'sche Kritik' ('Zeitschrift für Philosophie und spekulative Theologie,' vol. XIII, pp. 258–97, 1844).

[9] See J. G. Mussmann (1798?–1833), 'Lehrbuch der Seelenwissenschaft' (Berlin, 1827); K. Rosenkranz (1805–1879), 'Psychologie, oder die Wissenschaft vom subjektiven Geist' (Königsberg, 1837; 3rd edn. 1863); Carl Daub (1765–1836), 'Vorlesungen über die philosophische Anthropologie' (Berlin, 1838); C. L. Michelet, (1801–1893) 'Anthropologie und Psychologie' (Berlin, 1840); J. E. Erdmann, (1805–1892) 'Grundriss der Psychologie' (Leipzig, 1840; 2nd ed. 1873); J. V. Snellman (1806–1881), 'Versuch einer speculativen Entwicklung der Idee der Persönlichkeit' (Tübingen, 1841).

the complementarity of the three basic forms of thinking within the central principle of the Idea, the supposedly orthodox exponents of Hegelianism during the 1830's and 1840's laid themselves wide open to the onesided but incisively constructive criticism of Feuerbach and Marx. Although one might argue with some justification, that such criticism was only effective because Hegel's manner of expression was so obscure and because his followers had such an imperfect grasp of his general philosophical system, the outsider was swayed mainly on account of the inability of Hegel's professed exponents to deal competently with empirical disciplines. The simple empiricists among his colleagues at Heidelberg and Berlin had certainly criticized and cold-shouldered him, but they had done so with caution and respect, feeling no doubt that there may well have been some point in the sort of philosophical analysis to which he was submitting their disciplines. To those who ridiculed his followers' attempts at philosophizing about empirical data, it was, however, quite clear that they had nothing to lose. Discouraged by this polemic from even attempting to deal with anything concrete, the philosophers of the Absolute tended to retire into academic sequestration, into lucubrating upon the esoteric doctrines of the Logic, the great majority of which appeared to have the inestimable merit of being empirically unfalsifiable, or embodying their philosophical principles in various political and religious attitudes. It was an unhealthy state of affairs, deplorably different from what might have developed had Hegel been more successful as a teacher, and we are still living with the outcome of it. Looking back upon the reactions it has given rise to, it can hardly be denied that in spite of their own lack of empirical insight, especially in such crucial fields as natural science and psychology, Feuerbach and Marx did manage to pioneer the most constructive and balanced of the early attempts at reform.

Feuerbach actually attended some of Hegel's lectures at Berlin. Sensing that there was something radically wrong with the Hegelian logicians, he came to the conclusion that what they were really concerning themselves with was a logicized theology, and that their manner of philosophizing was therefore perpetuating an unwarranted duality: "The beginning of philosophy is not God, not the Absolute, not Being as a *predicate* of the Absolute or of the Idea — the beginning of philosophy is the finite, the determinate, the actual. The infinite cannot be thought at all *without* the finite... True speculation or philosophy is nothing but *true* and *universal* empiricism." By progressing from the Phenomenology and the Logic to the philosophies of nature, anthropology, politics, history etc., from what is abstract to what is concrete, speculative philosophy, as Feuerbach saw it, had reversed the correct procedure, and was simply engaged in realizing its own abstractions by violating the true, hard, material, factual nature of things. Its fundamental fault was that it was confusing subject and predicate: "The true relationship of thought to being is simply this: being is *subject*, *thought* is *predicate*. It is thought that derives from being, not being from thought."[10] Marx took up this point, and used it as the basic weapon in his critique of Hegel's 'Philosophy of Right' (1843). Postulating the concrete reality of the state as the true subject, he accused Hegel of having committed the elementary logical error of attempting to deduce property, capital, class-divisions, the police, the monarchy, the legislature etc. from a logic which is, in reality, on Marx's analysis, a predicate of what is ostensibly being deduced from it. He was evidently dissatisfied with the scholasticism of this distinction however, and he may well have suspected that his overriding interest in economic and social reform had led him into a certain amount of misrepresentation. In any case, he subsequently (1845) criticized Feuerbach for being content with a *theoretical* intuition of objectivity, instead of progressing to the subjective practicality of "sensuous human activity".[11]

Although this fateful inability to handle empiricism effectively only became the generally recognized characteristic of Hegelianism after the master's death, the change in the popular assessment of the movement during the early 1830's seems to have taken place extremely rapidly. In 1821 Michelet was speaking unabashedly of his teacher's materialism.[12] Schelling, who had studied with Hegel, worked together with him at the outset of his career, and had good reasons for following his later development closely, in the lectures he delivered at the University of Munich in 1827, actually accused him of doing the *opposite* of what Feuerbach and Marx were to regard as his cardinal sin: "Hegel has attempted to erect his abstract logic *over* the philosophy of nature. He has imported into it the method of the philosophy of nature however; it is not difficult to imagine what had to be perpetrated in order that a method having nothing but nature as its content and the perception of nature as its guide might be raised into something *simply* logical; he was forced to reject and violate these forms of intuition and yet proceed on the basis of them, and it is therefore by no means difficult to make the perfectly correct observation, that even in the first step of his logic Hegel presupposes intuition, and that without relying upon it, he could make no progress at all."[13]

Taking remarks such as these together with what records we have of the effect of Hegel's lectures upon his students, it seems reasonable to conclude that his empiricism must have constituted the one unmistakable and perfectly obvious link between his philosophical system and the everyday

---

[10] 'Vorläufige Thesen zur Reform der Philosophie' (1842), 'Sämtliche Werke' (ed. F. Jodl, 10 vols. Stuttgart-Bad Cannstatt, 1959–1960), II.230, 231, 239. Cf. 'Zur Kritik der Hegelschen Philosophie' (1839), 'Ueber „den Anfang der Philosophie"' (1841), II.158–215.
[11] 'Thesen über Feuerbach' (March, 1845). Cf. Enc. §§ 469–480.
[12] Nicolin, op. cit., no. 358.
[13] 'Werke' (ed. Schröter, 12 vols. Munich, 1927–1954), vol. 5, p. 208.

concerns of his contemporaries. In any case, the fame he enjoyed during his lifetime cannot possibly have been the outcome of any widespread comprehension of what he was writing, lecturing or talking about, for his manner of expression seems to have struck most of those who came into contact with him as elliptical and obscure in the extreme. Even his classical learning failed to provide him with a satisfactory mode of communication. Although he was in the habit of emphasizing the debt he owed to the Greeks, those trained in classical philosophy were frequently irritated at finding his language and manner of thinking so utterly incomprehensible. Rosenkranz, subsequently professor of philosophy at Königsberg and one of the most distinguished and accomplished of his interpreters, when he first attended his lectures, was amazed that the other students should have been able to follow anything at all of what was going on. Heinrich Heine suggested that he may not have wanted to be understood, and one of the unfortunates who attended some of the early lectures at Jena openly admitted that he, "could make absolutely nothing of them, had no idea what was being discussed, ducks or geese". A young Estonian nobleman told Rosenkranz that soon after he arrived at the University of Heidelberg in 1817, he went, "to the first good bookshop, bought those of Hegel's works available, and that evening settled comfortably on the sofa, intending to read them through." He continues, "But the more I read, the more strenuously I applied myself to what I was reading, the less I understood, and after I had struggled in vain for a couple of hours with one of the sentences, I was quite out of temper, and put the book aside. Out of curiosity, I later attended the lectures, but I must confess that I was unable to understand the notes I had taken." An acute observer of Hegel's effect upon his Berlin audiences subsequently (1847) analyzed it as follows: "I could not help feeling that what he was doing in Berlin had no real foundation. Physically and morally he was building on sand. When I think of the Poles who sat there with me, of the note-taking lieutenants, of the emaciated and elderly major, who sat by the master's side looking for all the world like a Germanic Don Quixote, I cannot rid myself of the uncomfortable impression that although Hegel certainly had enthusiastic admirers in Berlin, he had few followers who understood him."[14]

In the introduction to these lectures on the Philosophy of Subjective Spirit Hegel does what we might have expected him to do, and reminds us that the interrelationships established, "certainly have to be specified empirically" (I.83). The commentary to the present edition should make clear the extent to which this empirical aspect of his work has to be taken into consideration if we are to reach a balanced assessment of it. Philosophical must take precedence over empirical thinking however, and he warns against losing sight of the necessary presuppositional relationships: "In empirical psychology, it is the particularizations into which spirit is divided

[14] Nicolin, op. cit., nos. 269, 421, 362, 92, 246, 435.

which are regarded as being rigidly distinct, so that spirit is treated as a mere aggregate of independent powers, each of which stands only in reciprocal and therefore external relation to the other" (I.13–15). Empiricism is concerned with facts, propositions and laws, not with the more comprehensive connections elicited by philosophy, and we must therefore reject, "as being insufficient for and inadequate to science the whole finite manner of consideration of our ordinary consciousness and reflecting thought, — the empirical cognitions and appearances or the so-called facts of consciousness, the raising of these to the status of genera and classification, the abstract determinations of the understanding" (I.99). It is therefore the task of intelligence, "to dissolve the empirical connectedness of the manifold determinations of the general object" (III.167), in order that philosophy may re-expound it in its systematic or necessary interrelatedness: "Empirical statements... are of no help in cognizing the implicit nature of memory itself however; to grasp the placing and significance of memory and to comprehend its organic connection with thought in the systematization of intelligence, is one... of the most difficult points in the doctrine of spirit" (III.211). Whatever is empirical lacks certainty (II.373), and we are certainly incapable of explaining or acquiring an exact foreknowledge of our fate (II.281). We are not, however, incapable of distinguishing the exceptions from the rule, nor are we unable to bring out the main points of a topic (II.189; 287), and the final task of philosophy is to co-ordinate the results of such appositely selective thinking: "The Philosophy of Spirit can be neither empirical nor metaphysical, but has to consider the Notion of Spirit in its immanent, necessary development from out of itself into a system of its activity" (I.103). In order that both these kinds of thinking may in fact complement one another: "The material of the empirical sciences is taken up from without as it is provided by experience, and brought into external connection by being ordered in accordance with the precept of a general rule. Speculative thought, however, has to exhibit each of its general objects and their development, in their absolute necessity. This is brought about in that each particular Notion is derived from the self-producing and self-actualizing universal Notion, from the logical Idea. Philosophy has therefore to grasp spirit as a necessary development of the eternal Idea, and to allow that which constitutes the particular parts of the science of spirit to unfold itself purely from the Notion of spirit" (I.17).

Hegel's immediate followers may not have been very successful in solving the problem of the relationship in which the Phenomenology stands to the mature system, but they were right to concentrate upon the Philosophy of Subjective Spirit when attempting to do so. The circularity or self-sustaining nature of the situation lies in the fact that although the Idea, the co-ordination of the three modes of thinking, is as fundamental to the methodology of this section of the Encyclopaedia as it is to that of any other, it is in these lectures that Hegel is involved in the *comprehensive systematic analysis* of the commonsense realism on which it is based. It is essential to realize that his expositions involve the dual factor of the Idea providing the basic material by means of which commonsense realism is being analyzed.[15] In the treatment of Subjective Spirit, Hegel is not simply referring back to, restating or elaborating upon his original characterization of this realism at the beginning and end of the Logic. There, he was merely making a general statement about one of the basic ways in which we think about objectivity. Here, he is involved in analyzing this thinking into its constituents, and indicating precisely how they are interrelated.

It may be helpful to call attention to some specific examples of this. When pursuing the implications of the distinction between sensation and feeling, he notes that: "The sensuous presence of things has so light a hold upon objective consciousness, that I can also know of what is not sensuously present to me. I can, for example, be familiar with a distant country merely through what has been written about it' (II.205). When discussing the significance of attention as an aspect of intuition, he observes that, "It requires an effort, for if a person wants to apprehend one general object, he has to abstract from everything else, from all the thousand things going on in his head, from his other interests, even from his own person, and while suppressing the conceit which leads him into rashly prejudging the matter rather than allowing it to speak for itself, to work himself doggedly into it, to allow it free play, to fasten upon it without obtruding his own reflections" (III.127). He places a very high value indeed upon intelligent intuition, taking it to be the essential hall-mark of any distinguished intellect: "A talented historian for example, when describing circumstances and events, has before him a lively intuition of them as a whole, whereas a person with no talent for the presentation of history overlooks the substance of it and gets no further than the details. It has therefore rightly been insisted upon that in all branches of knowledge and especially in philosophy, what is said should arise out of an intuition of the matter... Only when thought is firmly based on the intuition of the substance of the general object can one avoid abandoning what is true when going on to consider the particular, which, though rooted in that substance, becomes mere chaff when separated from it. If genuine intuition of the general object is lacking from the outset however, or if it lapses, reflective thinking loses itself in consideration of the multitude of single determinations and relationships occurring in the object. By means of its tendency to separate, its onesided and finite categories of cause and effect, external purpose and means etc., the understanding then tears apart the general object, even when this is a living being such as a plant or an animal, and in spite of its extensive know-how, it therefore fails

---

[15] Hence the opening injunction, "Know thyself" (I.3).

to grasp the concrete nature of the general object, to recognize the spiritual bond holding together all the singularities" (III.139).

On Hegel's analysis, the relationships with objectivity in sensation, feeling, attention and intuition are the constituents of thought in that it could not be what it is unless it included or presupposed them, that is to say, unless it were more complex than they are and could therefore be analyzed into them. Since they are not its only constituents however, a more precise and comprehensive analysis will be necessary in order to clarify the exact presuppositional relationships in which they stand to one another. It may be of interest to note that the stages by which Hegel arrived at his final exposition of the interrelationships between these constituents of our thinking may be traced in some detail (II.485; 494).

Summing up, we may say therefore that the overall objective of the Philosophy of Subjective Spirit is to carry out as accurate and comprehensive a survey as is possible of the immediate constituent factors involved in our ordinary activity as conscious beings. As Hegel's contemporaries realized, this may well have practical implications in respect of the climate of opinion prevailing in the *empirical disciplines* involved, and it is of particular interest to anyone looking for an introduction to the Hegelian system, since the commonsense realism it analyzes is the fundamental factor in the methodology basic to the whole Encyclopaedia.[16]

b. STRUCTURE

> You need two beginnings for science: new myths, and a new tradition
> of changing them critically. — K. R. POPPER.

No empiricism is unstructured, and in eliciting a critically formulated or philosophical structure from the empirical content of the Encyclopaedia, Hegel makes use of certain general principles. If a modern philosopher is to come to grips with this structuring however, it is essential to remember, that at least in a part of the Encyclopaedia such as this, Hegel is not primarily concerned with the epistemological or methodological foundations of empiricism, with demonstrating the validity of a particular theory of knowledge or working out a logic of scientific discovery. On account of his basic realism, he accepts the empirical disciplines of his day in much the same way as they might have been accepted by any well-educated and capable person. This is not to say that he is uncritical of the prejudices or the incompetence of the professional practitioners, but simply that the examina-

---

[16] Puntel, op. cit., p. 132, has claimed that Hegel's Phenomenology and Psychology are as basic to his system as his Logic.

tion of such shortcomings is not his primary concern. What interests him are the relationships between what we might call the various 'branches' of any particular discipline, and the more comprehensive relationships between what the pure empiricist might regard as wholly unrelated fields of research.

It has often been assumed that the general principles he employs in establishing these inter-disciplinary relationships derive their ultimate validity solely and exclusively from the Logic. They are, however, as fundamental to the critical interrelating of logical categories as they are to inter-disciplinary work in the natural sciences and the philosophy of spirit, so that the Logic is only a help in grasping their significance in so far as it differs from these other fields by not being burdened with an historically contingent subject-matter. In this chapter I hope to show that these principles arise out of realism and empiricism, and that their structuring of empiricism forms the foundation of the fully elaborated philosophical system. It is certainly a temptation to translate Hegel's 'stages', 'developments' and 'spheres' into levels, hierarchies and sets, and to re-interpret his 'sublation' and 'double negation' in terms of asymmetrical relationships and presuppositional negation, but although it is of course desirable that contemporary logicians should find common ground with him, such an immediate link-up with modern logic would almost certainly give rise to as many misunderstandings as it removed. By and large, contemporary logicians are by no means as involved in empiricism as he was, and even the sections of the Encyclopaedia devoted to formal logic have failed to attract much attention.

The categories constituting the subject-matter of Hegel's Logic differ from the subject-matters of the Philosophies of Nature and Spirit on account of their greater abstraction and generality. Although they are certainly an integral part of empirical knowledge, they are treated as categories and submitted to a separate analysis in the Logic because a comprehension of any one of the various natural and spiritual phenomena in which they are involved does not provide a complete understanding of the part they play in comprehension as a whole; because they are, in fact, more general or universal than any of the particular instances in which they may occur. On account of their abstraction, they may certainly be regarded as the absolute presuppositions of both the natural sciences and the subject-matter of the Philosophy of Spirit. There could, however, be no critical analysis of the presuppositional relationships between them unless they had already been discovered in empiricism. Although there is, therefore, a certain empirical content to Hegel's Logic, Trendelenburg and others have not been justified in inferring that its categories are nothing but the natural or spiritual contexts from which they have been elicited.[1] The confusion here has its roots

[1] F. A. Trendelenburg (1802–1872) 'Geschichte der Kategorienlehre' (Berlin, 1846), 'Logische Untersuchungen' (2 vols. Leipzig 1862) pp. 38–42.

in an ultimate circularity of exposition, for although logical categories may be the most abstract and universal presuppositions of the Philosophy of Spirit, they can only be interrelated in the Logic in that they themselves presuppose spirit. Hegel calls attention to this when dealing with the precise placing of thought within the Psychology, implying, incidentally, that it is here rather than to consciousness that we should look for the basic presupposition of the Logic: "Thought is in its primary implicitness in the logic, and develops reason for itself within this oppositionless element" (III.223).

As we have seen, the immediate presupposition and final outcome of the Logic, the ultimate presupposition of the whole Encyclopaedia, is the Idea. Since all other logical categories fall short of it in degree of comprehensiveness, Hegel will often criticize the inadequacy of those used by empiricism in its attempt to grasp or interpret whatever it may be dealing with. In these lectures on Subjective Spirit he points out, for example, that empiricism fails to comprehend the nature of the soul on account of its ingrained tendency to treat it as a thing, limited in respect of other things, and that it employs similarly inadequate categories when attempting to come to grips with animal magnetism, — that it can make nothing of intelligence because it approaches it by means of such categories as entity, multiplicity and relation, and that the association of ideas must remain unintelligible to it in that it interprets it in terms of equivalence and disparity.[2]

It is important to note that such criticism is not only based upon the analysis of the relationships between these categories carried out in the Logic, but that it also involves a conception of the nature of the subject-matter being provided for empiricism by commonsense realism: "This manner of interpretation is entirely confined to mere reflection, and we are already raised above it by means of speculative logic, which demonstrates that in their abstract conception, all the determinations applied to soul, — such as thing, simplicity, individuality, unit, pass over into their opposite, and are therefore devoid of truth. This proof of the lack of verity in these categories of the understanding is borne out by the philosophy of spirit, which demonstrates that all fixed determinations are sublated within spirit, by means of its ideality" (II.11).

In our case, the *general principles* basic to the Logic and to every other part of the Encyclopaedia, can best be approached by examining the broad layout of the Philosophy of Spirit. If we bear in mind the final dialectical progression in this sphere from politics and history to art, religion and philosophy, it is not difficult to see why Hegel should have regarded psychic subjectivity as its foundation. The analytical judgement involved is simply that our physical existence, our being conscious of subject-object disparities and our psychology, constitute the immediate presupposition of our political,

[2] I.13; II.243; III.107; 159. See also II.129, 141, 253; III.19, 27, 91, 139, 221, 223, 279.

historical, artistic, religious and philosophical activity. Nor is it difficult to see that this presuppositional relationship is logically asymmetrical in that although it is possible to analyze what is psychically subjective into its constituent factors without presupposing politics, history, art etc., it is not possible to analyze these higher spiritual activities without presupposing psychic subjectivity. — The *level*, the *hierarchy* and the *sphere* are all involved in this general structuring of the empirical subject-matter of the Philosophy of Spirit. Although we shall now proceed to illustrate and analyze them with reference to Subjective Spirit, they are to be understood as being completely universal in their implications. As illustrated, they are the general principles of Hegel's critique of empiricism, the means by which he elicits philosophical form from an empirical content. It should not be forgotten, however, that as employed in the Logic they also constitute the philosophical structure elicited from the categories by analyzing the relationships in which they stand to one another. The categories are the subject-matter of the Logic, and although the commonsense realism and empiricism of the Idea are irrelevant to their immediate exposition, its philosophical aspect is certainly not.

The most fundamental of these principles is the *level*, which is still widely employed in the philosophy of science as well as the philosophy of mind. Hegel almost certainly took it over from the natural sciences of his day, probably well before the Jena period, since he makes good use of it in his inaugural dissertation of 1801.[3] Since he always sees it in the context of its relationship to other levels, and usually makes use of it when delineating an extensive or complex field of philosophical enquiry, he refers to it not as a level but as a stage. When introducing a particularly complicated part of the Anthropology for example, he distinguishes between the stages of the dreaming, deranged and actualized soul (II.213–215). When sketching the means by which the subject-object antithesis is overcome in the Phenomenology, he outlines the conscious, self-conscious and rational stages (III.15–19), and at similar junctures in the Psychology he distinguishes between the various stages of intelligence and presentation (III.115; 145).[4]

Not to recognize, or to confuse levels of enquiry, not to distinguish between distinct disciplines, is to fail one's first examination, to be disqualified right at the start from making any real advance beyond empiricism, and, indeed, a pretty confused and muddled empiricism at that. Hegel's own treatment of the here and now in the Phenomenology of 1807 (III.21), the prevailing approach to the relationship between language, psychology and logic (III.177), the naïve mathematicism of Eschenmayer and the fantastic romanticism of Steffens (I.101–103), are all criticized for perpetrating such

[3] P. Tibbetts, 'The "Levels of Experience" Doctrine in Modern Philosophy of Mind' ('Dialectica', vol. 25, fasc. 2, pp. 131–151, 1971). Phil. Nat. I.21–40; 372–373.
[4] Cf. II.331; III.71.

elementary confusions. The principal practical significance of the level is, therefore, that it enables the empiricist to exclude from his field of research material which is irrelevant in that it is intrinsically more or less complex than what is being investigated. Steffens, for example, is criticized for having mixed, "geology with anthropology to such an extent, that only about a tenth or twelfth of his book is concerned with the latter" (I.103). The physical and psychological aspects of sensation, though actually or empirically closely interrelated, have to be separated out into levels differing very greatly in degree of complexity, that is to say, in what they presuppose: "It is more on account of its relation to spiritual inwardness than through this peculiar measure of sensitivity, that exterior sensation becomes something which is peculiarly anthropological. Now although this relation has a multiplicity of aspects, not all of them have already to be brought under consideration at this juncture. For example, this is not the place for considering sensation's being determined as either pleasant or unpleasant, in which instance there is a comparing of exterior sensation, more or less interwoven with reflection, with our inherently self-determined nature, the satisfaction or non-satisfaction of which makes the sensation either pleasant or unpleasant" (II.177). The linguistic confusions resulting from an inadequate awareness of the levels implicit in a field of enquiry or exposition, can themselves breed and perpetuate further muddles: "Intuition is not to be confused either with presentation proper, which is first to be considered at a subsequent juncture, or with the merely phenomenological consciousness already discussed. In respect of intuition's relationship to consciousness, we have to observe that if one used the word 'intuition' in its broadest sense, one could of course apply it to the immediate or sensuous consciousness already considered. Rational procedure demands that the name should be given its proper meaning however, and one has therefore to draw the essential distinction between such consciousness and intuition" (III.137). The sciences of psychology and logic in particular are in need of a comprehensive reform of this basic aspect of their methodology, and in pointing this out Hegel touches upon the whole lay-out of the Philosophy of Subjective Spirit: "Psychology, like logic, is one of those sciences which have profited least from the more general cultivation of spirit and the profounder Notion of reason distinguishing more recent times, and it is still in a highly deplorable condition. Although more importance has certainly been attached to it on account of the direction given to the Kantian philosophy, this has actually resulted in its being proffered as the basis of a metaphysics, even in its empirical condition, the science here consisting of nothing other than the facts of human consciousness, taken up empirically simply as facts, as they are given, and analyzed. Through being so assessed, psychology is mixed with forms from the standpoint of consciousness and with anthropology, nothing having changed in respect of its own condition. The outcome of this has simply been the abandonment of the cognition of the necessity of that which is in and for itself, of the Notion and truth, not only in respect of spirit as such, but also in respect of metaphysics and philosophy in general" (III.99).[5]

Drüe, in his study of the relationship between empiricism and philosophy in Hegel's Psychology, has suggested that establishing levels of enquiry might hamper the cross-referencing which is so essential to progressive empirical work of every kind. The initial analytical procedure by means of which empiricism is raised to the level of philosophical thinking is not, however, to be regarded as a substitute for the actual investigation of the subject-matter of empirical disciplines. Thinking in terms of levels should provide a critical framework for empirical research, not replace or interfere with it. For example, although Hegel treats racial variety (II.45–83) and intelligence (III.103–229) as widely separated levels of enquiry, one could not say that he was "dealing with phenomena at the wrong juncture"[6] if he anticipated and made mention of intelligence while discussing racial characteristics. To be concerned with the systematic placing of a relatively simple or basic field of research is not to be prevented from referring to, or taking into consideration more complex levels of enquiry. In fact cross-references and the comparing of the levels distinguished are an integral part of Hegel's general philosophical procedure. When defining mental derangement for example, he compares this relatively primitive psychic state with the rational subject-object relationship finally achieved in the Phenomenology and presupposed by the Psychology: "When someone speaks in a deranged manner, one should always begin by reminding him of his overall situation, his concrete actuality. If, when he is brought to consider and to be aware of this objective context, he still fails to relinquish his false presentation, there can be no doubt that he is in a state of derangement. — It follows from this exposition that a presentation may be said to be deranged when the deranged person regards an empty abstraction and a mere possibility as something concrete and actual, for we have established that the precise nature of such a presentation lies in the deranged person's abstracting from his concrete actuality" (II.343). His definition of consciousness, which, incidentally, is analyzed in the Phenomenology as being only one remove from mental derangement, depends upon comparing it with the most primitive level of the soul, and judging it, once again, from the standpoint of the Psychology: "Although I am now conscious of these general objects, they also disappear, others come before my consciousness. Just as the form of feeling is the most insignificant form of the soul, so this simply immediate objective knowing by means of consciousness, immediate knowledge, is the lowest level of knowledge. The general object therefore has being for me, I do not know

[5] Cf. II.219, 38; 227, 15; 293, 15; III.239, 18.
[6] Drüe, op. cit., p. 55.

how, nor do I know where it comes from, but I find both to be so, the general objects being given to me with such and such a content" (III.303–305). He criticizes Condillac, not for regarding what is sensuous as being basic to what is spiritual, but for assuming, "that what is sensuous is not only the empirical prius but also persists as such, so constituting the true and substantial foundation" (III.91), that is to say, for *not* comparing the levels of complexity in the subject-matter he was dealing with.[7]

This comparing of the levels established by analyzing the results of empirical enquiry into their constituents gives rise to the second of Hegel's fundamental principles, that of development. In modern logic and philosophy of science, levels interrelating in degree of complexity are more often referred to as a *hierarchy*, and the term is preferable in that it emphasizes the systematic and structural as opposed to the natural and temporal aspect of what is being dealt with.[8] On occasions, however, Hegel does appear to lapse into treating this essentially structural or systematic relationship as a natural or temporal sequence (I.67; II.147; III.235), and there also seems to have been some uncertainty in his mind about the relationship between the philosopher and his subject-matter in respect of these interrelated levels. He sometimes speaks as if it is we who, "progress from the imperfect forms of the revelation of spirit to the highest form of it" (I.63), or begin with, "only the entirely universal, undeveloped determination of spirit" (I.67).[9] On other occasions the levels would appear to be distinguishing or inter-relating themselves: "Spirit distinguishes itself from nature, we compare spirit, as it is determined for us, with nature as we know it. *We* do this, we distinguish spirit, although it is related to nature. It is not only we who distinguish spirit from nature however, for it is of the essence of spirit to distinguish itself from it" (I.35).[10] The root of this apparent uncertainty may lie in Hegel's basic realism, or may simply be the result of his students' note-taking or Boumann's editing, but whatever he may have said about 'development' of this kind, or whatever his students may have thought he said, there can be no doubt that the philosophical significance of the relationships between the levels is not a natural but a systematic one. Analysis and comparison bring out either the inadequacy and limitedness of a certain level of enquiry, — in which case they make us aware of how it is related to more complex levels, or its complexity, — in which case they bring to our notice the relationship in which it stands to more limited or subordinate levels. The relationships between the levels may certainly coincide with a natural development in either of these cases, but such a development will not constitute their essential nature. When dealing with the significance of

---

[7] Cf. II.133, 30; 395, 31.
[8] Phil. Nat. I.21–40.
[9] Cf. III.349.
[10] Cf. I.45; II.203.

perception, Hegel indicates very clearly how our awareness of them arises out of our empiricism: "Perception starts with the observation of sensuous material. It does not remain confined to smelling, tasting, seeing, hearing and feeling however, but necessarily proceeds to relate what is sensuous to a universal which is not a matter of immediate observation, to cognize each singularization as in itself a connectedness, to comprehend force in all its expressions for example, and to search things for the relations and mediations occurring between them. Simple sensuous consciousness merely knows things, simply indicates them in their immediacy. Perception grasps their connection however, and by showing that the presence of certain conditions has a certain consequence, begins to demonstrate the truth of things. The demonstration is not final however, but still deficient. Since that by means of which something is here supposed to be demonstrated is presupposed, it is itself in need of demonstration. In this field one therefore enters into the infinite progression of moving from presuppositions to presuppositions. — This is the standpoint of experience. Everything has to be experienced. If this is to be a matter of philosophy however, one has to raise oneself above the demonstrations of empiricism, which remain bound to presuppositions, into proof of the absolute necessity of things" (III.27–29).

Hegel refers to the 'necessity' which philosophy introduces into a natural development by re-interpreting it as a systematic hierarchy, as being 'logical' or 'Notionally determined', and comments as follows upon the general significance of this coincidence of what is natural and systematic: "Naturally, our interpretation of derangement as a form or stage occurring necessarily in the development of the soul, is not to be taken to imply that every spirit, every soul, must pass through this stage of extreme disruption. To assert that it must, would be as senseless as assuming that since crime is treated as a necessary manifestation of the human will in the Philosophy of Right, it is an unavoidable necessity that every individual should be guilty of it" (II.331–333).[11] Looked at positively instead of negatively, not as a searching out of presuppositions but as a progression to increasingly complex considerations, these hierarchical relationships may be regarded as depending solely upon the finitude of what is being considered at a certain level: "The determination of finitude has been elucidated and examined long since at its place in the Logic. The further import of this exposition for the more fully determined but always still simple thoughtforms of finitude, like that of the rest of philosophy for the concrete forms of the same, is merely that the finite is not, i.e. that it is not that which is true, but simply a mere passing over which passes beyond itself. — That which is finite in the preceding spheres constitutes the dialectic, in that it passes away by means of an other and into an other" (I.71–73).[12]

[11] Cf. II.125, 131, 399; III.21, 89.
[12] Cf. III.31.

If we are to get a rounded view of the ways in which Hegel makes use of the hierarchical principle, we shall therefore have to consider it in both its natural and its systematic aspects. The soul can really relapse from consciousness into derangement (II.221, 5), and when examining the systematic relation between the two states, we also have to take this natural relation into consideration. On the other hand, it is we who interpret derangement as being "the second of the three stages of development through which the feeling soul passes as it struggles with the immediacy of its substantial content in order to complete its mastery and consciousness of itself by raising itself to the simple self-relating subjectivity present within the ego" (II.331). The analytical and philosophical significance of this development is, however, simply that of a progression from the most abstract to the most concrete or complex of the interrelated levels: "In order that the progression from something abstract to the concrete which contains it as a possibility may not appear to be somewhat singular and therefore suspect, it might be helpful to remember that a similar progression has to take place in the Philosophy of Right. Derangement consists of an abstraction to which the deranged person holds fast in the face of concrete objective consciousness, and this is therefore the reason for our having had to deal with it prior to objective consciousness, in the anthropological field" (II.351–353).

If a philosopher is to expound the significance of a hierarchy in which the various levels of an empirically ascertained subject-matter form a progression in degree of complexity, he will have to anticipate or be aware of levels which he will only be dealing with systematically later in his exposition. Analysis which brings out the relative complexity of a field by showing that it consists of subordinate levels, is certainly not identical with that which brings out the relative simplicity of a level by showing it to be a constituent of a more comprehensive field. Nevertheless, both procedures involve thinking in terms of a relationship closely resembling that which subsists between a whole and its parts, and it was evidently this relationship that led Hegel to formulate the third and most complex of his fundamental principles, that of the *sphere*.

To the extent that the sphere is the result of *anticipating* levels of complexity not yet included within the subject-matter under consideration, it is simply the outcome of distinguishing between levels and expounding them hierarchically, and this aspect of it has attracted some attention from recent commentators.[13] Hegel points out that we have to anticipate when expound-

---

[13] I. Fetscher, 'Hegels Lehre vom Menschen' (Stuttgart-Bad Cannstatt, 1970), pp. 30–31, is of the opinion that Hegel's anticipations disrupt the basic progression in degree of complexity, and Drüe, op. cit., pp. 54–56, agrees with him. Düsing, op. cit., p. 339, defends the principle of anticipation, as Hegel employs it in working out the relationships between *logical categories*, against the charge of being an essentially circular procedure.

ing systematically: "The determinations and stages of spirit occur in the higher stages of its development essentially only as moments, conditions, determinations, so that what is higher already shows itself to be empirically present in a lower and more abstract determination, all higher spirituality, for example, being already in evidence as content or determinateness within sensation. If lower stages are regarded with reference to their empirical existence however, higher stages will have to be simultaneously recollected. Since they are only present within these higher stages as forms, this procedure gives rise to the anticipation of a content which only presents itself later in the development. Consciousness is anticipated in natural awakening for example, the understanding in derangement etc." (I.24–25).[14] He will often make use of anticipation in order to indicate the general lay-out of an exposition: "As with the delineation of racial variety, a more specific and concrete account of what is spiritual throughout the natural course of the stages of its life, requires that knowledge of a more concrete spirituality than that which has yet been grasped scientifically from the standpoint of anthropology, should be anticipated and made use of in the distinguishing of the levels" (I.121).[15] And he frequently calls attention to the fact that he is anticipating in order to avoid possible confusion between the level under consideration and the more comprehensive context within which it occurs. When dealing with sensation in the Anthropology for example, he also reviews its simpler and more complex aspects: "Just as we have taken up the content of exterior sensations as demonstrated in its rational necessity within the Philosophy of Nature, which at this juncture lies behind us, so, now, it is to some extent necessary that we should anticipate the content of inner sensations, which first finds its proper place in the third part of the doctrine of subjective spirit. Our subject-matter at this juncture is simply the embodying of inner sensations, and more specifically the embodying of my sensations by means of gestures, which takes place involuntarily, for it is not dependent upon my will" (II.185–187).[16]

It is important to remember that a sphere is not only the outcome but also the prerequisite of a hierarchy of levels. Consequently, although Hegel chose to expound the empirical subject-matter of the Encyclopaedia philosophically by progressing from what is most abstract to what is most complex, he might just as well have reversed the procedure. He refers to the progressive inclusion of the levels within a sphere as *sublation*, but he could also have used this word for the converse but essentially identical procedure of

---

[14] Cf. II.177. The situation in respect of the *subject-matter* being dealt with is not the same in the Philosophy of Nature, in which it is possible to deal with a basic level such as planetary motion without referring to a complex level such as the animal organism.

[15] Cf. I.25; II.215, 327; III.115, 147, 161, 165.

[16] Cf. II.201, 233, 389; III.167, 179, 181.

analyzing a sphere into a sequence of increasingly abstract constituents. He does include a certain variety of relationships under the general concept of sublation however, and since the precise meaning of the word has been much debated, it may be of value to examine them in some detail.

In one respect sublation is real, since complex entities really do include less complex entities within themselves. In the natural world, for example, the eternal sublates the temporal (II.281), tone sublates vibration (II.171), living being sublates what is inorganic (II.5; III.35). It is apparent, however, that when Hegel speaks of the sentient soul sublating "the opposition between its being-for-itself or subjectivity, and its immediacy or substantial and implicit being" (II.149), of habit sublating corporeity (II.389), of the soul sublating itself in order to become consciousness (II.425) etc., he also means that he has analyzed the complexity relationship in which these levels stand to one another. Realism and empiricism guarantee that what is actual is reproduced in the analysis, but it has to be acknowledged that two aspects are involved. This correspondence between what is real and what is analyzed becomes particularly crucial when it is involved in the self-sustaining nature of the whole encyclopaedic system: "There is nothing less satisfactory than the expositions of divers relationships and combinations, encountered in materialistic writings, which are supposed to produce such a result as thought. They completely overlook that thought sublates that of which it is supposed to be the result, just as cause sublates itself in effect, and the means in the completed end, and that thought as such is not brought forth through an other, but brings itself forth from its implicitness into being-for-self, from its Notion into actuality, and itself posits that by which it is supposed to be posited" (II.19).

The most varied and interesting use of the concept of sublation is to be found in the treatment of consciousness. Evidently in order to help his students to follow a particularly abstract and difficult section of the Phenomenology, Hegel illustrates his exposition of the levels of recognitive self-consciousness by means of the master-servant dialectic (III.372; 377). This can certainly be interpreted as involving the real sublation of actual human situations, — of the death of a contestant, the submission of a servant, the communal provision of needs etc. (III.55; 59; 63; 67; 69). It is not a systematic treatment of such situations however, but an illustration of a level of consciousness which has sensuousness, perception and understanding as its immediate presupposition, and is itself the immediate presupposition of the sublation of the subject-object antithesis in reason. The implications of this sublation are therefore more universal, more purely epistemological in nature, than any of the actual human situations by means of which Hegel chose to illustrate them: "As this self-certainty with regard to the object, abstract self-consciousness therefore constitutes the drive to posit what it is implicitly, i.e. give content and objectivity to the abstract knowledge of itself, and conversely, to free itself from its sensuousness, to sublate the given objectivity, and to posit the identity of this objectivity with itself" (III.39).[17]

It is this second kind of sublation, uncomplicated by being associated with illustrative material, which is by far the most common throughout the other sections of the analysis of consciousness.[18] But although one level really sublates another, in the Phenomenology one also encounters the curious procedure of consciousness analyzing consciousness: "It is infinite spirit itself which, while making itself finite by presupposing itself as soul and as consciousness, also posits as sublated this self-made presupposition or finitude of the implicitly sublated opposition between consciousness and the soul on the one hand and consciousness and an external object on the other. The form of this sublation in free spirit is not identical with that in consciousness. For consciousness, the progressive determination of the ego assumes the appearance of an alteration of the object which is independent of the ego's activity, so that in the case of consciousness the logical consideration of this alteration still fell in us alone. For free spirit, however, it is free spirit itself which brings forth from itself the self-developing and altering determinations of the object, making objectivity subjective and subjectivity objective" (III.87–89). Like the rest of the Encyclopaedia, the Phenomenology involves sublation which derives its validity from the general principles of the system, not solely from the subject-matter being dealt with, and certainly not from the material used in illustrating it: "We comprehend consciousness, we know of it, have the Notion of it before us, and it is thus that we speak of it with its determinations before us. Consciousness as such, as what is empirical, does not have these determinations before it, does not know of them. That which lies in the Notion of consciousness does not pertain to consciousness as such, to empirical or everyday consciousness. Comprehending consciousness is certainly superior to empirical, reflecting consciousness, and what we comprehend of consciousness must certainly find itself in every comprehending consciousness. Consequently, when empirical consciousness objects to comprehending consciousness, the immediate refutation of the objection is in consciousness itself. Ordinary consciousness therefore has something which is superior to it, and this is the object, the negative of it, the beyond, the behind, the above, — which is a negative, an other than consciousness. For us, consciousness itself is object, is comprehended within our comprehending consciousness, and since it is therefore more than it is in empirical consciousness, we do get behind it" (III.283).

In comparison with the Phenomenology, the use of the principle of sublation in the Psychology, or the sphere of free spirit, is relatively straightforward. Intelligence is presented as sublating sensation, the sign intuition happiness impulse etc. (III.125; 143; 179; 263), because analysis makes it

[17] Cf. III.47, 293, 321, 323, 327.
[18] III.3, 31, 271.

apparent that each of these levels has that which it sublates as its presupposition. Since the sphere of Psychology is the initiation of spirit as such however, it has the sublation of the subject-object antithesis in reason as its immediate presupposition (III.347–357). The exposition of its general significance has, therefore, to take into consideration the general sublation of which spirit is capable in respect of the subject-matters of Logic and Nature: "Psychology is not an arbitrary abstraction; the Notion of spirit has made it apparent that spirit is precisely this elevation above nature and natural determinateness, above involvement with a general external object, above material being in general. All it now has to do is to realize this Notion of its freedom, that is to say, sublate the form of immediacy with which it starts once again. The content, which is raised into intuitions, consists of its sensations, intuitions and so on changed into presentations, its presentations changed into thoughts etc." (III.79).

On account of its having to rely upon commonsense realism and empiricism in sublating the subject-matter of the whole Encyclopaedia, the complete system, there is an undeniable element of finitude in spirit: "The finitude of spirit consists of the failure of knowledge to apprehend its reason as that which is in and for itself, or to the same extent, of its reason's failure to make itself fully manifest in knowledge. At the same time, it is only in so far as it is absolute freedom that reason is that which is infinite, that is to say, in so far as by rendering itself finite through taking itself to be the presupposition of its knowledge, it is the everlasting movement of sublating this immediacy, grasping itself and constituting rational knowledge" (III.85).[19] If the principles by means of which Hegel structures empiricism are most certainly universal, the empirical material which gives these principles a local habitation and a name is just as certainly not so. The extravagant nonsense promulgated by some would-be Hegelians on the basis of the perfectly valid claim that to know one's limits is to know of one's unlimitedness (I.75), is, therefore, not to be confused with Hegel's own treatment of the general powers of sublation implicit in the nature of thought or spirit.[20]

It is the principle of sublation within a sphere which provides us with the key to Hegel's conception of *negation*. One would not, in fact, be violating any aspect of his thought or language if one simply took the examples from the Psychology just considered, and said that sensation is negated by intelligence, intuition by presentation, impulse by happiness etc. Sublation is negation, what is negated being that which is presupposed, and what negates being that which sublates. If the philosopher is to use the concept of negation meaningfully therefore, he must already have analyzed the presuppositional relationships between the various levels in a certain sphere. If the level sublated is to be treated as being positive, then the level which

[19] Cf. I.53; 59; III.87; 91; 111.
[20] Cf. I.37; 47; 109; 154; III.107.

sublates it is to be regarded as negative, and vice versa. Since Hegel makes this perfectly clear on innumerable occasions, and since the logical structure involved here is by no means difficult to understand, one must draw one's own conclusions as to the significance of the controversies which his use of the concept has given rise to. The most primitive levels of spirit are contradicted by their immediate presupposition, nature, and it is this negation or contradiction which motivates the general power of sublation characteristic of spirit: "The Notion of spirit necessarily progresses into this development of its reality, since it is contradicted by the immediacy and indeterminateness of the initial form of its reality. That which appears as being immediately present in spirit is implicitly mediated, a positedness, not a true immediacy. This contradiction drives spirit to sublate the immediacy, the otherness it attributes to that which it presupposes. It is through this sublation that it first comes to itself, emerging as spirit" (I.67).

Since the spiritual sphere is the most complex of the three major levels of the whole hierarchy analyzed within the Encyclopaedia, the level within which the other two are sublated, Hegel often speaks of the negating power of spirit, and, indeed, of its absolute or infinite negativity.[21] This relationship between Logic, Nature and Spirit within the whole system is the most comprehensive expression of the identification of sublation with negation, and the general principle of it therefore enters into the detailed exposition of the presuppositional relationships between all the subordinate levels. In the Anthropology, for example, Hegel observes that, "since the substance of the soul is not a being or an immediacy but universal, it is in itself the subsisting of what is manifold, not a mere traversing of sensations which are, but a preserving of sensations which are posited as being of an ideal nature. This is because in the soul, the naked and abstract negation of that which has being is preserved and so sublated" (I.131). A similar observation is to be found in the Phenomenology: "Since abstract self-consciousness is the initial negation of consciousness, it is still burdened with an external object, with the formal negation of what pertains to it. At the same time it is therefore also consciousness, the preceding stage, and is the contradiction of itself as both consciousness and self-consciousness" (III.30).[22]

The structure of the whole Encyclopaedia of the Philosophical Sciences is *triadic* because the analysis of its subject-matter carried out in the course of its exposition brings to light the comprehensive triadicity of the distinctions between abstract logical categories, the subject-matter of the natural sciences, and spirit. Just as Nature negates or sublates the Logic, so Spirit negates or sublates Nature, and this central triadicity is, therefore, a matter of double negation. If we were to explain this triadicity fully, we should have

[21] Negating power — I.41; 43; 109; II.15; III.91. Absolute negativity — I.25; II.5; III.93; 221; 223; 271.
[22] Cf. II.217; 399; III.3; 9; 41; 91; 321.

to take into consideration the history of Hegel's development, and especially his preoccupation with the syllogism during the Jena period. Here we are only concerned with his structuring of empiricism, the foundations of his systematic work, and looked at simply from this point of view, one has to admit that the justification for the reproduction of this central triadicity within the general structure of levels, hierarchies and spheres is somewhat difficult to pinpoint. Remembering the logical significance of the sphere, one can certainly agree with the main part of his statement when he tells us that: "It is hardly necessary to observe that reason, which in our exposition appears as being third and last, is not simply a final term, a result proceeding from that which is somewhat alien to it. It is of course primary, since it is that which lies at the basis of consciousness and self-consciousness, and through the sublation of both these onesided forms shows itself to be their original unity and their truth" (III.19). It is the necessity of the triadicity here which is difficult to swallow. Why should the levels in the hierarchies not be grouped into fours or fives? Sir Thomas Browne made out a good case for thinking in terms of quincunxes. Why should we not prefer his pattern to Hegel's? Hegel certainly tells us that the second stage in any sphere is always the truth of the first (III.293), that self-consciousness is not yet absolute negativity in that it is not the negation of the negation of consciousness (III.41), that thought rounds off the sphere of intelligence because it so sublates intuition and presentation that "the end recurves to the beginning" (III.219), but when he raises our hopes by observing that "we have still to indicate the rational necessity of there being precisely these three forms" (II.93), all he provides us with is a justification for analyzing the subject-matter into a structured sphere.[23]

The truth of the matter is that the triadicity of the Hegelian structure derives from its final telos, not from its empiricism. It is certainly possible to grasp the significance of most of what he has to say about any particular subject without bothering very much about his triads, although it is quite impossible to follow anything essential to his expositions unless one has grasped the significance of the level and the hierarchy, and of sublation and negation within the sphere. There is even evidence in these lectures of a certain laxity in his attitude toward the triadicity of his expositions (III.442; 452). It would be unwise to attempt to draw any general conclusions from this however, since he quite evidently put a lot of effort into co-ordinating spheres and triads, and making double sublation or negation coincide with the rounding off of his expositions. What he almost certainly has in mind while he is doing this is the much discussed conclusion to the whole Encyclopaedia (§§ 574–577), in which he deals with the three major spheres of the whole system in terms of their syllogistic interdependence. In the Philosophy of Nature, he pictures the connection between this central triadicity and its

[23] Cf. II.167; III.71; 117; 149; 163 etc.

subordinate moments for us as follows: "The eternal life of nature consists in the Idea displaying itself in each sphere to whatever extent finitude makes possible, just as every drop of water yields an image of the sun" (I.220).

Based as it is upon a structure rooted in empiricism and commonsense realism, the sphere is central to Hegel's whole conception of philosophy as an inter-disciplinary procedure. While discussing the significance of the Idea at the end of the Longer Logic, he observes that: "By virtue of the nature of the method indicated, the science exhibits itself as a cycle returning upon itself, the end winding back into the beginning, the simple ground, through the mediation; this cycle is moreover a cycle of cycles, for each single member, as ensouled by the method, is so reflected into itself, that in returning into the beginning, it is at the same time the beginning of a new member. The individual sciences are the links in this chain, each of which has an antecedent and a sequent, — or, expressed more accurately, *has* only the antecedent and indicates its sequent in its conclusion."[24] And it is with this conception of a cycle that he introduces the Philosophy of Spirit: "Although our beginning with spirit involves the philosophy of spirit as a single philosophic science, it is in the nature of philosophy alone to be a whole, a cycle, the periphery of which consists of various cycles. Here we are considering one particular cycle of the whole, and it has a presupposition, its beginning being the product of what precedes it" (I.27).

Summing up, we may say therefore that the philosophical *structure* which Hegel elicits from his basic realism and empiricism consists of certain principles, and that these principles are consistently employed throughout the whole of the Encyclopaedia. The *subject-matter* of the Philosophy of Subjective Spirit certainly has to be mastered if we are to grasp the full significance of the specific levels, hierarchies and spheres employed in expounding this section of the Encyclopaedia, but this should not be taken to imply that they are as historically and culturally contingent as this subject-matter, or that they derive their validity wholly from the abstractions of the Logic.

c. SYSTEM

> Organisms are, therefore, the only beings in nature that, considered
> in their separate existence and apart from any relation to other things,
> cannot be thought possible except as ends of nature. — KANT.

We have already noticed that there are two aspects to the general principles involved in the structure which Hegel elicits from his empiricism. Although the logical categories are pure abstractions, an "oppositionless element within which thought develops reason for itself", they are also an

---

[24] L. Log. 842; cf. Enc. § 15.

integral part of the ways in which we interpret the soul, animal magnetism, intelligence, the association of ideas etc., and have to be assessed accordingly. Although the level has to be recognized as essential to clear thinking and constructive research in any logical, natural or spiritual context, it is always specific and concrete. It is we who progress from one level to another while expounding a hierarchy, but the levels also relate of their own accord, and sometimes even coincide with natural developments. And the sphere also, together with its subordinate principle of sublation or negation, has its general or universal significance, as well as its concrete or specific aspect. — It is important to note that what we are concerned with here is not the Kantian subject-object dualism analyzed in the Phenomenology, but what might be characterized as an abstract-concrete dualism. This is not a case of consciousness structuring sense-perceptions, as it would have been had Hegel's empiricism not been based upon realism, but of abstract principles structuring concrete disciplines, or of concrete disciplines giving specific significance to abstract principles. To raise the question of how it is that the philosopher has any knowledge of the concrete disciplines is to abandon the realism and adopt the Kantian standpoint. Hegel does not deny that Kant was concerned with an important problem here, in fact he devotes the greater part of the Phenomenology to finding a solution to it, but he does maintain that philosophy is not unavoidably committed to concerning itself with nothing else.

Levels, hierarchies and spheres may have been essential to the structuring of Hegel's empiricism, but it was the empiricism itself, his preoccupation with the concrete disciplines of the time, that provided him with the teleological principle which drew together the concrete and abstract aspects of this empirical structure into a philosophical *system*. Kant himself had already brought out the importance of thinking in terms of ends by analyzing the difference between a mechanism and an organism. In the third of his Critiques, he pointed out that an organism cannot be interpreted in terms of the mere motion of a machine, since it also has the power to impart form to material which is inherently devoid of it. He then proceeded to talk of the body politic in terms of the organism, and to ascribe a theological significance to teleological thinking: "Physico-teleological theology can serve as a propaedeutic to theology proper, since by means of the study of physical ends, of which it presents a rich supply, it awakens us to the idea of a final end which nature cannot exhibit. Consequently, it can make us alive to the need of a theology which should define the conception of God sufficiently for the highest practical employment of reason, though it cannot produce a theology or final evidences adequate for its support."[1] In the

---

[1] 'Critique of Judgement' (1790; tr. J. C. Meredith, Oxford, 1961), pt. II, p. 163. Cf. "Über den Gebrauch teleologischer Principien in der Philosophie' (1788; 'Werke' Academy edition VIII.157–184). For interesting contemporary discussions

lectures on the History of Philosophy, Hegel comments upon this part of the Kantian system at some length, noting the similarity between Kant and Aristotle in respect of their views on the purposefulness of nature, and criticizing Kant for having taken up only the subjective aspect of the matter: "Although he gives expression to the unity, once again he emphasizes the subjective side, the Notion."[2]

In Hegel's Philosophy of Nature, it is the subjectivity of the organism which sublates or negates the objectivity of Mechanics and Physics: "The sentience of individuality is to the same extent immediately exclusive however, and maintains a state of tension with an inorganic nature to which it is opposed as to its external condition and material. This basic division, or expulsion of the Sun and everything else, constitutes the precise standpoint of animation. The Idea of life is in itself this unconscious creativeness, it is an expansion of nature, which in animation has returned into its truth. For the individual however, inorganic nature is a presupposition with which it is confronted, and it is this which gives rise to the finitude of living being. The individual is for itself, but as the organic being has this negativity within itself, the connection here is absolute, indivisible, internal and essential. Externality is determined only as having being for organic being; organic being is that which maintains itself in opposition to it" (III.136). A close analysis of the historical background to the subject-matter of this section of the Encyclopaedia makes it abundantly apparent that the mechanized world-picture of Galileo, Descartes and Newton was so out of tune with progressive research in so many fields by the middle of the eighteenth century, especially in the organic sciences of course, but even in mechanics and physics, that a return to some kind of Aristotelian teleology was almost inevitable. Kant's concentration upon the philosophical significance of the organism, though not specifically related to such a comprehensive range of empirical disciplines as that analyzed by Hegel in the Philosophy of Nature, was, therefore, in complete harmony with general developments in the natural science of the time.

We now know from the manuscripts recently discovered in Berlin, that the idea of the organism and of subjectivity was central to Hegel's conception of a philosophical system from the very beginning of his university career.[3] As is so often the case, the Berlin writings and lectures simply reproduce in a fuller and more balanced form a basic conception the philosophical im-

---

of the logical structure of teleological explanations, see L. Wright, 'Teleological Explanations' (Berkeley, 1976), A. Woodfield 'Teleology' (Cambridge, 1976).

[2] Jub. vol. 19, pp. 605–606. The passage is omitted from the English translation.

[3] This might have been deduced long ago from the inaugural dissertation on the planets. These manuscripts, which relate to the lectures Hegel delivered at Jena during the Winter Term of 1801/2, are soon to be published by the Felix Meiner Verlag, Hamburg, in the 'Gesammelte Werke', vol. 5, pp. 3–16.

plications of which had already been worked out when he first arrived at Jena. In the mature Logic,[4] subjectivity is associated with the third major sphere, just as it is in the Philosophy of Nature. When we remember how well developed the Philosophy of Nature was during the Jena period, and how fragmentary the Logic remained until the Nuremberg period, this would certainly appear to bear out Schelling's view that "the abstract Logic was erected over the philosophy of nature." In the Encyclopaedia, subjectivity tends to be more confined to the first level of this sphere (§§ 163–193), but in both mature versions, the initial level of the Idea is life, the implication of this being that animation is the immediate presupposition of cognition, and so of commonsense realism and empiricism. As it is analyzed in this final sphere of the Logic, life presupposes the more abstract and general levels of mechanism, chemism and teleology, and sublates within itself the living individual, the life process and the genus.[5] The general principles implicit in the organism and subjectivity therefore constitute an integral part of the final outcome of the Logic, and the central principle of the whole Encyclopaedia.

This generalization in the Logic of the significance of the organism in the Philosophy of Nature has analogies or counterparts in many sections of the system. In the Philosophy of Right, for example, the subjectivity of the monarch unifies and co-ordinates the organic complexity of abstract right and ethical life into a sovereign state (§§ 275–279). In the Philosophy of History, the subjectivity of the historian unifies and inwardizes world events into a rational pattern.[6] In the Aesthetics, artistic expression culminates in the inwardness and subjectivity of romanticism (tr. Knox, p. 519). In these lectures on Subjective Spirit, while dealing with the levels involved in presentation, Hegel reminds us that, "a truly philosophical comprehension of this sphere consists in grasping the rational connection present between these forms, in recognizing the sequence of the organic development of intelligence within them" (III.147). After mentioning certain random facts relating to the nature of memory, he observes that: "Empirical statements such as this are of no help in cognizing the implicit nature of memory itself however; to grasp the placing and significance of memory and to comprehend its organic connection with thought in the systematization of intelligence, is one of the hitherto wholly unconsidered and in fact one of the most difficult points in the doctrine of spirit" (III.211–213). In the Phenomenology, the transition from consciousness to self-consciousness involves the ego's ability to recognize the full complexity of living being: "The transition is made from necessity to subjectivity, or what is implicitly

[4] L. Log. 575–844; Enc. §§ 163–193.
[5] L. Log. 761–774; Enc. §§ 216–222.
[6] World Hist., pp. 11–24; xv–xxv. Cf. G. D. O'Brien, 'Hegel on Reason and History' (Chicago and London, 1975).

Idea, animation. The sensuous object was first, secondly there was the object of perception as it is in the relationships of reflection, the point here being necessity, which is what is true in the connectedness, the internality of which is law. The higher factor is now living being, which is single, immediate, and is therefore contingent and not determined by the connectedness. It is, however, the inner source of activity, of motion, not within the connectedness of necessity, but free for itself. Secondly, it holds necessity within itself, is inner diremption, distinguishes itself within itself in its systems, in the moments of its animation, and is at the same time absolute ideality, the unity of these differences, single, not as a single sensuous being, but as subject. Within itself it is the resumption of the differences, which it sublates and permeates, and it therefore constitutes the third moment of animation, consciousness of life" (III.311–313).

The assessment of subjectivity or animation in contexts such as these is intelligible in its own right, and arises naturally out of an analytical structuring of the empirical subject-matter being dealt with. The corresponding assessment in the Logic is certainly a necessary universalization of these particular instances, and has a validity and significance of its own within the general context of abstract categories. This does not mean, however, that the assessment of the empirical subject-matter is simply an extension or projection of the corresponding categorial relationships worked out in the Logic. The Logic is more basic and universal than the natural sciences and spirit, and precedes them within the Encyclopaedia because its subject-matter is more abstract. It is not a substitute for the structuring of empirical research however, and as Hegel points out, could never have been worked out at all without presupposing realism, empiricism and philosophical insight. It is quite evident from the case in question, that Hegel associated the connecting or systematizing of levels, hierarchies and spheres with the organism. Kant's interpretation of it had evidently led him to the conclusion that just as in nature, mechanical, physical and chemical factors all contribute to the functioning of the animal body, so in many other fields of philosophical enquiry, complex entities hold series of less complex factors within themselves. The fundamental principle of the organism, that which constitutes the general significance of its comparative complexity, is its *subjectivity*. It is also subjectivity which is operative in all these other contexts, and the general significance of it has, therefore, to be worked out in the Logic.

The Anthropology, beginning as this sphere does with the close connection between nature and primitive psychic states, the immersion of psychic phenomena in the objectivity of nature, and progressing as it does to the consciousness which sets all that is natural over against itself, to the all-excluding subjectivity of the ego, is a good example of subjectivity functioning as a telos and so drawing a whole series of levels, hierarchies and spheres into an expositional unity. The importance of this particular sphere in the general history of modern philosophy is that it indicates the natural context of the post-Cartesian treatment of subjectivity, and shows it to be anything but presuppositionless. Hegel has no objection to Humean scepticism as long as it makes no attempt to progress beyond sensation: "The subjectivity of sensation must be sought not indeterminately in man's positing something within himself through sensing, for he also does this in thinking, but more precisely in his positing something not in his free, spiritual, universal subjectivity, but in his natural, immediate, singular subjectivity. This natural subjectivity is not yet a self-determining one, pursuing its own laws, activating itself in a necessary manner, but a subjectivity determined from without, bound to this space and this time, dependent upon contingent circumstances. Through being transposed into this subjectivity therefore, all content becomes contingent, and is endowed with determinations pertaining only to this single subject" (II.157–159). The ego which Descartes, Kant and Fichte took to be the bedrock of any wellfounded philosophy, is in fact a highly complex entity: "Ego, the universal, the being-for-self of the soul, proceeds from nowhere but the sphere of feeling. It is by this that it is conditioned, this is the ego's other, its feeling only, and it determines itself as such, it being only through the negation of the form or mode of feeling that the ego is for itself. Ego is for itself only as negation of its feelings, the determinations of its sensation. It has being in so far as it posits them as the negative of itself. It is only in that it relates itself to an object, and this object is the feeling it contains" (III.285). Descartes was saved by his interest in the natural sciences, Hume by his commonsense and his History, Kant by the thing-in-itself: "Error and stupidity only become derangement when a person believes his simply subjective presentation to be objectively present, and clings to it in spite of the actual objectivity by which it is contradicted. To the deranged, what is simply subjective is just as much a matter of certainty as what is objective. Their being centres upon the simply subjective presentation from which they derive their self-certainty" (II.343). The distinction between mental derangement and subjective idealism is, however, a narrow one: "If we have said that what is sensed derives the form of what is spatial and temporal from the intuiting spirit however, this statement must not be taken to mean that space and time are only subjective forms, which is what Kant wanted to make of them. The truth is that the things in themselves are spatial and temporal, this dual form of extrinsicality not being onesidedly imparted to them by our intuition, but in origin already communicated to them by the implicit, infinite spirit, by the eternally creative Idea. Our intuiting spirit therefore bestows upon the determinations of sensation the honour of endowing them with the abstract form of space and time and so assimilating as well as making proper general objects of them. What happens here is in no respect what subjective idealism takes it to be however, for we do not receive only the subjective mode of our determining to the exclusion of the object's own determinations" (III.135).[7]

In the systematic treatment of intelligence, the first major sphere of the Psychology, there is another progression in degree of subjectivity. A beginning is made with the all-pervasiveness of reason into which the subject-object antithesis of the Phenomenology has resolved itself, and a conclusion is reached in the subjectivity of thought (III.102–229). In the second and third major spheres of the Psychology however, there is a progression in degree of *objectivity*, from a feeling for what is practical to free social and political activity: "Intelligence has shown itself to us as spirit going into itself from out of the object, recollecting itself within the object and recognizing its inwardness to be what is objective. Now, conversely, will goes out into the objectification of its inwardness, which is still burdened with the form of subjectivity. Here in the sphere of subjective spirit however, we have to pursue this externalizing only to the point at which volitional intelligence becomes objective spirit, that is, to where the product of will ceases merely to be enjoyment, and begins to constitute deed and action" (III.231–233).

Within the Philosophy of Nature, the subjectivity of the organism objectifies in that it assimilates its environment and reproduces itself (III.136–213). In the Philosophy of Right, the monarchical subjectivity which unifies and co-ordinates the subordinate moments of a sovereign state, objectifies itself in the executive, the legislature and foreign policy (§§ 287–329). In the Philosophy of History, the subjectivity of the historian, which unifies and inwardizes world events, objectifies itself in that it interprets world history in terms of human freedom. In the Aesthetics, the comic spirit, "the good humour, the assured and careless gaiety, the exuberance and audacity of a fundamentally happy craziness, the folly and the idiosyncrasy" (tr. Knox, p. 1235) which conclude the whole exposition, initiate the earnest objectivity of the world religions surveyed in the succeeding sphere. The constant attention which Hegel paid to the general significance of this transition from subjectivity to objectivity is apparent in the mature Logic. As we have already noticed, in the earlier version, the whole of the third major sphere is presented as a subjective logic, whereas in the later version, subjectivity proper is confined to the initial level of the sphere.[8] In that subjectivity is the teleological principle systematizing the exposition of a sphere, Hegel speaks of the *Notion*: "Implicitly the soul is absolute ideality, that which overreaches all its determinateness, and it is the implication of its Notion that through the sublation of the particularities which have become fixed within it, it should make its unlimited power over them evident, that it should reduce to a mere property, a mere moment, that within it which still retains the

---

[7] Cf. Phil. Nat. I.315–316.
[8] L. Log. 575–844; Enc. §§ 163–193.

immediacy of being, in order to assume through this absolute negation the being-for-self of free individuality" (II.399). Self-consciousness also only sublates effectively in a merely subjective manner: "In its interior aspect, or in accordance with the Notion, self-consciousness has negated its own immediacy, the standpoint of desire, by the sublation of both its subjectivity and the general external object" (III.53).[9] In that objectivity is also included in the teleological principle systematizing the exposition of a sphere, he speaks of the *Idea*, which is more complex in that it includes the Notion. Aware, perhaps, of the difficulty his students would have in grasping the general implications of the relationship between the Notion and the Idea, he often refers back to the basic treatment of it in the Logic: "In its truth, that which in the preceding Paragraph we have called universal self-consciousness, is the Notion of reason. It is the Notion in so far as this exists not simply as the logical Idea, but as the Idea developed into self-consciousness, for as we know from the Logic, the Idea consists of the unity of what is subjective or of the Notion, and objectivity" (III.75). He never fails to emphasize, however, that the purely logical aspect of the Idea is a mere abstraction, which is simply implicit in the philosophies of Nature and Spirit: "In this development, spirit is preceded not only by the logical Idea, but also by external nature. For the cognition already contained in the simply logical Idea is not cognition present for itself, but merely the Notion of cognition thought by us; it is not actual spirit, but merely the possibility of it. In the science of spirit we have only actual spirit as our general object, which has nature as its proximate and the logical Idea as its primary presupposition" (I.27).[10] Although structured empiricism is systematized under the telos of the Idea, the Idea, provides no justification for violating the differences and distinctions out of which it arises. On the contrary, —: "In the two preceding Paragraphs we have dealt with the soul as formative within its corporeity in a way which is not absolute because it does not completely sublate the difference between the soul and the body. The nature of the logical Idea which develops everything out of itself demands that this difference should retain its significance" (II.427).[11]

The other major progression from subjectivity to objectivity within the sphere of Subjective Spirit, is the Phenomenology. Here, a beginning is made with the all-excluding subjectivity of the ego, and a conclusion is reached in the sublation of the subject-object antithesis in the all-pervasiveness of reason. The subject-matter of this sphere is simply consciousness, so we should not be misled by the way in which Hegel illustrates his expositions into ascribing more significance to it than it really possesses: "Ego is a completely simple, universal being. When we speak of it, we are certainly

[9] Cf. I.15–23; 79–81.
[10] Cf. I.3; 29; 53; 59.
[11] Cf. I.79; Enc. §237 Add.

referring to an individual being, but since every individual being is ego, we are merely referring to something extremely universal. It is on account of its universality that the ego is able to abstract from everything, even its life. Spirit is not merely this abstract and simple being however" (I.39). The final transition within this sphere of consciousness from subjectivity to objectivity is summarized by Hegel as follows: "Self-consciousness therefore necessarily progresses into repelling itself from itself and so placing another self-consciousness over against itself, thus giving itself an object which is identical with it and yet at the same time independent. Initially, this object is an immediate and singular ego. In that this ego is freed from the onesided form of subjectivity which thus still clings to it however, and is grasped as Idea, that is as a reality pervaded by the subjectivity of the Notion, self-consciousness advances out of its opposition to consciousness, into mediated unity with it. By this means it becomes the concrete being-for-self of the ego, absolutely free reason recognizing itself in the objective world" (III.17). Once again, however, we are merely dealing with an abstraction: "It is thus that we have progressed from consciousness to reason... The ego is the Notion, although not as the indwelling Notion of the Sun, the animal, the plant, which is inseparable from external reality. In self-consciousness the Notion is mine, abstractly for itself, and the reality opposed to self-consciousness is consciousness, ego as relating itself to an object. We have here the unity of consciousness as objectivity, i.e. the Notion within reality as self-consciousness" (III.349–351).

We have already noticed, in respect of the kinds of sublation evident in the Phenomenology, that in the interest of avoiding misunderstandings it is sometimes convenient to distinguish between the real aspect of Hegel's expositions, and the validity they derive from the general principles of the system. In respect of the teleological significance of subjectivity and objectivity in the various spheres of Subjective Spirit, it is also convenient to draw a distinction between the reality of the ego, reason, thought, free spirit etc., and the general system within which they are being expounded: "We have called the first form of spirit we have to consider subjective, since spirit here has not yet objectivized its Notion, is still Notionally undeveloped. In this its subjectivity, spirit is at the same time objective however, for it has an immediate reality, in the sublation of which it initiates its being-for-self by coming to itself, apprehending its Notion, achieving its subjectivity. It is therefore as true to say that spirit is initially objective and has to become subjective as it is to say the opposite, i.e. that it is at first subjective and has to objectivize itself. Consequently, the difference between subjective and objective spirit is not to be regarded as a rigid one. We have to grasp spirit at the outset not as mere Notion, as mere subjectivity, but as Idea, as a unity of subjective and objective, and each progression from this goes beyond the initial simple subjectivity of spirit, since it is an advance in the development of its reality or objectivity" (I.81–83). Each particular level or sphere of spirit is therefore subordinate to the full complexity of the whole, the syllogistic interdependence of the three major spheres of the whole system (Enc. §§ 574–577), and in that each is limited or determined by being sublated or negated by another, Hegel speaks of the progression involved in the interrelationships between them as the *dialectic*: "That which is finite in the preceding spheres constitutes the dialectic, in that it passes away by means of an other and into an other" (I. 73).[12] Although the final validity of these dialectical progressions therefore lies in the whole sphere of philosophy sublating or negating these subordinate levels, Hegel also pays close attention to justifying the details of specific dialectical progressions: "The necessity of the dialectical progression from the sex-relationship to the awakening of the soul lies in each of the sexually related individuals, on account of their implicit unity, finding itself in the other, so that the soul passes from its implicitness to being-for-self, which is precisely the passage from its sleep to waking" (II.131).[13] One result of this is, that in an explicitly dialectical exposition, it is sometimes difficult to distinguish what is real from what is formal, what is rooted in the empirical structure from what derives from the triadic telos: "Intelligence necessarily progresses from the subjective proof of the general presentation present within the symbol, through that mediated by means of the image, to its objective being in and for self. For since the content of the image serving as a symbol only coincides with itself in the content of the general presentation to be proved, the form of this proof's being mediated, this unity of subjective and objective, switches over into the form of immediacy. Through this dialectical motion, the general presentation, in that it is immediately valid, proved in and for itself, gets as far as no longer requiring the content of the image for its proof" (III.175).

Since sublation is negation, the dialectic is sometimes associated with the switchover into what is *really* an opposite, a complete contrast: "The content of sensuous consciousness is in itself dialectical. It is supposed to be the singular, but in that it is, it is not one but all singleness. Thus, since the single content excludes all other, it relates itself to another and shows that rather than being confined to itself, it depends upon and is mediated by the other, which it involves. The proximate truth of the immediately singular is therefore its being related to another. The determinations of this relation are known as the determinations of reflection, and it is the perceptive consciousness that takes them up" (III.25).[14] As if by some prescience of the subsequent history of the philosophical movements he was initiating, Hegel also takes the dialectic to be of the essence of what is comic: "Someone who is strutting about conceitedly trips up. If we laugh in such a case, it is

12 Cf. III.11; 37; 135; 327; Enc. §§ 81, 82.
13 Cf. II.147; 399; III.143; 291.
14 Cf. III.35.

because the character involved experiences in his own person the simple dialectic of undergoing the opposite of what he had reckoned with. True comedies are no different, for the essence of what gives rise to laughter is the immediate conversion of an intrinsically idle purpose into its opposite" (II.193-195).

To master the *empirical* disciplines included within the Encyclopaedia, to elicit the required *structure* from them, and to grasp the full significance of the final telos in order that this structure may be expounded as the philosophical *system*, was what Hegel required of his followers. As we have seen, not one of them managed to grasp the full significance of this programme, even in principle, and the movement he initiated disintegrated soon after his death. Feuerbach did not diagnose the malaise correctly when he said that Hegel himself had committed the elementary logical error of confusing subject and predicate, of treating what is real as the predicate of the Logic instead of regarding the Logic as the predicate of what is real. Nevertheless, he did call attention to what, for better or for worse, has turned out to be the main preoccupation of Hegel *scholarship*, and it may, therefore, be of some value to conclude this section of the Introduction by saying a word or two about this academic activity.

The Logic is, of course, an essential part of the Hegelian system. Logical categories are not simply the natural or spiritual contexts in which they are employed, and Hegel could never have criticized the use being made of them in these contexts in any really satisfactory manner if he had never undertaken the task of submitting them to a rigorous and comprehensive analysis, and showing how they all fall short of the Idea. The categories, which constitute the subject-matter of the Logic, differ from the subject-matters of Nature and Spirit on account of their greater abstraction: "Logic, as the formal science, cannot and should not contain that reality which is the content of the further parts of philosophy, namely, the philosophical sciences of nature and of spirit. These concrete sciences do, of course, present themselves in a more real form of the Idea than logic does. As contrasted with these concrete sciences logic is of course a formal science; but it is the science of the absolute form which is within itself a totality and contains the pure Idea of truth itself" (L. Logic 592). In that the categories are completely universal, and in that Hegel often makes use of his analytical assessment of their interrelationship in order to criticize the interpretations being foisted upon empirical research,[15] he is perfectly justified in claiming that, "in each particular philosophical science, what is logical is presupposed as the purely universal science, and so as the scientific factor in all science" (I.103).[16] What is more, since the structuring of Logic, Nature and Spirit presupposes

---

[15] I.55; 71; II.11; III.7; 307. Cf. I.13; II.11; 129; 243; 253; III.19; 27; 91; 107; 139; 159; 221; 223.
[16] Cf. II.87.

thought,[17] and the systematization of this structure involves the same principles throughout, the Idea being as central to the exposition of the Logic
as it is to that of the more concrete levels, it is not perhaps surprising that
comparison should bring out certain analogies between the three spheres.
In these lectures, for example, Hegel refers to the transition from being to
essence in the Logic, when dealing with the analogical transitions from
Anthropology to Phenomenology (II.427, 12) and from sensuous consciousness to perception (III.303, 25), and to the transitions from being to determinate being etc. when dealing with the soul (II.23, 1) and the ego (III.21,
16; 271, 8).[18] Such analogical comparisons are so rare and so obviously
casual however, that one would have to be awfully hard up for something
to say about Hegel's method, to attempt to concoct a theory of the regulative
nature of the Logic in empirical contexts on the basis of them. And if such
a theory is served up, one might well ask why such analogical comparisons
as those between sight and light (II.169), hearing and time (II.171), envisioning and space (II.281), nature and derangement (II.349), memory and
space (III.211) etc., should not be used in order to ascribe a similar rôle
to the Philosophy of Nature. Bearing in mind the syllogistic interrelationship
of the whole encyclopaedic system, one readily understands that although
the syllogism constitutes the subject-matter of a particular sphere of the
Logic (Enc. §§ 181–193), it should also be referred to when discussing the
dialectical system (II.7; 151; III.161). Hegel never suggests, however, that
the structures, let alone the subject-matters of the philosophies of Nature
and Spirit, are simply restatements of the Being (Enc. §§ 84–111), Essence
(§§ 112–159), Notion (§§ 160–244) sequence worked out in the Logic. The
analogies between the three major spheres of the Encyclopaedia are obvious
of course, and it is evidently quite easy in certain quarters to keep undergraduates and trusting colleagues fairly happy by pointing out, for example,
that the Anthropology shares certain characteristics with the sphere of
Being, the Phenomenology with that of Essence, the Psychology with that
of the Notion etc.,[19] or maintaining that the subject-matters of Nature and
Spirit are merely exemplifications of the Logic.[20] It should by now be pretty
clear however, that to do so is to grossly misrepresent the philosophy one is
professing to interpret. In this particular respect, Hegel evidently foresaw

---

[17] Cf. III.25; 185; 223.
[18] Cf. the use made of the syllogistic interrelationship between the universal and the
particular when dealing with the body-soul problem (II.15). Also II.438; 470.
[19] See, for example, Fetscher, op. cit., p. 237.
[20] Puntel, op. cit., p. 82. M. Clark, 'Logic and System' (The Hague, 1971), working
on the broadly correct assumption that Hegel's system is "a wide factual experience
seeking understanding in a logic and a logic seeking understanding in this experience" (p. 19), attempts to throw light upon Phil. Sub. Sp. III.144–229 by discussing the logic of essence. It is not surprising perhaps, that he should reach the conclusion that Hegel's general procedure is "obscure and ambiguous" (p. 200).

the possible abuse of the system, and when lecturing upon the absolute Idea, observed that, "When mention is made of the absolute Idea, it looks as though this is it, and everything is about to be revealed. But one can hold forth interminably about the absolute Idea without ever touching upon its content, and the true content is none other than the whole system, the development of which we have been considering" (Enc. § 237 Add.).[21]

Summing up, we may say therefore that the system developed from the empirical structure of the Encyclopaedia is teleological, and was almost certainly influenced by Kant's distinction between the mechanism and the organism. The general progression, which complements and never violates this structure, is one of increasing degrees of subjectivity, although in certain subsidiary spheres, notably the Phenomenology and Practical Spirit, it also involves increasing degrees of objectivity. By generalizing the sublation or negation of the basic structure, the dialectic exhibits the limitedness of the individual levels interrelated within these sequences. The teleological subjectivity of the Logic, like every other aspect of this sphere, provides us with a necessary universalization of the great variety of similar structures and sequences throughout the Philosophies of Nature and Spirit. We should not conclude from this, however, that the system, as expounded in these more complex spheres, is simply a projection or illustration of the Logic.

---

[21] Even some of the very best of Hegel's recent German interpreters have tended to overlook the importance of investigating the background and significance of his empiricism, the *subject-matter* dealt with in the Philosophies of Nature and Spirit, although the theoretical foundations for doing so have now been laid, and the publication of reliable editions of the Jena works is bringing to light the range and depth of his early empirical knowledge. Fulda (op. cit.) called attention to the crucial problem of the relationship between the Phenomenology and the Logic as early as 1965, and discussion of the central question of the significance of the concept of subjectivity in the early development of the system was opened up by R.-P. Horstmann at about the same time, 'Hegels Vorphänomenologische Entwürfe zu einer Philosophie der Subjectivität' (Heidelberg, 1968), but there is still no general agreement on the way in which the Phenomenology and the Logic relate to the mature system as a whole. Düsing (op. cit.), in a thoroughly convincing analysis of the rôle played by subjectivity in synthesizing and co-ordinating the whole range of categories surveyed within the *Logic*, rejects the possibility of there being any ontological, empirical or pragmatic foundation for what is logical (p. 344), regards the Logic as 'fundamental' to the explication of Nature and Spirit (pp. 294, 334, 344) and the dialectic as being a matter of 'purely logical' contradiction (pp. 324, 335), and simply dismisses the possibility of philosophy's playing any part in our researches into the neurological, physiological and biochemical foundations of subjectivity (p. 11). Cf. II.433; III.438.

d. ANTHROPOLOGY

The new philosophy makes man, including nature as the basis of man, the sole, universal and highest object of philosophy. It therefore makes anthropology, including physiology, the universal science. — FEUERBACH.

Kant had come to the conclusion that since philosophically coherent knowledge is an impossibility in a field such as anthropology, the pragmatic approach is the only reasonable one: "Anthropology is the systematically worked out doctrine of our knowledge of man, and can be either physiological or pragmatic. — Physiological knowledge of man involves investigating what nature makes of him: pragmatic, what man as a free agent makes, or can and should make, of himself. If we ponder the possible natural causes behind the power of recollection for example, we can speculate interminably, as Descartes did, about the traces of impressions left by what we have experienced, and persisting in the brain. But since we neither know the cerebral nerves and fibres, nor how to use them for our purposes, we shall have to admit that we are mere observers in this play of our ideas, and have to let nature have its way. Consequently, all this theoretical speculation is a sheer waste of time. — But when we use our observations of what has been found to hinder or stimulate memory in order to increase its scope or efficiency, and employ knowledge of man to this end, this is part of pragmatic anthropology; and that is precisely what concerns us here."[1] In the course of the work, he touches upon many of the topics dealt with by Hegel under the general heading of Anthropology, but there appears to be no particular significance in the lay-out he adopts. No attempt is made to indicate any systematic connection, and such fields of enquiry as the imagination, the poetic faculty, memory, wit etc., which Hegel has good reasons for including within the general sphere of Psychology, are discussed together with such topics as witchcraft, prophecy, derangement, national characteristics etc., which Hegel has equally good reasons for including within the general sphere of Anthropology proper. Nevertheless, the work makes fascinating reading. We can hardly fail to be delighted by the wealth of detail and concrete observation with which its pragmatic approach is worked out, and it is very easy to understand why it should have had such an effect upon Kant's contemporaries and given rise to a whole genre of imitative publications. It is, however, a good example of the inhibiting and restricting effect of his unrealistically narrow and academic theory of knowledge. From a philo-

---

[1] 'Anthropologie in pragmatischer Hinsicht' (Königsberg, 1798), 'Werke', Academy edition VII.117–333, foreword. Cf. I. Kant 'Anthropology from a pragmatic point of view' (tr. M. J. Gregor, The Hague, 1974).

sophical point of view, many of his non-philosophical followers produced more satisfactory surveys of the field, some of which were also more pragmatically effective. It is perhaps of interest to note that although Kant begins his Anthropology with a consideration of the ego, which is where Hegel ends his (II.633), he concludes the work with a discussion of the social, constitutional and international implications of Anthropology, in a manner which may well have influenced the formulation of Hegel's transition from Subjective to Objective Spirit (III.264–269).

Many of those who published works on Anthropology in Germany during the early part of the last century, Pölitz, Funk, Liebsch, Goldbeck, Ennemoser (1828) for example,[2] quoted and praised Kant, and as Hegel notes (III.169), Fries and his followers even attempted to interpret the whole Kantian system in anthropological or psychological terms. Their interest in the material and physiological foundations of psychology, in the subject-matter dialectically analysed by Hegel in the section of the Encyclopaedia (§§ 350–376) immediately preceding Subjective Spirit, was quite marked. Neumann, for example, began his work with a consideration of matter, and Loder, Bartels, Carus (1808), Gruithuisen (1811), Masius, von Baer and Hartmann all treated Anthropology as a predominantly physiological subject. Many university lecturers taught it as a general introduction or background to medicine, and the books by Bartels, Liebsch, Gruithuisen (1810), Masius and Hartmann were evidently intended to meet the need among medical students for broad surveys of the ground usually covered by such courses. Hegel's treatment of mental derangement as a level of Anthropology was, therefore, a perfectly natural extension of the ordinary academic attitudes of the time, and the medico-anthropological works published by Bischoff, Ideler and Nasse (1830) also included quite a lot of material on the subject. Like Hegel, Abicht, Wenzel, Liebsch, Gruithuisen (1810), Herbart (1813), Steffens, Bischoff and Ideler all treated Anthropology as requiring a particularly sensitive and balanced approach in that it involves fields of research in which it is difficult to categorize what is being dealt with as either clearly physical or clearly psychic. The rather more explicit treatment of the body-soul problem preoccupied Vering, Hillebrand, Ideler and Choulant, and in Carus (1808), Steffens, Choulant and Suabedissen (1829), one notices an attention to geography and racial characteristics which anticipates later works on Anthropology. Ennemoser (1825), Bischoff and Weber emphasized the importance of the subject to criminologists and lawyers, and Kant's ethical idealism found expression in attempts, notably by Funk, Gruithuisen (1810), Weber (1810), Herbart (1816), Salat, Bischoff, Ennemoser (1828), Suabedissen (1829), and later, of course, by Feuerbach and the Marxists, to treat anthropology either as pointing beyond itself to social life and religion,

---

[2] See the list of contemporary German works on Anthropology at the end of this chapter. It is not exhaustive, I have only listed works I have been able to consult.

or as providing the foundation for the future liberation or ennoblement of mankind.

Schmid's publication resembled earlier works on Psychology in that it seemed to go out of its way to avoid imposing any order at all upon the presentation of its material. Quite a number of writers, — Wenzel, Pockels, Weber (1810), Masius, Vering, Eschenmayer, Heinroth, Nasse, Schubert and Carus (1831) in particular, dealt with the same subject-matter as Hegel, illustrated it with the same instances and examples, but differed from him in not classifying it in the same way. We know from the draft of the projected book on Subjective Spirit (I.101–103), as well as the other records from the Berlin period, that it was this lack of a well-founded system of classification, this elementary confusion of levels, which Hegel objected to most strongly in the writings of his contemporaries (III.386; 388; 392). Pölitz, who was deeply influenced by Kant, divided Anthropology into consideration of what is external in origin, — the physical aspect of the body, stimulation, the five senses etc., and what is internal in origin, — language, memory, feelings, will etc. This distinction corresponds very roughly to that between Hegel's Anthropology and Psychology, and certain writers, notably Liebsch, Herbart (1816), Kiesewetter and Schulze, not only drew it, but classified the spheres on the basis of it. In many other cases however, — Abicht, Carus (1808), Fries (1821), Heinroth for example, we either find the distinction drawn without the classification, or the classification without the distinction. One disastrous result of this lack of method was general confusion in respect of terminology, and Gruithuisen (1811), Suabedissen (1818) and Salat in particular devoted a lot of attention to the definition of terms. In fact this approach was so fashionable about 1820, that Fries (1820, pp. 12–13) felt obliged to point out that there was not much point in simply treating Anthropology as a language game, reducing it to lexicography, and that what was required was a new critical approach to the theories in which the terminology was rooted. He cited chemistry as a good example of a science which had minimized its terminological confusions on account of its having worked out an effective theory of classification. Kiesewetter and Eschenmayer even went so far as to include a consideration of logic in their anthropological or psychological writings (I.101; III.177).

A few of these contemporary writers, notably, Goldbeck, Hillebrand, Berger (III.360), dealt with what they called 'phenomenology', but there is very little resemblance between their use of this word and Hegel's. Hegel alone seems to have grasped the precise systematic relationship between the subject-matter of physiology, Anthropology and Psychology, and the subject-object problems raised by Kant and Fichte. The idea of a broad progression from what is most simple to what is most complex as being basic to the rational exposition of a field like this, is, however, fairly common in the writings of the time. It may not have been worked out with Hegel's rigour and thoroughness, but it is, nevertheless, clearly discernible in Liebsch, Goldbeck, Gruithuisen (1810), Weber, Herbart (1813), Neumann, Kiese-wetter, Hillebrand, Steffens, Berger, Schulze, Mussmann, Bonstetten, Schubert and Carus (1831). As in the Philosophy of Nature therefore, this basic feature of Hegel's methodology is simply a more consequential applica-tion of a principle the effectiveness of which had already been proved in superabundance by ordinary empiricism. Towards the end of the 1820's Hegel's own publications and lectures were beginning to have some effect upon the work being done in these fields, but although Keyserlingk (p. 6) and Mussmann (pp. v–vi) make mention of him, their works are of no great importance.

   Although Hegel was extremely interested in the general subject-matter of the Philosophy of Subjective Spirit as early as 1786/7,[3] the philosophical background to his early reading in this field was that of the Leibniz–Wolff tradition. At the Tübingen Seminary in the summer term of 1790 however, he attended a class given by J. F. Flatt (1759–1821) in which he was intro-duced to the relationship between Kantianism and empirical psychology. The extensive notes on the subject which he wrote out in 1794 are to a great extent based on this course, and most of the subject-matter they deal with was later included in the sphere of Psychology as defined in the Berlin Encyclopaedia. What is more, the general lay-out of this material already involves a broad progression in degree of complexity, — from basic cogni-tion, to phantasy, memory and recollection, and finally to the understanding, language and judgement. There are, however, some important differences between these notes and the dialectical classification of the same material in the mature system, — sensation and feeling (pp. 170–172), dreaming, somnambulism and derangement (pp. 179–184), are subsequently treated as part of Anthropology, and the logic of cognition (pp. 186–187; 190–192)[4] is partly included in the first major sphere of the Encyclopaedia. Since Hegel quite evidently revised his conception of this material pretty carefully after he had written out these notes, it is rather surprising that there should be such a dearth of evidence concerning the development of his Anthropology during the Jena period. The 'System der Sittlichkeit' (1802) deals with the psychological presuppositions of practical activity but not with the anthro-pological presuppositions of consciousness, the first sketch of a philosophical system (1803/4) makes the transition from the Philosophy of Nature to the

---

[3] 'Dokumente zu Hegels Entwicklung' (ed. J. Hoffmeister, Stuttgart, 1936), pp. 87–104, 110, 115–136. Cf. H. S. Harris 'Hegel's Development' (Oxford, 1972), p. 175.
[4] 'Zur Psychologie und Transzendentalphilosophie. Ein Manuskript. 1794'. The pagination here is that of the forthcoming critical edition: 'Gesammelte Werke' I.167–192, Felix Meiner Verlag, Hamburg.

Philosophy of Spirit by means of consciousness (pp. 273–281),[5] and the second and third sketches of 1804/5 and 1805/6 contain nothing that might be regarded as anticipating the mature treatment of the sphere. This may be due in part to the incompleteness of the records that have survived, we know from the correspondence with van Ghert, for example, that Hegel dealt with animal magnetism (II.242–323) during the lectures of 1805/6 (II.560), but it seems unlikely that the absence of any record would have been so complete had he spent much time on the subject in the lecture-room. One can only assume that during this early period he was uncertain about the way in which the relationship between the Kantian and Fichtean ego and empirical psychology was to be worked out. He had evidently grasped the significance of the ego as the foundation of consciousness, cognition and practical activity, but had not yet explored its *presuppositions*, the transition from the Philosophy of Nature to consciousness.

As we have seen, contemporary works on Anthropology were of no great help to him in this respect, since although many of them drew attention to the general transition from what is physiological to what is psychic, none of them dealt with the systematic placing of the ego within the transition. In fact the first evidence we have of his mature distinction between Anthropology, Phenomenology and Psychology is to be found in the 'Philosophical Encyclopaedia' he prepared for the sixth formers at the Nuremberg Grammar School in 1808.[6] Some development must have taken place during the Nuremberg period, for in the Longer Logic of 1816 he is more specific about the subject-matter of the first of these spheres: "Anthropology is concerned only with the obscure region in which spirit is subject to what used to be called sidereal and terrestrial influences, and lives as a natural spirit in sympathy with nature, becoming aware of nature's changes in dreams and premonitions, and inhabiting the brain, the heart, the ganglia, the liver and so on" (781). In the Heidelberg Encyclopaedia (1817) it finally assumed a form similar to that worked out in volume two of this edition, — a general progression being made from the objectivity of psychic states closely related to natural environment, to the subjectivity of the ego which excludes from itself all such involvement with nature, and a fairly detailed triadic structure being apparent (§§ 308–328). It was, however, not until the Berlin period that Hegel began to pay particular attention to this part of the system, and to rethink, revise and expand it into one of the most elaborately researched and carefully structured sections of the whole Encyclopaedia. The transition from the organic sciences to consciousness was quite evidently one of his main interests during these years. The draft of the proposed book (1822/5) is con-

[5] 'Jenaer Systementwürfe I' (ed. K. Düsing and H. Kimmerle, Hamburg, 1975). See the extraordinarily condensed reference to the subject-matter of Anthropology on p. 285.
[6] 'Nürnberger Schriften' (ed. J. Hoffmeister, Leipzig, 1938), p. 267.

cerned with the initial stages of it (I.90–139), and the lectures show very clearly how his preoccupation with the subject-matter involved increased steadily after 1820. The summer term began at the end of April and ended at the end of August: in 1820 he finished lecturing on Anthropology on July 7th, in 1822 on July 22nd, in 1825 on July 28th. In the last instance he had devoted thirty-five lectures to the subject, and had to deal with Phenomenology and Psychology in only ten (III.455).

When Hegel refers to the subject-matter of Anthropology as the soul, he is not using this word, as Aristotle did, simply to mean animation, since at this level he has already dealt with the predominantly physical aspect of animal life (§§ 350–376). Nor is he using it, as did many of his contemporaries, with reference to the subject-matter of psychology, which he defines as presupposing the rationality of self-consciousness. For him, the soul is the subject-matter of Anthropology in that this science is concerned with psychic states, closely dependent upon but more complex than purely physical ones, and not yet involving the full self-awareness of consciousness. It might, therefore, be quite accurately defined as the generic term for the various factors involved in what is sub-conscious. On his analysis there is therefore no hard and fast line dividing animal and psychic life, let alone matter and soul or body and mind, and the sort of either-or dichotomy perpetrated by the Cartesians could only have been blown up into a central philosophical problem by those who were ignorant of the relevant empirical research being carried out on the physiological foundations of psychology (II.432). Body and mind are distinct but complementary, and shade into one another through a whole complexity of levels, brought to light by empirical research, and philosophically surveyed in the Anthropology. Hegel cannot resist treating the pretentiousness of the philosophers' approach to this pseudo-problem of their own making with a certain amount of irony: "A cognate question is that of the communion of soul and body. It was assumed that the union was a fact and that the only problem was the way in which it was to be comprehended. To deem this an incomprehensible mystery might be regarded as the ordinary answer here, for if both are presupposed as absolutely independent of each other, they are as mutually impenetrable as any two matters, and it is to be presumed that each occurs only in the non-being, i.e. the pores of the other. When Epicurus assigned to the gods their residence in the pores, he was therefore consistent in sparing them any communion with the world. — There is however no squaring this answer with that given by every philosopher since this relationship has been a matter of enquiry. Descartes, Malebranche, Spinoza, Leibnitz have all proffered God as constituting the connection" (II.5–7).

In order not to draw false conclusions from this, and so misunderstand the precise nature of the relationship between the physiology dealt with in the final sphere of the Philosophy of Nature, and the levels of the soul distinguished in the Anthropology, it is important to remember that Hegel is only able to comment in this way upon the shading of body into mind and the vacuity of the traditional philosophical approach to the transition, because he is engaged in philosophically systematizing its *empirical* structure. Remove the empirical basis, discredit the value of the close attention he pays to the relevant empirical research of his time, and his criticism becomes an empty assertiveness, in no respect superior to that of the kind of philosophy he was calling in question. The empiricism, the structuring and the systematization are the work of spirit, and are, as we have seen, basic to the whole cycle of the Encyclopaedia. What they deal with in the Philosophy of Nature is not, however, identical with what they deal with in the Philosophy of Spirit, and when surveying spirit they are also concerned with what they are themselves: "Spirit's coming into being means therefore that nature of its own accord sublates itself as being inadequate to truth, and consequently that spirit no longer presupposes itself as this self-externality of corporal singularity, but in the concretion and totality of its simple universality" (II.3).

To make the transition directly from organic being to consciousness, as Hegel did in his first major sketch of a philosophical system,[7] was to come very close to reproducing Cartesian, Kantian and Fichtean dualism. By developing his Anthropology, he brought the broad development of post Cartesian philosophy into line with progressive empirical research of the kind he found in superabundance in the ordinary anthropological publications of the time. Having rid himself of the unrealistic and restricting dualism of the prevailing philosophy, he was then at liberty to draw attention to the physical foundations of many of the psychic phenomena which the more uninformedly idealistic of his contemporaries felt obliged to exploit in the interest of asserting the superior significance of spirituality. His treatment of dreams, for example, is notable mainly on account of its sobriety and matter-of-factness, his attitude having much more in common with that of those who were attempting to cure people of the nightmare by means of carbonate of soda, than with the elaborate and fantastic theorizing then getting under way in central Europe (II.143).[8] By classifying water-divining and pendulation within the general sphere of feeling (II.263), he dissociates from the Schellingians, who had theorized exuberantly on the basis of the discovery that inorganic factors such as electricity and magnetism could have an effect upon such organic phenomena as frogs' legs and human corpses (II.517). The most explicit example of the way in which he calls attention to the importance of the physical foundation of what is psychic is, however, his criticism of the spiritualized medicine being advocated by his friend Windischmann: "He requires of a doctor that he should be a devout person,

---

[7] 1803/4: op. cit., pp. 265–273.
[8] Cf. II.482; Olga König-Flachsenfeld 'Wandlungen des Traumproblems von der Romantik bis zur Gegenwart' (Stuttgart, 1935).

a good Catholic Christian. He is thinking here of what is physical, and maintains that since bodily illness is not something external to spirit, a doctor has to take account of the soul if he is to cure the body, has to cultivate a relationship with what is mightiest in spirit, with what is religious in man, and that he can only do this in that he himself becomes religious. This relationship is therefore established at a level of culture at which spirit has not yet reflected itself into itself in this way, has not yet liberated itself. The subject in our time is more reflected, illnesses are more corporeal, more a matter of the body, and it is therefore not surprising that, as Windischmann would say, the manner of healing should be less spiritual, less godly, concerned with bodily modes of operation. One can say that this sharper division between body and soul is no reason for accusing medicine of an inappropriate approach" (II.321).[9]

As we have seen, the telos of the whole sphere of Anthropology is the ego, the extreme subjectivity of which excludes from itself all that is natural. The broad dialectical progression is, therefore, that of increasing degrees or levels of subjectivity: "All the activities of spirit are nothing but the various modes in which that which is external is led back into internality, to what spirit is itself, and it is only by means of this leading back, this idealizing or assimilation of that which is external, that spirit becomes and is spirit. — If we consider it somewhat more closely, we find that the ego is its primary and simplest determination. Ego is a completely simple, universal being. When we speak of it we are certainly referring to an individual being, but since every individual being is ego, we are merely referring to something extremely universal. It is on account of its universality that the ego is able to abstract from everything, even its life. Spirit is not merely this abstract and simple being however, it does not resemble light, as it was thought to when reference was made to the simplicity of the soul as opposed to the compositeness of the body... The ideality of the ego first gives proof of itself in the relation of the ego to the infinite multiplicity of the material confronting it" (I.37–39). The extreme form of subjectivity which concludes the sphere, is therefore anticipated, during the course of the exposition, by its more or less natural counterparts. To awaken from sleep, for example, is to leave behind a more natural state and enter into a more subjective one: "The truly spiritual activity of will and intelligence occurs in waking of course. We do not yet have to consider waking in this concrete sense however, for here we are only concerned with it as a state and consequently as something essentially different from will and intelligence. Spirit, which is to be grasped in its truth as pure activity, has the states of sleeping and waking implicit within it however, for it is also soul, and as such reduces itself to the form of a natural or immediate being. In this shape spirit merely endures its becoming being-for-self. One can therefore say that awakening is brought about by the lightning stroke of

[9] Cf. II.541.

subjectivity breaking through the form of spirit's immediacy" (II.133). The whole sphere of the feeling soul (II.215—409), as finally worked out by Hegel, is presented as an advance in degree of subjectivity beyond the mere receiving of stimulation from without characteristic of sensation: "The feeling individual is simple ideality, subjectivity of sensation. Consequently, the task here is for it to posit the simply implicit filling of its substantiality as subjectivity, to take possession of itself and assume the being-for-self of self-mastery. In that it feels, the soul is inwardly and no longer merely naturally individualized" (II.215). It is rounded off or concluded by habit, because in habitual activity subjectivity has mastered the body, although it is not yet conscious of this corporeal objectivity: "This particular being of the soul is the moment of its corporeity. Here it breaks with this corporeity, distinguishing itself from it as its simple being, and so constituting the ideal nature of its subjective substantiality. This abstract being-for-self of the soul in its corporeity is not yet ego, not yet the existence of the universal which is for the universal... In habit, the soul makes an abstract universal being of itself and reduces what is particular in feelings and consciousness to a mere determination of its being. It is thus that it possesses content, and it so contains it that in such determinations it is not sentient, but possesses and moves within them without sensation or consciousness, — standing in re-lationship to them, but neither distinguishing itself from nor being immersed within them" (II.389–391). The culmination of the whole sphere, that which makes it the immediate presupposition of consciousness, is the ego: "In so far as the soul has being for abstract universality, this being-for-self of free universality is its higher awakening as ego, or abstract universality. For itself, the soul is therefore thought and subject, and is indeed specifically the subject of its judgement. In this judgement the ego excludes from itself the natural totality of its determinations as an object or world external to it, and so relates itself to this totality that it is immediately reflected into itself within it. This is consciousness" (II.425).

Consciousness, in that it involves an antithesis in which the subject relates itself to the totality of its objectivity, is the predominating characteristic of the philosophical standpoint of Descartes, Kant and Fichte. These thinkers were essentially phenomenologists, in that although they had not seen that this totality of objectivity was also the presupposition of their subjectivity, and had therefore failed to work out a systematic philosophy of nature and anthropology, they had at least grasped it as a totality (III.11–13). On Hegel's analysis however, phenomenological consciousness is not very differ-ent from derangement, and the latter has to be dealt with in the general sphere of Anthropology and not that of Phenomenology, solely on account of its distorting its awareness of objectivity by fixing upon one particular aspect, and so failing to relate to it as to a totality: "It follows from this exposition that a presentation may be said to be deranged when the deranged person regards an empty abstraction and a mere possibility as something concrete and actual, for we have established that the precise nature of such a presentation lies in the deranged person's abstracting from his concrete actuality" (II.343). In fact, however, it is not always possible to distinguish quite clearly between the two levels: "The word foolishness involves spirit's being obsessed by a single and merely subjective presentation, which it regards as objective. For the most part the soul gets into this state when the person is dissatisfied with actuality, and so confines himself to his subjectivity. The passion of vanity and pride is the main reason for the soul's spinning this cocoon about itself. Spirit which nestles within itself in this manner easily loses touch with actuality, and finds that it is only at home in its subjective presentations. Such an attitude can soon give rise to complete foolishness, for if this solitary consciousness still has any vitality, it will readily turn to creating some sort of content out of itself, regarding what is merely subjective as objective, and fixing upon it. We have seen that the soul in a state of imbecility or desipience does not possess the power to hold fast to anything definite. Foolishness proper does possess this faculty however, and it is precisely by means of it that it shows that since it is still consciousness, it still involves a distinction between the soul and its fixed content. There are therefore two aspects here, for although a fool's consciousness has fused with this content, its universal nature also enables it to transcend the particular content of the deranged presentation. Consequently, together with their distorted view of one point, fools also have a sound and consistent consciousness, a correct conception of things, and the ability to act in an understanding manner" (II.361–363).

Within this general progression in degree of subjectivity, Hegel distinguishes three subordinate spheres, each of which exhibits a similar structure and is in its turn analyzed into further subordinate progressions (II.21–25). Since the triadic pattern is strictly maintained, even throughout the details of the exposition, this is one of the most elaborately 'dialectical' sections of the whole Encyclopaedia.

As one might expect from the growing interest Hegel showed in Anthropology during the Berlin period, the published Paragraphs on which his lectures were based were extensively revised in the 1827 and 1830 editions of the Encyclopaedia. The section on Natural Changes (§§ 396–398), for example, was very largely re-written, a new Paragraph on the Sex-relationship being inserted in order to indicate a revised triadic progression which, curiously enough, seems never to have been discussed in the lecture-room (II.477). The most important development was, however, the establishment of a clear distinction between sensation and feeling, and the insertion of §§ 399–402, which have no counterpart in the 1817 edition (§ 318). It looks as though Hegel's friction with Schleiermacher on matters of religion led him to reconsider his already well-formulated exposition of sensation in the

Philosophy of Nature (III.103, 136), and then to the realization that it was relevant to a more carefully worked out treatment of this section of the Anthropology (II.485). Recently published lecture material seems to indicate that he had grasped the importance of distinguishing between sensation and feeling in this context by 1822, although it was not until 1830 that he standardized his terminology (II.494). On his final analysis, sensation is more completely involved with objectivity by means of the senses, whereas in feeling there is a greater degree of inwardness or subjectivity. The great advantage of this distinction was that it enabled him to elicit a much more satisfactory unity from the subject-matter dealt with in §§ 403–410, than would have been possible had he simply kept to his original characterization of it as 'the subjective soul's opposition to its substantiality' (1817 ed. p. 215). All the Paragraphs relating to this second major sphere of the Anthropology (§§ 403–410) were re-written and greatly expanded in the second and third editions of the Encyclopaedia, although the broad triadic pattern remained much the same as in the first (§§ 319–325). The extraordinary wealth of empirical material dealt with in the lectures on this sphere was evidently presented in a variety of ways, and in certain cases it is difficult to pick up the thread of the chronological development. There are, for example, considerable differences between the lay-out of the treatment of animal magnetism (§ 406) in the recorded lectures and the subdivisions of the subject in Boumann's text (II.557), and the precise development lying behind the final version of the treatment of habit (§§ 409–410) is also difficult to trace (II.624). The fact that the last major sphere of the Anthropology (§§ 411–412), that dealing with the ego as the telos of the whole development, undergoes the least change, would seem to confirm the supposition that after the systematic significance of the sphere was indicated in 1808, there was no substantial alteration in Hegel's conception of the importance of this focal point.

Much of the subject-matter of the Anthropology is so much of a commonplace in the literature of the time, that there is no point in attempting to indicate specific sources or influences. The treatment of racial and national characteristics for example (§§ 393–394), can with some justification be regarded as part of a widespread vogue of relaxed philosophizing on the subject which was not above drawing its inspiration from such literary sources as Goldsmith's 'Traveller'. The progression in degree of subjectivity from natural and geographical environment to race, nationality and individuality is Hegel's, but the subject-matter itself was simply part of the common intellectual property of the time. The same is true of the treatment of the stages of life (§ 396), which were dealt with in nearly all the contemporary textbooks on anthropology, and from the most varied points of view (II.471). Hegel alone treated them as part of the transition from natural qualities to sensation, and he did so primarily on account of the general principles of his philosophical system. His personal experience, especially as a schoolmaster, evidently coloured the empirical aspect of his exposition however, and in respect of its empirical content he might have applied the general principles of his system to this field, and indeed to any other such field of enquiry, in any one of a whole variety of ways. Since he simply relies upon the ordinary professional knowledge of the time in order to endow his expositions with objectivity, even what appears to be the outcome of his own direct observation, such as the fascinating treatment of deportment, gesture and physiognomy in § 411, will often be found to derive from the general interests and preoccupations of the time (II.631). The main importance of the expositions has therefore to be sought in their cultural context and systematic placing rather than the idiosyncrasies of Hegel's empirical knowledge. This can be well illustrated by his treatment of race. Since the idea of an evolutionary progression from what is most primitive to what is most developed was fairly widely accepted at that time, one might have expected him to do what he did in § 396 with the stages of life, and simply reproduce a natural and temporal progression in systematic terms. Curiously enough, he adopted instead the now almost forgotten theory according to which racial variety is the result of an original white race having degenerated under the influence of the climatic differences between the regions into which it has migrated (II.450). He seems to have had only a limited interest in the subject, probably because the research of the time had failed to establish anything that might have been regarded as a body of basically uncontroversial knowledge, but he evidently thought it worthwhile to forego a neat correspondence in the analogy between empiricism and system, in the interest of keeping to what he considered to be the more likely of the two empirical theories. Although the systematic aspect of the exposition is therefore universal in its import, its empirical aspect is of purely historical significance, and Hegel's personal assessment of this empirical aspect of interest to no one but the antiquarian. Bearing in mind the general principles of his system, we may very well agree with him when he implies that an investigation of the effects drugs can have upon psychic states ought to be so classified that it involves natural and easy reference to both physiology and psychiatry (II.297; 305), but we are not likely to find his discussion of the actual empirical state of this field of enquiry very satisfactory (II.547; 564).

The treatment of animal magnetism (II.243–323) is the most extensive and detailed exposition of any one topic in the Philosophy of Subjective Spirit, and one of the most extensive expositions of the whole Encyclopaedia. It is not certain whether Hegel associated it with the organic sciences or the Philosophy of Spirit during the Jena period (II.560), but it is clear that he regarded it as of particular interest on account of its involving such a close and intimate interrelationship of inorganic and organic factors. On October 15th, 1810 he replied as follows to his friend van Ghert's request for a precise classification of the field: "This obscure region of the organic relationship seems to me to be particularly worthy of attention on account of its not being open to ordinary physiological interpretations; it is precisely its simplicity which I regard as being its most remarkable characteristic, for what is simple is always said to be obscure. The instance in which you applied magnetism also consisted of a fixation in the higher systems of the vital processes. I might summarize my view as follows: in general magnetism seems to me to be active in cases in which a morbid isolation occurs in respect of sensibility, as also in the case of rheumatism for example, and its effect to be a matter of the sympathy which one animal individuality is able to enter into with another in so far as its sympathy with itself, its inner fluidity, is interrupted and hindered. This union leads life back again into its general and pervasive stream. The general idea I have of magnetism is that it pertains to life in its simplicity and generality, and that in it life relates and manifests itself as does the breath of life in general, not divided into particular systems organs and their special activity, but as a simple soul to which somnambulism and the general expressions are connected; they are usually bound up with certain organs, but here they may be exercised by others almost promiscuously. It is up to you, as your experience provides you with the intuition of the matter, to examine and define these thoughts more closely" (Letters, I.329). This classification is a good example of the effectiveness of the dialectic in avoiding the sort of conceptual confusion which hinders constructive empirical work. Some of Hegel's contemporaries had overvalued animal magnetism and regarded it as capable of providing direct revelation of religious or philosophical truth, others had been encouraged by these overvaluations to dismiss it outright as a bogus phenomenon, unworthy of serious attention. In that his assessment of it implied that it ought to be carefully investigated, it was in full accord with the informed and progressive views of the 1820's. In that it also indicated the precise significance of the phenomenon in the anthropological hierarchy however, it was considerably in advance of these views. In respect of the quality of the critical judgments involved, there is moreover nothing even remotely comparable to Hegel's exposition in the anthropological literature of the time, most of which makes some mention of this field (II.504; 511).

In the introduction to the section on animal magnetism, Hegel calls attention to the fact that if we are content with simply treating such mutually contradictory interpretations of a phenomenon as irreconcilable, there is no particular reason why we should look to empiricism to provide us with a foundation for a more constructive philosophical approach: "Confirmation of the factual aspect could appear to be the primary need. For those for whom it might be required it would be superfluous however, since they simplify their consideration of the matter by dismissing accounts of it, infinitely numerous though they are, and accredited by the professionalism and character etc. of the witnesses, as delusion and imposture. They are so set in their a priori understanding, that it is not only immune to all evidence, but they have even denied what they have seen with their own eyes. In this field, even believing what one sees, let alone comprehending it, has freedom from the categories of the understanding as its basic condition" (II.243).

Since the dialectic is not only founded upon empiricism, but also involves the clarification of the conceptual contexts within which we can most effectively employ it, it is relevant not only to our understanding of disciplines, but also to our applying or making use of them. If we knew nothing of the historical background to Hegel's exposition of animal magnetism for example, we might be tempted to argue that his interest in it was purely academic. Once we discover that the phenomenon was not only being used by the medical practitioners of the time, but that one of Hegel's main reasons for taking an interest in it was the hope that their effectiveness and efficiency might be improved by a philosophical critique capable of providing them with greater conceptual clarity concerning the precise significance of what they were exploiting (II.319), his exposition appears in a new light. One has to admit that the pragmatic effectiveness of philosophy was not his primary concern. On the other hand, it can hardly be maintained that the dialectic is irrelevant to the applied sciences. Those who have overriding political, religious or philosophical concerns, and have never engaged in constructive intellectual work in any empirical discipline, may well demand more of philosophy than simply a clarification of conceptual contexts. They should not, however, underrate the pragmatic potential of such clarifications, and whatever they may expect of Anthropology, they would do well to consider that section of Hegel's treatment of it devoted to mental derangement (II.327–387).

Hegel could have done much worse than base his conception of such a theoretically complicated and practically important field upon the writings of Pinel, for although there were certain inauspiciously formal elements in his thinking, his practical experience had taught him the futility of 'metaphysical discussions and ideological ramblings', as well as the very limited efficacy of the purely physical treatment of the insane. He was, in fact, the first really influential psychiatrist to advocate gentleness, understanding and goodwill in dealing with mental derangement. Hegel was particularly impressed by his putting forward the revolutionary view that in certain cases the psychiatrist can transform the life of a patient by allowing his latent moral sense to develop, and he notes with approval (II.371) the following observation in Pinel's main work: "I cannot here avoid giving my most decided suffrage in favour of the moral qualities of maniacs. I have no where met, excepting in romances, with fonder husbands, more affectionate parents, more impassioned lovers, more pure and exalted patriots, than in the lunatic asylum, during their intervals of calmness and reason. A man of sensibility may go there every day of his life, and witness scenes of indescribable tenderness associated with most estimable virtue" (II.602).

Hegel's dialectical exposition of mental derangement is in many respects a reproduction at the psychic level of his treatment of bodily disease at the organic level (Phil. Nat. III.193–209). Just as in bodily disease one part of the organism establishes itself in opposition to the activity of the whole, so in mental disease, "the subject which has developed an understanding consciousness is still subject to disease in that it remains engrossed in a particularity of its self-awareness which it is unable to work up into ideality and overcome" (II.327). Since the mental activity of the individual presupposes its functioning as an organism, its mental aberrations are often rooted in the malfunctions of its body. On the other hand, self-awareness is the presupposition of morality, and emphasis upon the potential moral and social capabilities of the deranged can therefore play an important part in rehabilitating them. The psychiatrists of Hegel's day tended to lay emphasis upon one or the other of these two aspects of mental derangement, the somatic school concentrating upon the organic presuppositions of derangement, the psychic school upon the higher spiritual activities of which the mentally disturbed are potentially capable (II.578). Hegel brings out the complementarity of these differences of approach by applying the general principles of his system, — attempting to give every aspect of the phenomenon brought to light by practical experience and empirical research its systematic placing, and so indicating the specific relevance which each has to an overall understanding of it. In respect of the various forms of derangement, this systematic placing involves a classificatory progression from those which are predominantly physical, such as cretinism, to highly intellectual cases involving moral and ethical idealism. The simplicity and effectiveness of his procedure at this juncture contrasts sharply with the elaborate artificiality of the other attempts at classifying derangement current at the time.[10]

One might call attention to his assessment of the second sight as another good example of the effectiveness of his philosophical system in indicating a balanced and highly plausible interpretation of a phenomenon which had given rise to contradictory assertions among contemporary philosophers (II.541). Homesickness (II.247) and St. Vitus's dance (II.257) were also subjects of controversy in that physicians were not agreed as to whether they were predominantly physical or psychic, and Hegel's classification of these maladies is therefore to be assessed in the light of the mutually contradictory

---

[10] II.622–624. Cf. D. von Engelhardt, 'Hegels philosophisches Verständnis der Krankheit' ('Sudhoffs Archiv', vol. 59, no. 3, pp. 225–246, 1975); E. Fischer-Homberger, 'Eighteenth century Nosology and its Survivors' ('Medical History', vol. XIV, no. 4, pp. 397–403, October, 1970). Hegel's criticism of the random empiricism of the physicians of the time (II.373) is also justified (II.604).

theories they were then applying when attempting to cure them (II.507; 513).

In this detailed and close interrelationship between practice, empirical theory and dialectic, it is the *level* rather than the hierarchy, the sphere or the triad which has to be borne in mind if we are to grasp the immediate significance of Hegel's expositions. It is essential, for example, that the treatment of waking and sleep at the level of the natural changes of the soul (§ 398) should be kept quite distinct from their treatment at the natural and phenomenological levels (II.478). Hegel may conceivably have sympathized with Dupuis when he admitted that, "the genius of a man capable of explaining religion seems to me to be of a higher order than that of a founder of religion" (II.443), but when drawing attention to his explanation of Christianity he is careful to point out that it is unwarrantedly reductionist, and depends upon a blatant confusion of levels (II.35). It is the precise level at which the dialectic deals with natural disposition in relation to temperament and character, that enables Hegel to give a preciser meaning and clearer significance to it than did any of his contemporaries, — the major transition from nature to the soul being exactly paralleled or reproduced in this minor one (II.85; 464). The five senses are to be ranged in a different sequence according to the level at which they are being considered. At the physiological level (Phil. Nat. III.138–140) they are considered in accordance with the extent to which their external equivalents approximate, through them, to the inwardness and expressiveness of animal being, — touch in this case being the most general and abstract, and sight and hearing the most expressive of this inwardness. At the anthropological level however, they are considered in the light of the transition from the relative abstraction of sensation to the relative concreteness of feeling, — sight, in this respect, being the most general and abstract, and touch the most specific and concrete (II.167; 486). Even laughter, "although it is certainly an anthropological phenomenon in that it pertains to the natural soul, ranges from the loud, vulgar, rollicking guffaw of an empty-headed or boorish person, to the gentle smile of a noble mind, to smiling through tears, falling into a series of gradations through which it frees itself to an ever greater extent from its naturality, until in smiling it becomes a gesture, a matter of free will" (II.195).[11]

It will be apparent from this survey of the background to Hegel's Anthropology and the range of empirical disciplines it deals with, that it is not easily related to the subsequent intellectual developments that have taken place under this general heading. Had Boumann published it a few years earlier, Feuerbach might have had second thoughts about criticizing Hegel

---

[11] It should perhaps be noted that Hegel's immediate followers, and even those who specialized in Anthropology, had some difficulty in following the analytical and synthetic judgements involved in formulating these levels (II.477).

as he did, and emphasizing the novelty of his anthropocentric philosophy. On the other hand, there are many real and radical differences between Feuerbach's conception of the subject and that of his teacher. Hegel certainly bases his Anthropology upon physiology, just as he bases his Philosophy of Spirit upon Anthropology, but he would undoubtedly have regarded Feuerbach's conception of the central importance of what is natural or physical as naïvely reductionist. To indicate the immediate presupposition of a discipline, though certainly a necessary procedure in the working out of the dialectic, is by no means the same thing as attempting to interpret or comprehend it in the light of what it presupposes.[12]

[12] See 'Zur Kritik der Hegelschen Philosophie' (1839, Werke II.158–204).

Contemporary German Works on Anthropology

1.	J. C. Loder	'Anfangsgründe der physiologischen Anthropologie' (Weimar, 1800).
2.	K. H. L. Pölitz	'Populäre Anthropologie, oder Kunde von dem Menschen nach seinen sinnlichen und geistigen Anlagen' (Leipzig, 1800).
3.	J. H. Abicht	'Psychologische Anthropologie' (Erlangen, 1801).
4.	G. I. Wenzel	'Menschenlehre, oder System einer Anthropologie nach den neuesten Beobachtungen, Versuchen und Grundsätzen der Physik und Philosophie' (Linz and Leipzig, 1802).
5.	C. L. Funk	'Versuch einer praktischen Anthropologie, oder Anleitung zur Kenntniß des Menschen, und zur Vervollkommnung seiner Seelenkräfte, als Vorbereitung zur Sitten- und Religionslehre' (Leipzig, 1803).
6.	C. C. E. Schmid	'Anthropologisches Journal' (4 vols. Jena, 1803/4).
7.	C. F. Pockels	'Der Mann. Ein anthropologisches Charaktergemälde seines Geschlechts' (4 vols. Hanover, 1805/8).
8.	E. D. A. Bartels	'Anthropologische Bemerkungen über das Gehirn und den Schädel des Menschen. Mit beständiger Beziehung auf die Gall'schen Entdeckungen' (Berlin, 1806).
9.	W. Liebsch	'Grundriß der Anthropologie physiologisch und nach einem neuen Plane bearbeitet' (2 vols. Göttingen, 1806/8).
10.	F. A. Carus	'Psychologie' (2 vols. Leipzig, 1808).

11. J. C. Goldbeck   'Grundlinien der organischen Natur' (Altona, 1808).

12. F. von P. Gruithuisen   'Anthropologie, oder von der Natur des menschlichen Lebens und Denkens; für angehende Philosophen und Aerzte' (Munich, 1810).

13. H. B. Weber   'Anthropologische Versuche zur Beförderung einer gründlichen und umfassenden Menschenkunde für Wissenschaft und Leben' (Heidelberg, 1810).

14. F. von P. Gruithuisen   'Organozoonomie, oder: Ueber das niedrige Lebensverhältniß, als Propädeutik zur Anthropologie' (Munich, 1811).

15. G. H. Masius   'Grundriss anthropologischer Vorlesungen für Aerzte und Nichtärzte' (Altona, 1812).

16. J. F. Herbart   'Lehrbuch zur Einleitung in die Philosophie' (Königsberg, 1813).

17. K. G. Neumann   'Von der Natur des Menschen' (Berlin, 1815).

18. D. T. A. Suabedissen   'Die Betrachtung des Menschen' (3 vols. Cassel and Leipzig, 1815/18).

19. J. F. Herbart   'Lehrbuch der Psychologie' (Königsberg and Leipzig, 1816).

20. J. G. C. Kiesewetter   'Faßliche Darstellung der Erfahrungsseelenlehre, zur Selbstbelehrung für Nichtstudierende' (2 pts. Vienna, 1817).

21. A. M. Vering   'Psychische Heilkunde' (3 pts. Leipzig, 1817/21).

22. J. Salat   'Lehrbuch der höheren Seelenkunde. Oder: Die psychische Anthropologie' (Munich, 1820).

23. J. F. Fries   'Handbuch der psychischen Anthropologie, oder der Lehre von der Natur des menschlichen Geistes' (2 vols. Jena, 1820/21).

24. A. C. A. Eschenmayer   'Psychologie in drei Theilen als empirische, reine und angewandte' (2nd ed. Stuttgart and Tübingen, 1822).

25. J. C. A. Heinroth   'Lehrbuch der Anthropologie' (Leipzig, 1822).

26. H. Steffens   'Anthropologie' (2 vols. Breslau, 1822).

27. J. Hillebrand   'Die Anthropologie als Wissenschaft' (Mainz, 1823).

28. F. Nasse   'Zeitschrift für die Anthropologie' (4 vols. Leipzig, 1823/6).

29. K. E. von Baer   'Vorlesungen über Anthropologie' (Königsberg, 1824).

30. J. E. von Berger	'Grundzüge der Anthropologie und der Psychologie mit besonderer Rücksicht auf die Erkenntniß- und Denklehre' (Altona, 1824).
31. E. Stiedenroth	'Psychologie zur Erklärung der Seelenerscheinungen' (2 pts. Berlin, 1824/5).
32. J. Ennemoser	'Ueber die nähere Wechselwirkung des Leibes und der Seele, mit anthropologischen Untersuchungen über den Mörder Adolph Moll' (Bonn, 1825).
33. G. E. Schulze	'Psychische Anthropologie' (Göttingen, 1826).
34. C. H. E. Bischoff	'Grundriß einer anthropologischen Propädeutik zum Studio der gerichtlichen Medicin für Rechts – Beflissene' (Bonn, 1827).
35. K. W. Ideler	'Anthropologie für Ärzte' (Berlin and Landsberg, 1827).
36. H. von Keyserlingk	'Hauptpunkte zu einer wissenschaftlichen Begründung der Menschen – Kenntniß oder Anthropologie' (Berlin, 1827).
37. J. G. Mussmann	'Lehrbuch der Seelenwissenschaft oder rationalen und empirischen Psychologie, als Versuch einer wissenschaftlichen Begründung derselben, zu akademischen Vorlesungen bestimmt' (Berlin, 1827).
38. K. V. von Bonstetten	'Philosophie der Erfahrung oder Untersuchungen über den Menschen und sein Vermögen' (2 vols. Stuttgart and Tübingen, 1828).
39. L. Choulant	'Anthropologie oder Lehre von der Natur des Menschen für Nichtärzte' (2 vols. Dresden, 1828).
40. J. Ennemoser	'Anthropologische Ansichten oder Beiträge zur Kenntniß des Menschen' (Bonn, 1828).
41. J. F. Fries	'Neue oder anthropologische Kritik der Vernunft' (1807; 2nd ed. 3 vols. Heidelberg, 1828/31).
42. D. T. A. Suabedissen	'Die Grundzüge der Lehre von dem Menschen' (Marburg and Cassel, 1829).
43. H. B. von Weber	'Handbuch der psychischen Anthropologie mit vorzüglicher Rücksicht auf das Practische und die Strafrechtspflege' (Tübingen, 1829).
44. F. Nasse	'Jahrbücher für Anthropologie und zur Pathologie und Therapie des Irrseyns' (Leipzig, 1830).

45. G. H. Schubert   'Die Geschichte der Seele' (Stuttgart and Tübingen, 1830).

46. C. G. Carus   'Vorlesungen über Psychologie' (1831; ed. E. Michaelis, Erlenbach-Zürich and Leipzig, 1931).

47. P. C. Hartmann   'Der Geist des Menschen in seinen Verhältnissen zu einer Physiologie des Denkens. Für Ärzte, Philosophen und Menschen im höhern Sinne des Wortes' (1819; 2nd ed. Vienna, 1832).

ℓ. PHENOMENOLOGY

If the object has its ground solely in the acting of the Ego, and is completely determined through the Ego alone, it follows that, if there be distinctions amongst the objects, these their distinctions can arise only through different modes of acting on the part of the Ego. — FICHTE.

As we have seen, the ordinary non-philosophical works of the time from which Hegel drew the greater part of the empirical material structured and systematized in the Philosophy of Subjective Spirit, provided him with nothing even remotely resembling the Phenomenology. This part of the system is rooted almost exclusively in the more purely philosophical developments initiated by the Kantian revolution. The gap between the general and the philosophical culture of the time was too great for the anthropologists and psychologists to have assessed current philosophy very profoundly, or for the ordinary philosophers to have so mastered current empiricism as to have been capable of integrating their epistemological problems into it in any very satisfactory manner. Hegel's philosophical accomplishment is extraordinary in that he does manage to assimilate the empirical disciplines of the time with a fair degree of comprehensiveness. Since he was able to do so primarily on account of the general principles he adopted, it was only natural that at one juncture or another in the exposition of his philosophical system, he should comment upon the accomplishments of other philosophers. He does this mainly in the Logic, in which the categories they and the empiricists employ are submitted to a rigorous analysis, and in the lectures on the History of Philosophy, which do not constitute an entirely satisfactory part of the Encyclopaedia in that they are arranged chronologically rather than systematically. In this respect the Phenomenology may be classed with the Logic and the History of Philosophy, since although it is primarily concerned with a systematic analysis of consciousness, it also involves a critique of the epistemological implications of post-Kantian subjectivism. In its final systematic form it has a fairly close affinity with Fichte's 'Grundlage der gesamten Wissenschaftslehre', concerned as this is with the exploration of the ego — non-ego antithesis, "not as the object of knowledge, but solely as the form of the knowledge of all possible objects."[1] In the less systematic form in which it appeared in 1807, it bears a closer resemblance to Schelling's 'System des transzendentalen Idealismus', in which the fundamental ego — non-ego antithesis is developed into a history of self-consciousness involving the purely teleological assimilation of a whole range of theoretical and practical activity.[2]

In an advertisement put out by the publishers in 1807, Hegel summarized the significance of the original Phenomenology as follows: "It includes the various shapes of spirit within itself as stages in the progress through which spirit becomes pure knowledge or absolute spirit. Thus, the main divisions of this science, which fall into further subdivisions, include a consideration of consciousness, self-consciousness, observational and active reason, as well as spirit itself, — in its ethical, cultural and moral, and finally in its religious forms. The apparent chaos of the wealth of appearances in which spirit presents itself when first considered, is brought into a scientific order, which is exhibited in its necessity, in which the imperfect appearances resolve themselves and pass over into the higher ones constituting their proximate truth. They find their final truth first in religion and then in science, as the result of the whole."[3] It is clear, therefore, that although there is of course evidence of Hegel's having employed logical categories in working it out, and although our knowledge of the natural sciences is certainly touched upon in the course of its expositions, it was intended primarily to be neither a logic nor a philosophy of nature, but the first full-scale sketch of a Philosophy of Spirit.

Now that the lecture-notes from 1801/2 are extant, and the three attempts Hegel made at working out a system while teaching at the University of Jena have appeared in critical editions, there is plenty of scope for exploring the similarities between his systematic work prior to the publication of the Phenomenology and that of the mature Encyclopaedia.[4] The broad conception is there from the beginning, for although the system of 1803/4 lacks the overall symmetry of its Berlin counterpart, it already involves a comprehensive tripartite division into logic, nature and spirit (p. 268).[5] The empirical methodology is also much the same. Commonsense realism is basic, and provides the foundation for the elementary procedure of minimizing con-

---

[1] Jena and Leipzig, 1794; 2nd ed., 1802.
[2] Tübingen, 1800.
[3] 'Intelligenzblatt der Jenaischen Allgemeinen Litteraturzeitung', October 28th, 1807.
[4] 'Jenaer Systementwürfe' I (1803/4), II (1804/5), III (1805/6), in 'Gesammelte Werke', vols. 6, 7, 8 (Hamburg, 1975, 1971, 1976).
[5] For a much fuller statement of this conception, see the forthcoming edition of the 1801/2 notes: 'Gesammelte Werke', vol. 5, pp. 7–9.

ceptual confusions by the precise definition of levels of complexity. As during the Berlin period, the criticism of what Hegel considers to be unwarranted holism is often implied, while that of reductionism is nearly always explicit. Even the actual empirical contexts out of which such criticism arises are often identical in the earlier and later periods. In the system of 1805/6 for example, the students are warned against applying the concepts of finite mechanics to celestial motion (p. 24), regarding the galvanic process as organic (p. 107), misinterpreting the biochemistry of plant-life (p. 133), precisely as they were twenty years later. Refraction is treated as the immediate presupposition of colour (p. 85), chemistry as subordinate to organics (p. 110), the state as the sublation of the will, the law etc. (p. 253), much as they were in the Heidelberg and Berlin lectures. Had those who listened to Hegel mumbling and coughing in 1803/4 accompanied their sons to the University of Berlin a quarter of a century later, they would have been perfectly familiar with such memorable dialecticisms as the transition from nature to spirit by means of disease and death (pp. 241–265), the treatment of galvanism as an integral part of the chemical process (p. 146), the widely separated expositions of light and colour (pp. 7, 80) — although curiously enough, in several of such cases it was only toward the end of the Berlin period that he returned to his original conceptions. In the Jena sketches, as in the mature dialectic, the initiation of a major sphere often involves cross-references to its counterparts in the overall exposition. In 1805/6 for example, the treatment of space as the initiation of Mechanics is presented as having its parallel in that of the initial categories of the Logic (pp. 4–7), the exposition of life at the outset of the Organics as warranting reference to space and light (p. 108), the conception of intuition as basic to spirit as being in some sense a restatement of what has already been established in respect of being and space (p. 185).

There are, of course, differences, but it is difficult to see how any of them relate in any way whatever to what was accomplished in the Phenomenology. In 1803/4 for example, the rigorous procedure of the dialectic, with its set pattern of sublation comprehensively co-ordinated under a final telos, had not yet been developed, so that the expositional procedure of interrelating the established levels of complexity had to rely upon more or less unsatisfactory analogical procedures. Hegel rightly criticizes certain of his contemporaries for their insufficiently analytical approach to the natural sciences (pp. 126, 150, 154), but his own use of the four Empedoclean elements (pp. 111–114) in order to interrelate such widely diverse levels as meteorology (pp. 85–92), chemistry (pp. 144–146), fertilization (p. 130), botany (p. 196) and consciousness (p. 277), could obviously be criticized on the same grounds. In 1805/6, as in the Encyclopaedia, mass is presented as the unity of inertia and motion (p. 23), crime as the unity of recognition and contract (p. 232), philosophy as the unity of art and religion (p. 286) etc., but in comparison with the later system, the triadic structure of this early work is loose, and the comprehensive co-ordination of these triads with the basic structure of levels, hierarchies and spheres still very imperfect. In this connection, it is perhaps of interest to note that these lectures contain plenty of evidence that Hegel was thinking in terms of syllogisms rather than triads (pp. 74, 109, 218).[6]

The clearly empirically based Philosophy of Nature is by far the most developed part of the system during the Jena period. There is plenty of evidence, however, that Hegel had already planned the general outlines of the later Logic at the very beginning of his university career, for the publishing house of Cotta was announcing a book by him on the subject as early as June 1802, and the lectures show that by then he had already conceived of it as the first major sphere in a triadic encyclopaedic system.[7] The system of 1804/5 contains his most extensive exposition of the subject prior to the publication of the Longer Logic of 1812–1816, and since Bubner called attention to the coherence of its systematic structure, there has, in Germany at least, been more readiness to relativize the significance of the Phenomenology in the history of his development.[8]

Of the three major spheres of the system delineated in 1801/2, and later worked out in detail in the mature Encyclopaedia, that of Spirit remained the least developed for the longest period of time. The way in which this must have disturbed Hegel becomes apparent when we bear in mind that the subject-matter of Spirit — psychology, ethics, sociology, politics, history, art, religion etc., seems to have been his main concern during the 1790's, and that as the third and most comprehensive of the spheres, he was bound to regard it as the key to the self-sustaining circularity of the system as a whole, that is to say, as the immediate presupposition of, or introduction to, the Logic. There is, however, no evidence that he came anywhere near exhibiting it as such prior to the completion of the Phenomenology in 1807. The most developed treatment of it prior to this is the 'System der Sittlichkeit' (autumn, 1802), which deals systematically and in some detail with practical psychology and a wide range of social and political phenomena. The system of 1803/4 covers much the same ground, and is notable mainly on account of its also containing a treatment of consciousness as the transition from Nature to Spirit and the immediate presupposition of psychology (pp. 273–279). The 1804/5 manuscript contains a Logic and a Philosophy of Nature,

[6] See the thesis he defended for his habilitation in 1801: "Syllogismus est principium idealismi"; K. Rosenkranz, 'Hegels Leben' (Berlin, 1844), p. 157. Cf. Werke, vol. 7, pp. 94–105.
[7] H. Kimmerle, 'Dokumente zu Hegels Jenaer Dozententätigkeit' ('Hegel-Studien', vol. 4, pp. 53, 85, 1967); 'Werke', vol. 5, p. 7; vol. 6, p. 268.
[8] R. Bubner, 'Problemgeschichte und systematischer Sinn einer Phänomenologie' ('Hegel-Studien', vol. 5, pp. 129–159, 1969). Cf. Düsing, op. cit., pp. 76–108, 'Hegels Konzeption der Logik in der frühen Jenaer Zeit'.

but no third section, and that of 1805/6, though it provides us with a much more carefully structured Psychology (pp. 185–222) and a quite extensive treatment of social and political matters (pp. 222–277), constitutes no real advance upon its predecessors.

Unlike Fichte and Schelling, Hegel begins his treatment of consciousness in the Phenomenology of 1807[9] not with the ego, but with an exploration of the nature of sense-certainty. When regarded in the light of the mature system, the first main section of the work[10] might therefore be regarded as corresponding rather more closely to certain sections of the later Anthropology. It then progresses to a consideration of self-consciousness and reason in much the same way as the mature work. Had Hegel been in as full a command of the Philosophy of Spirit as he was when lecturing upon the Encyclopaedia, the treatment of consciousness, the Phenomenology proper, would have been concluded with the discussion of the identity of thinghood and reason (p. 372). From then on the subject-matter of the work, though presented in an extraordinarily impressionistic and turgid manner, emerges in a sequence corresponding pretty closely to that of the subsequent Philosophy of Spirit.

Hegel had written out the final version of the work with extreme rapidity, and in order to clear his mind in respect of a specific problem which he must have realized he had only half solved.[11] The reviewers pointed out, understandably enough, that the "scientific order" into which he claimed to have brought "the apparent chaos of the wealth of appearances in which spirit presents itself when first considered" was by no means self-evident.[12] Goethe was horrified by the reference to a botanical matter in the preface, and wrote as follows to his friend T. J. Seebeck about it: "To say anything more monstrous is indeed not possible. It seems to me quite unworthy of a rational man to want to annihilate the eternal reality of nature by means of a miserable sophistical joke... When a distinguished thinker manages to contradict and obliterate an idea by means of ingenious and mutually self-nullifying words and phrases, one knows not what to say."[13] Unlike so many of his would-be exponents, Hegel had enough commonsense to take the hint pretty quickly. In respect of the first section of the work, he saw that once a

---

[9] The original title was 'Wissenschaft der Erfahrung des Bewußtseins': see the excellent survey of the present state of scholarship concerning the work, by W. Bonsiepen of the Hegel Archive, Bochum, who is preparing the critical edition: 'Hegel-Einführung in seine Philosophie' (ed. O. Pöggeler, Freiburg and Munich, 1977), pp. 59–74.

[10] Tr. J. B. Baillie, 1949 ed., pp. 147–213.

[11] O. Pöggeler, 'Die Komposition der Phänomenologie des Geistes' (1966), now most conveniently available in 'Materialien zu Hegels "Phänomenologie des Geistes"' (ed. H. F. Fulda and D. Henrich, Frankfurt, 1973), pp. 329–390.

[12] Bonsiepen, loc. cit., p. 59.

[13] November 28th, 1812: 'Werke', IV Abtheilung 23 Band; Phil. Nat. I.83.

clear distinction had been established between the ego and its presuppositions, and a certain eccentric discursiveness concerning physiognomy, phrenology, urination etc. had been removed, there was no reason why it should not find its place in a systematic Philosophy of Spirit. In the 'Philosophical Encyclopaedia' he prepared for the sixth formers at the Nuremberg Grammar School in 1808, Phenomenology is therefore already confined to its later systematic placing as the transition from Anthropology to Psychology,[14] and there is a letter from him to Niethammer confirming that its triadic structure was meant to correspond to the first three major divisions of the published work.[15] It seems to have been quite clear to him after 1808, as indeed it should have been in May 1805 when he first began to excogitate what eventually appeared in 1807,[16] that the subject-object problems raised by Fichte and Schelling were, in their epistemological if not in their logical aspect, essentially a matter of consciousness or Phenomenology, and that although they certainly constituted a valid field of enquiry, they were not to be confused with those involved in the structuring and systematization of the Philosophy of Spirit as a whole. Since 1801, he had been presenting this sphere to his students as the transition from Nature to Logic, but it was not until he began to lecture on the basis of the Heidelberg Encyclopaedia that he was able to bring the general principles of his system to bear upon the task of demonstrating the necessity of regarding it as such.[17] The original Phenomenology was, therefore, no more than a voyage of discovery. When he left Jena, the well-ordered commerce and trade of empire still had to be established.

Although Hegel eventually ascribed a strictly subordinate role to Phenomenology within the mature Philosophy of Spirit, he continued to emphasize its importance as a critique of contemporary philosophical attitudes, and especially of the sharp distinction Kant had drawn between things in themselves and our knowledge of them: "In phenomenology, the soul now raises itself by means of the negation of its corporeity into the purely ideal nature of self-identity. It becomes consciousness, ego, has being-for-self in the face of its other. This primary being-for-self of spirit is however still conditioned by the other from which spirit emerges. The ego is still completely empty, an entirely abstract subjectivity, which posits the entire content of immediate spirit as external to itself and relates itself to it as to a world which it finds before it. Thus, although that which in the first instance was merely our general object certainly becomes the general object of spirit itself, the ego does not yet know that that which confronts it is itself natural spirit" (I.85–87). Hegel is making a systematic point here, in that he is characterizing the

---

[14] 'Nürnberger Schriften' (ed. J. Hoffmeister, Leipzig, 1938), p. 267.
[15] Hoffmeister, op. cit., p. xix. Cf. III.359.
[16] 'Werke', vol. 8, p. 358.
[17] Cf. I.93.

precise nature of consciousness, but it is a systematic point which, as he points out, is directly relevant to the whole philosophical development of post-Cartesian subjectivism (III.11). Descartes, Kant, Fichte etc. were essentially phenomenologists in that they had treated what is objective to the ego as being simply objective, and not also the presupposition of there being an ego. They had negated the subject-matter of the natural world and treated it as accessible to subjective reason only through deduction, mathematics, activity, pragmatism, and had therefore failed to place the ego in its context, work out a systematic philosophy of nature and anthropology.[18] The fundamental natural and subjective presupposition of working out such a systematic analysis of the presuppositions of the ego is the identity of subject and object in reason (III.346–357) or thought (III.216–229), and this is the immediate *outcome* of Phenomenology, not its starting point. In its broad conception, the mature Phenomenology exhibits a progression in degree of objectivity from the extreme subjectivity of the ego to the abstract but comprehensive sublation of the subject-object antithesis in reason. If we approach the work with only Kant's fundamental distinction between subject and object in mind, it must therefore remain completely unintelligible. But if any of the central issues of modern philosophy are of any concern to us, as soon as we grasp the fact that what Kant regarded as simply the negative of the ego is also its presupposition, the record of the way in which Hegel analysed the stages by which the ego approximates to the realization that what is objective to it is also what it is itself, can hardly fail to bear in upon us as one of the most fascinating and rewarding of all philosophical texts (III.270–357).

Hegel recognizes that Kantianism had made an essential philosophical advance upon Spinozism: "With regard to Spinozism however, it is to be observed that spirit emerges from substance as free subjectivity opposed to determinateness, in the judgement whereby it constitutes itself as ego, and that philosophy emerges from Spinozism in that it takes this judgement to be the absolute determination of spirit" (III.13). He emphasizes, however, the severe limitations it had imposed upon itself by its rejection of common-sense realism: "The precise stage of consciousness at which the Kantian philosophy grasps spirit is perception, which is in general the standpoint of our ordinary consciousness and to a greater or lesser extent of the sciences. It starts with the sensuous certainties of single apperceptions or observations, which are supposed to be raised into truth by being considered in their connection, reflected upon, and at the same time, turned by means of certain categories into something necessary and universal" (III.27). Although he mentions Reinhold (III.11), it looks very much as though it was primarily Fichte's version of Kantianism that he had in mind when working out his mature analysis of consciousness: "In the Fichtean exposition, the thought-

18 See, for example, III.289.

determinations of the progressive formation of the object are not merely expressed objectively, but in the form of subjective activity. It is not necessary for us to distinguish this activity as a particularity, and to give particular names to the subjective mode; it is thought in general which determines and progresses in its determinations, the absolute determination of the object being that the determinations of what is subjective and what is objective are identical. If we observe the determinations of the object therefore, we are also observing those of the subject; we do not need to distinguish them, for they are the same" (III.295). There is a nice irony about the way in which he illustrates the fundamental significance of Kant's Copernican revolution: "Certainty now expresses such a content's being identical with me as a quality of what I am, a determination of my ego's general objectivity, my reality. It is what is mine, what I hear, believe etc., that of which I am certain. It is fixed within me, it is in my ego, be it reason, immediate consciousness, intuition etc., and it is unseparated although not inseparable in my ego, since ego is the pure abstraction and can discard it once again, abstract from everything. I can take my own life, free myself from everything. All such content is also separable from the ego therefore, and it is this that constitutes the further difference between certainty and truth. People have been completely certain that the Sun moves about the Earth. Although they were unable to abstract from it, this is separable, and people today have other views. Man finds that he has deceived himself therefore, and abandons a view, is convinced of the untruth of his certainty" (III.277).

Once he had drawn the distinction between Phenomenology proper and the rest of the Philosophy of Spirit, Hegel seems never to have entertained any thought of treating his earlier work as an integral part of the mature system. It was not to be rejected outright, since for all its imperfections it could not be regarded as entirely at odds with systematic thinking, but it was quite obvious that there was no point in encouraging anyone to take it very seriously. In the introduction to the Encyclopaedia, as we have seen, the three ways in which we relate to objectivity are presented as basic to systematic thinking (§§ 26–78), and there was quite evidently no doubt at all in Hegel's mind that he was better employed making use of common-sense realism, empiricism and philosophical principles in working out a satisfactory dialectic of Anthropology, Politics, History, Art etc., than in phenomenologizing. Just as chemistry can never be completely separated from mineralogy, botany or even organic being (Phil. Nat. III.28, 54, 152), just as sensation enters into all that is spiritual (II.153), so: "Consciousness is in everything, in what is ethical, legal, religious; here, however, we are only considering what consciousness and what the relationship of consciousness is, and what is necessary to its being able to progress to spirit. The relationship of consciousness is exhibited once again in what is spiritual. In the Phenomenology therefore, the concrete formations of spirit are also developed in order to indicate what consciousness is within spirit, while at the same time the content too is developed. Here, however, we have to confine ourselves strictly to dealing with consciousness and its forms" (III.297).[19] Mineralogy, botany and organic being are not to be interpreted solely in chemical terms however, any more than spirit is to be interpreted solely in terms of sensation. Similarly, although consciousness certainly enters into morality, for example, and although to some extent it would therefore be true to say that morality can be understood in terms of consciousness, it is also very much more than consciousness, and if it is to be philosophically considered in a comprehensive and constructive manner, has also to be treated as a "concrete formation" of spirit in its own right, and not simply in order "to indicate what consciousness is within spirit" (Phil. Right §§ 105–141; cf. Phen. 611–679).

Although the main developments in Hegel's Philosophy of Spirit after 1808 are in other spheres than Phenomenology, he by no means entirely neglected this level once he had structured the first part of the 1807 work in a fairly satisfactory manner. The treatment of Reason has a fairly complicated history for example (III.380), the transition from Consciousness to Self-consciousness was altered considerably during the 1820's (III.463), the struggle for recognition was not always presented in precisely the same way (III.467), and every Paragraph of the exposition printed in the Encyclopaedia was revised in one respect or another in the second or third edition.

As Düsing has quite rightly pointed out, the presuppositional structure of the unity of subjectivity in Hegel's Logic, in that it is a purely logical principle, is not open to the criticism of involving an infinite regressional reiteration of itself.[20] Greene, in his pioneering work on the relationship between the Anthropology and the Phenomenology, noticing that Hegel is engaged in placing the ego within a context, claimed that he is "rejecting Kant's argument that the ego's knowledge of ego would entail a vitiating circularity".[21] This is not quite the case, in that as Düsing has shown, Hegel makes good use of Kant's 'transcendental' ego in giving teleological unity to the Logic. In the phenomenology he is not dealing with *logical* subjectivity, but subjectivity in so far as it is involved in being conscious of the objectivity of the natural world. Kant had equated the transcendental ego with the conscious unity of apperception,[22] and Hegel is therefore justified in criticizing him in the Phenomenology. Hegel, however, distinguishes logical subjectivity from the empirical ego, and it is of course only the empirical ego that is being dealt with as the telos of the Anthropology and the presupposition of consciousness. It should be noted however, that the philosophical procedure of structuring and systematizing the subject-matter of the Encyclo-

[19] Cf. Enc. § 25.
[20] 'Das Problem der Subjektivität in Hegels Logik' (Bonn, 1976), pp. 345–346.
[21] Murray Greene, 'Hegel on the Soul' (The Hague, 1972), pp. 16, 164.
[22] 'Critique of Pure Reason', A 107; B 143.

paedia is basic to the exposition of both logical subjectivity and phenomenological egoity. At this juncture, Hegel distinguishes between philosophical
and empirical 'consciousness' : "It is commonly asserted that each will find
in his empirical consciousness what proceeds forth from the Notion of consciousness, and this has also given rise to the dicta that one cannot get
behind consciousness, that it is the highest, that one cannot cognize what
lies behind it, that Fichte wanted to comprehend consciousness itself, but
that since it is the highest, one is unable to get over and beyond it. Although
this clearly implies that consciousness cannot be comprehended, comprehending consciousness is certainly superior to empirical, reflecting consciousness, and what we comprehend of consciousness must certainly find
itself in every comprehending consciousness. Consequently, when empirical
consciousness objects to comprehending consciousness, the immediate
refutation of the objection is in consciousness itself. Ordinary consciousness
therefore has something which is superior to it, and this is the object, the
negative of it, the beyond, the behind, the above — which is a negative,
an other than consciousness. For us, consciousness itself is object, is
comprehended within our comprehending consciousness, and since it is
therefore more than it is in empirical consciousness, we do get behind it"
(III.283).

Hegel evidently enjoyed the polemic involved in exhibiting Kant, Fichte
and Reinhold as subjectivists and phenomenologists whose theory of knowledge had brought them to the brink of mental derangement.[23] He should
have distinguished between the theoretical subjectivity of which he had made
such good use in his own Logic, and the natural subjectivity or ego he was
analysing in the Phenomenology, but he concentrated instead upon calling
attention to the Anthropology when defining the ego and its objectivity:
"Ego, the universal, the being-for-self of the soul, proceeds from nowhere
but the sphere of feeling. It is by this that it is conditioned, this is the ego's
other, its feeling only, and it determines itself as such, it being only through
the negation of the form or mode of feeling that the ego is for itself. Ego is
for itself only as negation of its feelings, the determinations of its sensation.
It has being in so far as it posits them as the negative of itself. It is only in
that it relates itself to an object, and this object is the feeling it contains"
(III.285). When dealing with the relatedness of self-consciousness, he emphasizes the organic and even purely physical nature of its presuppositions:
"The man has before himself a self-consciousness which is wholly abstract,

---

[23] Cf. Phil. Nat. I.200: "According to a metaphysics prevalent at the moment, we
cannot know things because they are uncompromisingly exterior to us. It might be
worth noticing that even the animals, which go out after things, grab, maul and
consume them, are not so stupid as these metaphysicians." See also Phil. Sub. Sp.
II.327, 19; III.9, 13; 456.

and in that they find themselves as such, these beings are mutually opposed. There is, therefore, a positing of the highest contradiction — that between the clear identity of both on one side and the complete independence of each on the other. Since each is a particular subject, a general corporeal object, they appear to me as being mutually opposed. The other differs from me as I differ from the tree, the stone etc." (III.329–331).

Since Phenomenology is the second level of Subjective Spirit, presupposing Anthropology and the presupposition of Psychology, it has the same systematic placing within this sphere as Essence has within the whole sphere of the Logic, and Hegel calls attention to this: "As ego, spirit is essence, but since reality is posited in the sphere of essence as immediate being, and at the same time as of an ideal nature, spirit as consciousness is only the appearance of spirit" (II.9; 363). A strict analogy between the two major spheres would have classified Phenomenology logically as corresponding to Quantity (Enc. §§ 99–106), but since Hegel does not expect us to read any more into such logical cross-references than we would normally read into his referring to light (III.3) as the initiation of the second major sphere of the Philosophy of Nature,[24] it would be pointlessly pedantic to pull him up on the matter. It is worth noting however, that there is a more general structural analogy here. Just as essence involves being, which is its immediate presupposition in the Logic, so ego involves the soul or feeling, which is its immediate presupposition in the Philosophy of Spirit. In the dialectical progression, essence presupposes, sublates or negates being, just as the ego presupposes, sublates or negates the soul. What is more, the content of the ego or consciousness is that which it negates: "Consciousness, as relationship, contains only those categories pertaining to the abstract ego or formal thinking, and it takes them to be determinations of the object. Of the object therefore, sensuous consciousness knows only that it is a being, something, an existing thing, a singular etc. Although this consciousness appears as the richest in content, it is the poorest in thought. The wealth with which it is filled consists of the determinations of feeling; they are the material of consciousness, what is substantial and qualitative, what the soul is and finds in itself in the anthropological sphere" (III.19).[25] Precisely the same structure is elicited from the relationship between self-consciousness and consciousness (III.41), and the struggle for recognition (III.59). This is, of course, a perfectly general structural interrelationship, basic to the system as a whole, and deriving no

[24] Phil. Nat. II.13: "Light corresponds to this identity of self-consciousness, and is the exact image of it. It is not the ego however, for in itself it is not self-dimming and refractive, but is merely abstract appearance. If the ego were able to maintain itself in a state of undisturbed equability, as the Indians would like it to, it would pass away into the abstract transparency of light."
[25] Cf. III.93, 25.

particular significance from our having become aware of it in Hegel's survey of categorial relationships or exposition of the levels of consciousness.[26]

If we remember that the categories of being constitute the beginning of the Logic as such (Enc. §§ 84–111), when Hegel comments as follows upon the nature of self-consciousness, he would appear to be implying that subjectivity is an even more extreme level of abstraction: "Since the ego here is still not determined at the same time as being generally objective, there is a lack of reality, of determinate being. This deficiency may also be said to be the initially wholly abstract nature of self-consciousness, of ego-ego, of my being for myself. The deficiency here consists of self-consciousness, the further forms of which now have to be considered, being wholly abstract. Such abstract self-consciousness is merely subjective, posited merely subjectively, still without being, not being the determinate being of self-consciousness" (III.317–319). The same inference might be drawn from the discussion of the logical categories that emerge in perceptive consciousness: "Perceptive consciousness involves seizing the general object, no longer immediately, but as mediated and as self-relating in the mediation. Through this there is a mediation of what is sensuous and of thought-determinations, and the thoughts, the categories emerge in the relation. The relating of that which as such is a manifold pertains to the unity of the ego, and these relations are categories, thought-determinations in general. We have, therefore, sensuous determinations and thought-determination" (III.307). The truth of the matter would appear to be, however, that Hegel is simply calling attention to the sort of categories employed in the extremely abstract cognition characteristic of consciousness, in the same way as he calls attention to those employed by the understanding (III.139), or implicit in the grammar of a language (III.181). In that the treatment of the natural ego in the Phenomenology constitutes a level in the whole sphere or hierarchy of Spirit, and Spirit is the immediate presupposition of the Logic, the ego can be regarded as one of the presuppositions of the Logic. It is not, however, to be confused with the subjectivity constituting the teleological unity of the Logic as such, and having the categories of Being and Essence as the immediate presuppositions of its own particular sphere (Enc. §§ 160–244).

The general progression in degree of objectivity throughout the Phenomenology is one in which that which confronts the ego approximates to it with an increasing degree of adequacy. In Sensuous consciousness, Perception and Understanding, the ego is confronted with what simply is (III.23), with what is connected (III.29), with what is lawlike (III.33). This is the epistemological world of the infinite revisability of Popperian science, and indeed, of the empirical foundation of Hegel's own philosophy of the natural

---

[26] L. B. Puntel, 'Darstellung, Methode und Struktur' (Bonn, 1973) thinks otherwise, and ascribes a fundamental systematic significance to the Phenomenology and Psychology.

sciences.[27] Through the discovery of natural laws, which are "the determinations of the understanding dwelling within the world itself... the understanding consciousness rediscovers itself and so becomes its own general
object" (III.33).[28] At the level of Self-consciousness, the ego is confronted
with living being, which is unlike the world of mechanical or chemical laws
in that the interdependence of the whole and the parts is unmistakable: "In
living being however, consciousness intuites the very process of the positing
and sublating of the different determinations, perceives that the difference
is not a difference, that is to say, that it is not absolutely fixed. For life is that
inner being which does not remain abstract, but enters wholly into its
expression" (III.35).[29] Finally, in Reason the ego recognizes itself, that is
to say, all that its own recognition presupposes, in that which confronts it.[30]
It has, therefore, the self-justifying or self-explaining circularity of the logical
category Hegel calls the Idea (Enc. §§ 213–244), which is the central principle of the whole philosophical system. In Reason the ego is no longer
confronted with anything inadequate or alien to itself, the subjectivity of the
Notion has resolved itself in the Idea, and Phenomenology exhibits itself as
the immediate presupposition of Psychology: "Reason is not the unity of the
object which occurs in consciousness however, nor is it the unity of consciousness which occurs in self-consciousness, it is the Idea, the actively
effective Idea, and it is therefore the unity of the Notion in general with
objectivity. Self-consciousness is as the Notion which is for itself, as free
Notion, which is ego as it is for itself. The ego is the Notion, although not as
the indwelling Notion of the Sun, the animal, the plant, which is inseparable
from the external reality. In self-consciousness the Notion is mine, abstractly
for itself, and the reality opposed to self-consciousness is consciousness, ego
as relating itself to an object. We have here the unity of consciousness as
objectivity, i.e. the Notion within reality as self-consciousness. At this
juncture, reality is self-consciousness, ego in relationship with an object
presented to it externally. Reason in general is the Idea, for the Idea is
reason: we do not possess the Idea, it possesses us, so that reason also possesses
us, being our substance" (III.349–351).[31]

It should not be overlooked that the exposition of this general progression
in degree of objectivity by means of which that which confronts the ego
approximates to it with an increasing degree of adequacy, is the outcome of
an analytical and synthetic procedure on Hegel's part. Taking the extreme
subjectivity of the ego as his starting point, and the rational resolution of

[27] Phil. Nat. I.193; 201.
[28] Cf. III.311.
[29] Cf. III.313.
[30] See III.371, note 47, 11.
[31] Strangely enough, Boumann failed to include anything of Hegel's lectures on
Reason in his edition of this section of the Encyclopaedia (III.77).

the subject-object antithesis as his telos, he analysed the intervening levels into degrees of complexity, and then synthesized the results of this analysis in the dialectical exposition. He refers to this as a 'logical' procedure: "For consciousness, the progressive determination of the ego assumes the appearance of an alteration of the object which is independent of the ego's activity, so that in the case of consciousness the logical consideration of the alteration still fell in us alone" (III.89). There is some evidence that on account of his own earlier preoccupation with Kantianism, the development of this procedure, which is simply the general one employed throughout the Encyclopaedia, caused him some difficulty in this sphere (III.457).

The best known part of the Phenomenology is the master-servant dialectic by means of which Hegel illustrates the relationship between certain levels of Self-consciousness (III.37–69).[32] This *illustrative* material serves the *didactic* purpose of making an abstract and difficult subject more easily understandable, and unlike the subject-matter of the rest of the Encyclopaedia, is not to be regarded as having been given its *systematic* placing in this sphere of Subjective Spirit. Anthropological observations, social contract theory, historical events, Defoe's classic, religious attitudes, are all touched upon, but merely in order to illustrate the main exposition, not in order to establish its significance. The 'desire' dealt with at the first level (III.43–53) for example, is not to be confused with the animal's instinctive drive to satisfy need (Phil. Nat. III.145), the intelligent drive of practical spirit (III.249), or the desire to acquire property (Phil. Right §§ 54–58). At this juncture, the state of nature (III.57–59) is more the natural soul being negated by the ego than the postulate of the political theorists (III.374).[33] Hegel does not say that the relationship between the two egos *is* a struggle for recognition which resolves itself in the relationship of mastery and servitude, but that it is to be *presented* as such (III.331), and Boumann confirms the accuracy of Griesheim and Kehlers' reporting here: "In order to avoid eventual misunderstandings of the point of view just presented, it has also to be observed that the struggle for recognition in the extreme form in which it is here presented can occur only in the state of nature, in which men are simply singular beings. It remains alien to both civil society and the state, within which the recognition constituting the result of this struggle is already present" (III.59).

What suggested the master-servant dialectic to Hegel in the first instance is not known. His earliest exposition of it is in the 'System der Sittlichkeit' (1802), where it occurs in the context of such economic and personal relationships as are dealt with later in the Philosophy of Right §§ 189–208. In the first sketch of a system (1803/4) it is given a rather more extended treatment in much the same sort of context, whereas in the third sketch (1805/6),

[32] Cf. III.314–345.
[33] Cf. III notes 53, 33; 57, 6.

although the context remains the same, the treatment is cursory in the extreme. In each of these early instances therefore, it is evidently conceived of as an analysis of the *actual* personal relationships that subsist within society. In the Phenomenology of 1807 however, as in all its subsequent occurrences, it is simply used in order to illustrate a level of consciousness. Since the expositions of it by Hegel's immediate followers and commentators were neither informed nor illuminating, it is not surprising that such an impatient and incisively constructive critic as Marx should have felt free to criticize him for having presented it as a dialectic of consciousness, and not as a real dialectic of socio-economic relationships. The *systematic* placing of Marx's fundamental concern is to be found not in the Phenomenology of 1807 or 1830 however, but in the Philosophy of Right, within which clear confines most of his main ideas are already formulated, and quite evidently pleading for practical realization. With regard to the *subject-matter* of the Encyclopaedia therefore, Hegel's initial assessment of the master-servant dialectic was without any shadow of doubt superior to what Marx thought he was criticizing (III.377–379).

f. PSYCHOLOGY

And so we may observe, how the Mind, by *degrees*, improves, and *advances* to the Exercise of those other Faculties of *Enlarging, Compounding*, and *Abstracting* its *Ideas* and of reasoning about them. — LOCKE.

Since Hegel devotes so much of the introduction to the Encyclopaedia to outlining the significance of basing philosophy upon commonsense realism and empiricism, it is understandable that when the Prussian Ministry of Education asked him to make suggestions as to the subjects most suitable for serving as an introduction to philosophy, he should have recommended Logic and Psychology.[1] Taken together as they are expounded in the Encyclopaedia, these two subjects not only lead on into a systematic consideration of the natural and political sciences, but also provide us with a radical and realistic analysis of the methodology basic to their own exposition. By analysing, structuring and systematizing logical categories and psychological phenomena, we bring philosophical insight to bear upon the initially uncritical adoption[2] of the commonsense realism and empiricism basic to the whole philosophic enterprise. Although these disciplines can hardly be regarded as constituting the *beginning* of an essentially self-sustaining system,

[1] April 16th, 1822: 'Berliner Schriften' (ed. J. Hoffmeister, Hamburg, 1956), pp. 541–556.
[2] In Enc. § 25 Rem., Hegel admits that it is, and apologizes, pleading practical necessity.

they should certainly be pointed out as providing us with the readiest and most well-founded *introduction* to it.[3]

The Logic is particularly well-suited for beginners in that the main dialectical thread of its exposition may be pursued without direct reference to anything concrete. It is true that its pure abstractions are in fact derived from the concrete sciences and the history of philosophy, but they may be understood in their systematic setting without any knowledge of their empirical origin or use. Deplorable though it may be that anyone should be idle, foolish or ignorant enough to pursue the study of Logic without reference to the concrete contexts within which logical categories and judgements occur, it has to be admitted that it is not only convenient to do so for didactic purposes, but that if a teacher can prevent the young and inexperienced from misassessing the significance of any skills they might acquire, the benefits they can derive from thinking clearly and constructively in abstract terms are very considerable: "Logic is in one respect the most difficult of the sciences, for whereas even the abstract presentations of geometry have a sensuous aspect, it is concerned not with intuitions but with pure abstractions, and demands a force and facility of withdrawing into pure thought, holding fast to and moving within it. On the other hand, it can also be regarded as the easiest of the sciences, since its content is nothing other than the simplest and most elementary of the familiar determinations of our own thought" (Enc. § 19). To introduce the young to the study of philosophy via Anthropology, would be to face the same problems as those encountered in every other sphere which requires a prior knowledge of the empiricism being structured and systematized. To introduce them to it via Phenomenology is to run the risk of their never finding their way out of the subject-object problems raised, it being so much easier to fossick about in epistemology than to get down to the hard grind of mastering empirical disciplines. In Psychology, however, as in Logic, one is dealing with a subject-matter which is immediately familiar, devoid of potentially unmanageable didactic dangers, and quite evidently relevant to a wide range of other disciplines.

It was, however, Psychology's involvement with other disciplines, which in Hegel's view had given rise to one of the main difficulties encountered when attempting to treat it systematically, expound it philosophically: "Psychology, like Logic, is one of those sciences which have profited least from the more general cultivation of spirit and the profounder Notion of reason distinguishing more recent times, and it is still in a highly deplorable condition. Although more importance has certainly been attached to it on

[3] The failure to distinguish between what Hegel might have regarded as the *beginning* of the system, and what he thought might constitute the best *introduction* to it, has reduced the usefulness of the very important debate opened up by H. F. Fulda's 'Das Problem einer Einleitung in Hegels Wissenschaft der Logik' (Frankfurt/M., 1965).

account of the direction given to the Kantian philosophy, this has actually resulted in its being proffered as the basis of a metaphysics, even in its empirical condition, the science here consisting of nothing other than the facts of human consciousness, taken up empirically simply as facts, as they are given, and analysed. Through being so assessed, psychology is mixed with forms from the standpoint of consciousness and with anthropology, nothing having changed in respect of its own condition. The outcome of this has simply been the abandonment of the cognition of the necessity of that which is in and for itself, of the Notion and truth, not only in respect of spirit as such, but also in respect of metaphysics and philosophy in general" (III.99).

The root of this problem in the immediate post-Kantian period lay in the failure to distinguish effectively between the subjectivity involved in unifying logical categories, and the natural ego constituting the subject-matter of the Phenomenology and the immediate presupposition of Psychology. Kant himself had certainly distinguished between the logical and natural aspects of subjectivity, but he had also equated the transcendental ego with the conscious unity of apperception.[4] Hegel criticizes him severely for this in the lectures on the History of Philosophy: "Since Kant shows that thought has synthetic judgements *a priori* which are not derived from perception, he shows that thought is so to speak concrete in itself. The idea which is present here is a great one, but, on the other hand, quite an ordinary signification is given it, for it is worked out from points of view which are inherently rude and empirical, and a scientific form is the last thing that can be claimed for it... Kant remains restricted and confined by his psychological point of view and empirical methods" (III.430–431).[5] Kant and his followers had confused Logic and Psychology because they had simply accepted the empirical fact of their being united in subjective cognition, and had failed to analyse the *systematic* relationship between them. In order to grasp the significance of Hegel's criticism of their work, it may be of value to spend a moment or two comparing it with his criticism of Newtonian Optics. — It is perfectly understandable that the physicists of Hegel's day, and indeed of our own, should have treated the study of light and the study of colours as closely related disciplines. Goethe had shown, however, that the circumstances in which we *perceive* colours cannot be explained simply by

---

[4] 'Critique of Pure Reason', B 407/8: "That the I or Ego of apperception, and consequently in all thought, is singular or simple, and cannot be resolved into a plurality of subjects, and therefore indicates a logically simple subject — this is self-evident from the very conception of an Ego, and is consequently an analytical proposition... The proposition of the identity of my Self amidst all the manifold representations of which I am conscious, is likewise a proposition lying in the conceptions themselves, and is consequently analytical." Cf. A 107; B 143.

[5] Cf. Phil. Sub. Sp. III.388.

means of the Newtonian theory of the composite nature of white light. He was wrong to conclude from this that the Newtonian theory was erroneous, but right to insist that it needed revision. As Heisenberg has pointed out, the Newtonian and Goethean theories of colour are in fact complementary, Newton's being concerned with its purely physical nature, and Goethe's with the more complex chemical, physiological and psychological factors involved in our actual perceptions of and reactions to chromatic phenomena.[6] Hegel, not as careful as he should have been in distinguising all the levels of complexity implicit in Goethe's field of empirical research, concluded from his experiments that it is the various *physical* circumstances in which light and darkness are combined which give rise to colour. Taking light itself, in its simplicity, to be the initial, the most basic, the most universal level of Physics (Phil. Nat. II.12–25), he did not, therefore, deal with colour until he had worked out the dialectical exposition of the further physical levels he considered to be involved in its production (Phil. Nat. II.135–160). What might reasonably be regarded, on account of their empirical occurrence, as closely related fields of research, are therefore presented within the overall dialectical exposition at widely separated levels.

Precisely the same basic procedure of resolving empirical confusions by distinguishing levels of complexity gave rise to the widely separated spheres of Logic and Psychology in the Encyclopaedia, and the criticism of Kant's general treatment of subjectivity. The logical unity of apperception, the focal point of Kant's categories, reappears, as Düsing has shown, as the teleologically unifying principle of subjectivity in Hegel's Logic. The conscious unity of apperception, the focal point of Kant's subjective cognition, reappears as the ego of Hegel's Phenomenology, and resolves itself, through the dialectical exposition of this sphere, into the rational presupposition of the sublated subject-object antithesis which initiates Psychology proper. On Hegel's view therefore, Psychology is certainly more complex than, and so presupposes, logical categories, just as it presupposes Phenomenology, Anthropology and the whole sphere of the natural sciences. On the other hand, the systematic exposition of the abstract categories of the Logic presupposes the History of Philosophy as well as Psychology, and is therefore to be *approached systematically* through the Philosophy of Spirit (Enc. §§ 440–577). The whole circularity or self-sustaining nature of the system is therefore apparent in the relationship between these two spheres.

Hegel himself had some difficulty in mastering the full implications of it. Although one might take the broad division between the faculty of cognition

---

[6] W. Heisenberg, 'Die Goethesche und die Newtonsche Farbenlehre im Lichte der modernen Physik', in 'Wandlungen in den Grundlagen der Naturwissenschaften' (4th ed., Leipzig, 1943), pp. 58–76.

and cosmology in the 1794 notes on the subject[7] as the forerunner of the later distinction between Theoretical (III.103–229) and Practical (III.231–265) Psychology, it is the two 'potencies', roughly corresponding to theory and practice, distinguished in the 'System der Sittlichkeit' (1802), which provide us with the first clear evidence of the mature structuring of the sphere. The precise determination of Psychology in respect of Phenomenology and Social or Political Philosophy took longer however, for although something resembling the later transition from Phenomenology to Psychology is already formulated in the system sketched in 1803/4,[8] material later dealt with in Psychology is classified with that subsequently included within the spheres of Anthropology, Phenomenology and Objective Spirit, throughout the whole of the Jena period. In fact the Psychology was the very last of the major spheres of the Encyclopaedia to be given a triadic structure, and Hegel only managed to elicit it from its subject-matter by revising the transition to Objective Spirit in the 1830 edition (III.452). The reform of the subject was certainly on his mind while he was working at the Longer Logic, and he seems to have planned a similar full-scale book on it, for we find him writing to his friend Niethammer in the October of 1811 that he hoped to be able to publish his Logic the following spring, and that this would then be followed by his Psychology.[9] Almost a decade later he mentions the same project in the introduction to the Philosophy of Right, calling attention to the importance of its purely systematic aspect, and its relevance to political philosophy.[10] Not only did he never manage to write such a work, but he also failed to devote very much attention to the subject in the lecture-room. Placed as it is at the end of the sphere of Subjective Spirit, it tended to be increasingly neglected as his interest in Anthropology grew throughout the 1820's. In fact of the forty-five lectures on the whole sphere given in the Summer Term of 1825, only three were devoted to Psychology. As a result of this, there has been considerable uncertainty among his followers as to the precise nature and significance of his proposed reform. The elementary terminological confusion among those of his friends and contemporaries who attempted to expound his views is little less than disastrous when one takes into consideration the labour he expended upon clarifying this basic aspect of the field. Carl Daub (1765–1836), for example, who gave sympathetic and detailed expositions of this part of the system in his lectures, treated it under the heading of Consciousness, and had his work published as an Anthropology, and J. E. Erdmann (1805–1892) in-

---

[7] 'Zur Psychologie', to be published in 'Gesammelte Werke', vol. I, pp. 167–192, but at present still most conveniently accessible in J. Hoffmeister, 'Dokumente zu Hegels Entwicklung' (Stuttgart, 1974).

[8] 'Jenaer Systementwürfe I', pp. 273–296.

[9] 'Briefe von und an Hegel' (ed. J. Hoffmeister, 4 vols., Hamburg, 1969), I.389.

[10] Phil. Right § 4.

cluded the whole doctrine of Subjective Spirit under the general heading of Psychology.[11] No one has seen the full significance of presenting the Phenomenology as the immediate presupposition of Psychology, and despite the rather obvious procedure of treating psychological phenomena as the immediate presuppositions of the law and political philosophy, the fact that Hegel refers to his Psychology in the opening Paragraphs of the Philosophy of Right can still be regarded as surprising, even by informed students of his works.[12]

When he criticizes the "direction given to the Kantian philosophy" in respect of Psychology, Hegel almost certainly has in mind thinkers such as J. F. Fries (1773–1843) and F. E. Beneke (1798–1854), who were attempting to interpret all philosophy, including not only logic but also the treatment of things in themselves, as a psychological science of experience, a psychic anthropology (III.389). In the introduction to the treatment of Intelligence however, he criticizes an older eighteenth century theory, which was still current in the 1820's: "One form favoured by reflection is that of the powers and faculties of the soul, intelligence or spirit. — Like power, faculty is represented as being the fixed determinateness of a content, as introreflection. Although power is certainly infinity of form, of inner and outer, its essential finitude involves the content's being indifferent to the form. In this lies the irrationality which, by means of this reflectional form, is introduced into spirit when it is considered as a multitude of powers, as it is into nature through the concept of forces. Whatever is distinguishable in the activity of spirit is defined as an independent determinateness, a procedure which results in spirit's being treated as an ossified and mechanical agglomeration. And it makes no difference whatever if activities are spoken of instead of faculties and powers, for the isolating of activities also involves treating spirit as nothing but an aggregation, and considering their relationship as an external and contingent relation" (III.107). C. A. Crusius (1715–1775) had distinguished between empirical psychology and noology, or the science of the powers of the human mind, which he thought could be treated as a kind of logic.[13] The distinction became a commonplace, and Kant in his 'Anthropology' also made good use of the power or faculty concept, but by the turn of the century it was being criticized from various standpoints, many of which, like Hegel's, involved emphasizing the unity of the soul. Herbart is interesting in this respect, in that although he also called in

[11] C. Daub, 'Vorlesungen über die philosophische Anthropologie' (Berlin, 1838); J. E. Erdmann, 'Grundriss der Psychologie' (Leipzig, 1840). For an interesting survey of later views, see H. Drüe, 'Psychologie aus dem Begriff' (Berlin, 1976), pp. 16–29.
[12] 'Hegel-Studien', vol. 1, p. 10, 1961.
[13] The terminology and the concept have recently been revived by L. B. Puntel, op. cit. (p. 145 et seq.), in an attempt to establish a connection between the Logic and the philosophies of Nature and Spirit (p. 31 et seq.).

question the reality of faculties by insisting upon the simplicity and unity of the soul, he introduced a very similar concept of "presentations which, in that they penetrate one another in the one soul, check one another in so far as they are opposed, and unite into a composite power in so far as they are not opposed", which he proceeded to expound mathematically in terms of statics and mechanics.[14] Beneke accepted this criticism of faculties, but rejected Herbart's presentations in favour of certain pre-dispositions or impulses, which are harmonized in the self. Hegel had no very high opinion of Beneke,[15] but he shared more common ground with Herbart than he perhaps realized (III.389), and we know from the Berlin University records that his attitude to these contemporary developments in the field was not quite so straightforward as one might assume from what he published in the Encyclopaedia. He comments as follows upon Stiedenroth's Psychology for example: "The author thinks too highly of himself for having rejected the form of psychic faculties formerly common in Psychology — and in any case it is a form which Fichte and others have already replaced by the category of activity. It is a matter of small importance whether one heads a chapter 'The Faculty of Recollection' or simply 'Recollection'."[16] It is clear, however, from his criticism of Stiedenroth as well as his commendation of such an entirely insignificant work as Mussmann's 'Lehrbuch der Seelenwissenschaft',[17] that it was the lack of a significant hierarchical structure in the arrangement of their subject-matter which he regarded as the primary fault of the text-books of the time. As he saw it, a philosophical exposition of Psychology could not dispense with the attempt to range its subject-matter in degree of complexity, ideally, as he had indicated, within the general sphere delimited by Phenomenology and the Law. He could, perhaps, hardly have expected there to be any general recognition of the significance of his formulation of the transition from Phenomenology to Psychology, but the significance of the Psychology-Law relationship, expounded as it was in quite a number of contemporary works (III.387), could well have been more widely recognized. The closest contemporary approximation to his philosophical Psychology is, however, to be found in such early and relatively obscure publications as G. A. Flemming's 'Lehrbuch der allgemeinen empirischen Psychologie' (Altona, 1796), or J. H. Abicht's 'Psychologische Anthropologie' (Erlangen, 1801), neither of which is in itself of any philosophical interest. In fact his work in this particular field tends to have some

[14] J. F. Herbart (1776–1841), 'Lehrbuch zur Einleitung in die Philosophie' (Königsberg, 1813; 2nd ed. 1821), § 158.
[15] 'Berliner Schriften' (ed. Hoffmeister), pp. 612–626.
[16] ibid., pp. 569–70; E. Stiedenroth (1794–1858), 'Psychologie' (2 pts. Berlin, 1824/5).
[17] ibid., pp. 570, 646: J. G. Mussmann (1798?–1833), 'Lehrbuch der Seelenwissenschaft' (Berlin 1827).

affinity with that of the eighteenth century Leibnizians, and there is evidence that he was aware of this.[18]

Before leaving the general topic of Hegel's assessment of the Psychology of his time, it should perhaps be noted, that although it was bound, on account of its origin in the overall philosophical system, to call many current developments in question, especially in so far as they were rooted in certain Kantian confusions, it also found important points of contact with contemporary views. We have already noticed that like Herbart and his school, Hegel emphasized the unity of the subject-matter being dealt with in Psychology. Interestingly enough, he evidently followed Kant in distinguishing between the pure or logical intuition of the Idea, and the empirical intuition to be dealt with in Psychology. As one might have expected from the Phenomenology, he differs from Kant in regarding empirical intuition not as simply mediating between matter and knowledge, but as also having matter as its presupposition. He saw that there could be no empirical or psychological intuition without it itself presupposing the material world, that a systematic or philosophical treatment of intuitive cognition must, therefore, involve recognition of the fact that in its 'other', such cognition also intuites this basic aspect of itself, and that unlike consciousness, it is therefore no longer merely confined to a subject-object antithesis. He realized in fact, that Kant's 'objects' do not simply 'affect' the 'faculty of representation', but that this faculty is what it is in that it also involves some awareness of these objects as the presuppositions of its own being (III.394). His treatment of Thought concludes the sphere of Intelligence initiated by the treatment of Intuition, and once again we find him referring to Kant, this time in order to praise his distinction between understanding and reason. He adds, however, that Kant had not grasped the "variegated forms of pure thought", the various subsidiary levels at which subject and object are identical. In that these levels are sublated in rational thought, the sharp distinction between an understanding "unifying phenomena by virtue of rules" and a reason giving unity to the understanding but "never applying directly to experience, or to any sensuous object"[19] loses something of its significance. Looking back over the general sphere of Intelligence as analysed by Hegel, one notices that the recollected image, "requires the determinate being of an intuition" (III.155), that the reproductive memory, "possesses and recognizes the matter in the name" (III.201), and both these levels can be rationally sublated within thought. Consequently, although such rational thought may not apply directly to experience, it does so through the intermediary of what it sublates, and unlike the understanding, does not "fall apart into form and content, universal and particular" (III.225).[20]

[18] ibid., p. 570. Cf. M. Dessoir, 'Geschichte der neuern deutschen Psychologie' (Berlin, 1902), p. 377 et seq.
[19] 'Critique of Pure Reason', B 355–B 359.
[20] Cf. note III.440.

In the lectures on the History of Philosophy, he criticizes Locke in the same way as he criticizes Fries and Beneke, for psychologizing philosophy in general, for making a theory of knowledge the be all and end all of it, for being, in his own terms, such an inveterate phenomenologizer: "From Locke a wide culture proceeds, influencing English philosophers more especially... It calls itself Philosophy, although the object of Philosophy is not to be met with here... When experience means that the Notion has objective actuality for consciousness, it is indeed a necessary element in the totality; but as this reflection appears in Locke, signifying as it does that we obtain truth by abstraction from experience and sensuous perception, it is utterly false, since, instead of being a moment it is made the essence of the truth" (III.295–296). His low opinion of our philosophical culture was not entirely justified however, for whatever our self-styled philosophers may have been up to, our empiricists had remained unimpressed, and during the opening decades of the nineteenth century many of our psychologists were also engaged in calling in question the whole Lockean and Humean conception of the association of ideas.[21] One has to admit that they were not doing so for entirely the same reasons as Hegel, but nevertheless their empirical work does link up in an interesting manner with his broader philosophical insight. It was not the concept of association as such that he objected to (III.201), but the loose definition of 'ideas' and the woolly conception of 'law' involved in the general talk about the laws of the association of ideas: "These have attracted a great deal of interest, particularly during the flourishing of empirical psychology which accompanied the decline of philosophy. Yet in the first place it is not ideas that are associated, and in the second place the modes of relation are not laws, the precise reason for this being that the same matter is subject to so many laws, that what occurs tends to be quite the opposite of a law in that it is capricious and contingent — it being a matter of chance whether the linking factor is an image or a category of the understanding such as equivalence and disparity, reason and consequence etc. In general, the sequence of images and presentations in the associative imagination consists of the play of a thoughtless presenting in which the determination of intelligence still constitutes formal universality in general, while the content is what is rendered in the images" (III.159). It is also interesting to note that English reviewers received John Millard's version of Gregor von Feinaigle's 'Mnemonik' in much the same way as Hegel (III.203), one of them condemning a similar home-grown product as, "about as valuable as a catalogue of past snow-storms".[22]

With regard to the basic structure of Hegel's Psychology, it is absolutely essential to grasp the fact that it is concerned with a progression in degree

---

[21] R. Hoeldtke, 'The History of Associationism and British Medical Psychology' ('Medical History', vol. 11, pp. 46–65, 1967).
[22] III.432: 'Quarterly Review', vol. IX, p. 135, March, 1813.

of complexity, and not with a natural development. When Locke, in the passage quoted at the head of this chapter,[23] spoke of our observing how the mind progresses by degrees from sensations to reasoning, he was thinking in terms of a temporal sequence. Such a sequence may in fact coincide with a systematic one, but this is by no means a matter of necessity. Hegel was therefore cautious about making too much of evolutionary theories in the Philosophy of Nature,[24] and he is equally cautious, although not, be it observed, dismissive, when he discusses the implications of Condillac's assessment of sensations. Condillac had been influenced by Locke's postulation of sensation and reflection as the two fundamental sources of ideas. In his 'Traité des sensations' (Paris and London, 1754), he made the point that all mental operations and functions are reducible to sensations, and being left with the problem of our knowledge of the external world, solved it to his own satisfaction by locating the external origins of sensations. His ideas were well known in Germany on account of the works of Michael Hissmann (1752–1784), professor at Göttingen, and were generally used in order to foster a physiological approach to psychology (III.384). Hegel, in the introduction to his Psychology, comments upon them as follows: "In this context one is not to think of the development of the individual, for this is involved in what is anthropological, and in accordance with it faculties and powers are observed to emerge in succession and to express themselves in existence. On account of Condillac's philosophy, there was a time when great importance was attached to the comprehension of this progression, it being assumed that such a conjectured natural emergence might demonstrate how these faculties arise and explain them. Although it is not to be denied that such an approach holds some promise of grasping the unity of spirit's multifarious modes of activity and indicating a necessary connectedness, the kind of category it employs is entirely inadequate to this. The overriding determination is broadly justified in that the sensuous is regarded as primary, as the initial foundation. It is assumed however, that the appearance of the further determinations proceeds forth from this starting point in a merely affirmative manner, that which is negative in the activity of spirit, that whereby this material is spiritualized and its sensuousness sublated, being misconstrued and overlooked. The approach assumes that what is sensuous is not only the empirical prius but also persists as such, so constituting the true and substantial foundation" (III.91).

The subject-matter of Psychology is to be treated within this sphere and not within Anthropology or Phenomenology on account of its no longer involving a subject-object antithesis. If we are concerned with the disparity between subjectivity and objectivity in intuition or language for example, we shall therefore have to classify our field of enquiry as anthropological or

[23] 'Human Understanding', bk. II, ch. i.
[24] Phil. Nat. I.25; III.229; 366.

phenomenological, not psychological. These forms of cognition become the subject-matter of Psychology only in so far as they constitute levels at which cognition is what is cognized and vice versa. It is in the Psychology, therefore, that Hegel analyses the commonsense realism basic to the whole philosophic enterprise, and it should be noted that although Anthropology and Phenomenology are included within the general sphere of Spirit, he takes Psychology to be the beginning of Spirit proper, the initial juncture at which what is natural and therefore to some extent apparently external to Spirit, is effectively sublated (III.381). In Psychology, therefore: "Spirit is no longer susceptible to natural change, no longer involved in the necessity of nature but in the law of freedom. It is no longer soul, influenced externally, and is concerned not with any general object but simply with its own determinations. It is self-relating" (III.382). In this sphere the general principles of philosophical systematization are therefore dealing with a subject-matter which itself exhibits their own freedom from subject-object disparities. After pointing out that this was not the case in the Phenomenology, Hegel goes on to characterize the situation in the Psychology as follows: "For free spirit however, it is free spirit itself which brings forth from itself the self-developing and altering determinations of the object, making objectivity subjective and subjectivity objective. The determinations of which it is conscious, while certainly dwelling within the object, are at the same time posited by itself. Nothing within it is mere immediacy. Consequently, when the 'facts of consciousness' are spoken of as if for spirit they were primary and must remain for it as an unmediated factor with which it is simply presented, it has to be observed that although a great deal of such presented material certainly occurs at the standpoint of *consciousness*, the function of free *spirit* is not to leave these facts as independent, presented factors, but to explain them by construing them as acts of spirit, as a content posited by itself" (II.89).[25] Psychology is therefore "the unity of the soul and of consciousness" (III.99), and just as the ego had the natural soul not only as its presupposition but also as its general object, so spirit or the psyche has consciousness as both its presupposition and its object, the difference being that this object is quite clearly what it is itself: "Just as consciousness has the preceding stage of the natural soul as its general object, so spirit has consciousness as its general object, or rather makes it so; i.e. in that it is only implicitly that consciousness is the identity of the ego with its other, spirit so posits their being-for-self that it now knows them to be this concrete unity. The productions of spirit accord with the rational determination that in that it is free, both the content and what is implicit belong to it" (III.93–95). This parallel structure between Phenomenology and Psychology gives rise to an interesting comparison between the Sensuous consciousness basic to the one (III.19–25), and the Intuition basic to the other (III.117–145). Hegel notes that the

[25] Cf. III.111, 30.

similarity could give rise to confusion, and draws attention to the qualitative difference between the two levels elucidated and established by means of the analytical and synthetic procedures of the dialectic: "In respect of intuition's relationship to consciousness however, we have to observe that if one used the word 'intuition' in its broadest sense, one could of course apply it to the immediate or sensuous consciousness already considered in § 418. Rational procedure demands that the name should be given its proper meaning however, and one has therefore to draw the essential distinction between such consciousness and intuition. The former, in unmediated and wholly abstract certainty of itself, relates itself to the immediate singularity of the object, which falls apart into a multiplicity of aspects. Intuition, on the contrary, is a consciousness which is filled with the certainty of reason, its general object having the determination of being a rationality, and so of constituting not a single being torn apart into various aspects, but a totality, a connected profusion of determinations" (III.137).

The precise systematic placing of Psychology is, therefore, the direct outcome of the presentation of the ego as the telos of the Anthropology, and the dialectical resolution of the subject-object antithesis in the Phenomenology: "Psychology is therefore concerned with the faculties or general modes of the activity of spirit as such, — intuiting, presenting, recollecting etc., desires etc. It is not concerned either with the content occurring after the appearance in empirical presenting and in thinking, in desiring and willing, nor is it concerned with the two forms here, that in which the soul is a natural determination, and that in which consciousness is itself present to consciousness as a general object which is for itself. Psychology is not an arbitrary abstraction however; the Notion of spirit has made it apparent that spirit is precisely this elevation above nature and natural determinateness, above involvement with a general external object, above material being in general. All it now has to do is to realize this Notion of its freedom, that is to say, sublate the form of immediacy with which it starts once again. The content, which is raised into intuitions, consists of its sensations, intuitions and so on changed into presentations, its presentations changed into thoughts etc." (III.79). To confuse sensuous consciousness with intuition, to raise subject-object problems in respect of recollection, language, memory or thought, is to become involved in the subject-matter of consciousness already dealt with in the Phenomenology. Psychology is a distinct discipline in that it presupposes the sublation of such problems in the subject-matter it deals with.

As early as the lectures of 1803, Hegel takes the essence of spirit to be its sublation of its otherness, or rather, its having sublated its otherness and recognized it as being what it is itself.[26] Since such a conception could only have been arrived at if he had already been in command of the main principles of his mature system, it is not perhaps surprising that although the

[26] See the forthcoming 'Gesammelte Werke', vol. 5, p. 24. Cf. Phil. Nat. III.213.

precise classification and critical assessment of the subject-matter of Psychology may not have matured very early, its broad outlines did. In the first sketch of a system (1803/4), the consciousness which organizes itself into language is taken to be the theoretical unity basic to the 'potencies' corresponding to the two major divisions of the mature sphere. In the third (1805/6), the motion of free intelligence generates that which is theoretical in the form of the image and recollection, neither of which involves the reception of anything alien.[27] The 'theoretical spirit' of the 1817 Encyclopaedia (§ 368) is intelligence, knowledge which presupposes the rationality of possessing what is given as its own. These earlier uses of the word theoretical are built into the mature exposition of the Psychology, 'theoretical spirit' being synonymous with 'intelligence' throughout these lectures. What is more, every level of the first major sphere of 'Theoretical Spirit' (III.103–229) presupposes the sublation of subject and object in precisely the same way as the exposition of spirit does in the lectures of 1803. As has already been observed, this does not imply that the image involved in recollection always corresponds to what has been intuited (III.149), that words are always used rationally (III.179), or that it is always the case that what is thought is (III.217), but rather that if there are disparities between intuition and recollection, reason and language, being and thought, as there most certainly are for example in dreams (II.233), mental derangement (II.327) or sensuous consciousness (III.19), then the 'recollection', 'language' and 'thought' involved have to be considered in anthropological or phenomenological, and not in psychological terms.

To approach the subject-matter of Psychology without having grasped the conception of spirit's having sublated its otherness is, therefore, to be disqualified at the outset from treating it philosophically: "It is not for philosophy to take such imperfections of determinate being and presentation for truth, to regard what is bad as the nature of the matter. The numerous other forms employed with reference to intelligence, — it is said that it receives or takes up impressions from without, that the operations of external things cause the emergence of presentations etc., belong to a standpoint involving categories which do not pertain to the standpoint of spirit and of philosophic consideration" (III.105–107). Within this sphere, therefore, the systematic progression is such that the *apparent* disparity between subject and object is sublated with an increasing degree of completeness: "The usual interpretation of this development of spirit from sensation is however, that since intelligence is originally entirely empty, all content is derived by it from without as something which is entirely alien to it. This is erroneous, for what intelligence appears to take up from without is in truth nothing other than that which is rational, and which is therefore identical with spirit and immanent within it. Consequently, the sole purpose of the activity of

[27] 'Systementwürfe' I.297–298; II.201.

spirit is to sublate the apparent self-externality of the implicitly rational object, and so refute even the apparency of the general object's being external to spirit" (III.123).[28] The central principle of this systematic progression in the sphere of 'Theoretical Spirit' is that of an increasing degree of inwardness. Attention is an inwardizing of intuition for example (III.125, 4), and presentation is to be interpreted from the same point of view (III.145, 15). The images of recollection are more directly related, through intuition and sensation, to the particularity of natural things, than are the more highly inwardized images of the imagination (III.404), and in imagination itself: "The interrelating of images is a higher activity then merely reproducing them. On account of its immediacy or sensuousness, the content of images has the form of finitude, or relation to another. In that at this juncture it is now I who am the general determining or positing factor, this relation is also posited by me. Through this positing, intelligence gives the images a subjective bond, replacing their objective one" (III.163–165). Thought, which rounds off the sphere, is presented as the inwardizing of memory, the immediately preceding level (III.211).

We have now examined enough instances of the working of the dialectic closely enough, to realize that this progressive sublation of the apparent disparity between subject and object, and this progression in degree of inwardness, are not to be interpreted in natural or temporal terms, but in the light of the analytical and synthetic procedures by which Hegel structures and systematizes his ultimately empirical material. In recollection, intuition is inwardized as the image (III.149). In memory, it is the name, which corresponds to that which is intuited as well as having a coherence of its own (III.415), which is inwardized or recollected. The two levels of inwardization differ in that the second presupposes, and is therefore more complex than the first. It is quite evident that such a neat and illuminating dialectical progression could only have been formulated by analysing the inherent and proximate levels of complexity involved in the subject-matter under consideration, and then synthesizing in the light of one's knowledge of the further levels of complexity still to be dealt with. The new element in the dialectic at the level of spirit proper is that the nisus or impulse motivating the transition from one level to the next is no longer primarily the result of the intellectual labour involved in analysing and synthesizing the interrelatedness of the subject-matter in the light of a telos, but also arises naturally and directly out of the subject-matter itself. In the Phenomenology, it was spirit which systematized consciousness, here in the Psychology spirit is systematizing itself: "Since consciousness is in immediate possession of the object, one cannot very well say that it possesses impulse. Spirit has to be grasped as impulse however, for it is essentially activity, and what is more, it is in the first instance the activity by which the apparently alien object receives,

[28] Cf. III.401 (note 143, 8).

instead of the shape of something given, singularized and contingent, the form of something recollected, subjective, universal, necessary and rational. By undertaking this alteration with the object, spirit reacts against the one-sidedness of consciousness, which rather than knowing objects, subjectively relates itself to them as to an immediate being. In this way it constitutes theoretical spirit, within which it is the impulse to knowledge, the drive toward cognition which is dominant. I know that the content of cognition has being or objectivity, and at the same time that it is within me and there-fore subjective. This standpoint therefore differs from that of consciousness in that the object is no longer determined as the negative of the ego" (III.95–97). Basically, it is a structured empiricism which co-ordinates the various fields of enquiry and makes explicit this coincidence of the real and the systematic aspects of a philosophic Psychology: "At the standpoint of presentation, even more than at the preceding stage of intelligence, the various forms of spirit tend to be regarded as singularized and mutually independent powers or faculties. One speaks not only of the general faculty of presentation, but also of the powers of imagination and memory, the mutual independence of these spiritual forms being taken entirely for granted. The precise nature of a truly philosophical comprehension consists however in the grasping of the rational connection present between these forms, in recognizing the sequence of the organic development of intelligence within them" (III.147).

A triadic structure is also imposed upon theoretical spirit of course, and Hegel introduces the third and final level of the sphere as follows: "Thought is the third and last main stage in the development of intelligence, for it is within it that the immediate, implicit unity of subjective and objective present in intuition is reconstituted, out of the subsequent opposition of these two aspects in presentation, as a unity which is in and for itself through its being enriched by this opposition. The end therefore recurves to the beginning" (III.219).[29] In that this simply implies that it is reasonable, in the light of the overall hierarchy of levels to be distinguished within Psycho-logy, to regard theoretical spirit as a qualitative unity, a distinct sphere, and that the subordinate spheres of Intuition, Presentation and Thought, Recollection, Imagination and Memory etc. are also brought to light by empirical research and dialectical analysis, there is no reason why we should take any exception to it. In that Hegel is also asking us to acknowledge the significance of the triadic divisions and sub-divisions however, we should at least be aware that he is bringing a further factor into play, and that it derives its validity from the overall telos of the system, rather than the ana-lytical and synthetic procedures we have been investigating here.

In the introduction to the Philosophy of Spirit, Hegel tells us that,

[29] Cf. III.227, 28, and the note (229, 27) on the casual revision of this triadicity (III.442).

"Aristotle's books on the soul, as well as his dissertations on its special aspects and conditions, are still by far the best or even the sole work of speculative interest on this general topic" (I.11).[30] In 'De Anima' (431a), after distinguishing between sensation and thinking, which correspond very roughly to Hegel's Anthropology and Theoretical Spirit, Aristotle goes on to deal with the practical intellect in operation, and this is almost certainly the ultimate origin of Hegel's differentiating the first and second major spheres of his Psychology. Although the difference between cognition and cosmology in the 1794 notes on Psychology might be regarded as prefiguring the mature distinction, the first clear evidence of it is in that between the first and second potencies in the 'System der Sittlichkeit' (1802). What is, perhaps, even more interesting about this early work, is that 'practical spirit' is already presented as the immediate presupposition of what might be regarded as an embryonic Philosophy of Right. In the first system of 1803/4 we find the mature transition from theoretical to practical spirit outlined, and in the third of 1805/6 not only its outlines but also many of its details.[31]

The general progression within this sphere is that of increasing degrees of objectification, and Hegel seems to have seen it as in certain respects a reproduction at the psychological level of self-consciousness at the phenomenological level: "Practical spirit has the opposite point of departure, for unlike theoretical spirit it starts not with the apparently independent object but with subjective determinations, with its own purposes and interests, and only then proceeds to objectify them. In doing so, it reacts against the one-sided subjectivity of self-enveloped self-consciousness to the same extent as theoretical spirit reacts against consciousness, which is dependent upon a given general object" (III.97).[32] Although he states at the outset that, "Here in the sphere of subjective spirit we have to pursue this externalizing only to the point at which volitional intelligence becomes objective spirit, that is, to where the product of will ceases merely to be enjoyment, and begins to constitute deed and action" (III.231–233), it was not until 1830 that he reached a clear conception of the immediate telos of the sphere, and of happiness (III.261–265) as the sublation of practical feeling, impulses and wilfulness (III.235–261). This section was extensively revised between 1827 and 1830, but the existing lecture notes do not shed much light upon the reasons for this. In 1817 and 1827 happiness was treated as subordinate to wilfulness, not as its sublation. It is not improbable that Hegel reversed this sequence on account of the analogous dialectical progression in the Philosophy

[30] There is an erudite discussion of Hegel's interpretation of Aristotle in Drüe, op. cit., pp. 336–352. G. R. G. Mure, 'An Introduction to Hegel' (Oxford, 1940), ch. VIII, is also quite useful.
[31] 'Jenaer Systementwürfe' I.282–300; III.185–222.
[32] Cf. III.231; 235.

of Right (§§ 90–104), in which crime or wilfulness is subordinate to the assertion of the law in legal punishment.³³ He mentioned eudemonism at this juncture in the 1825 lectures (III.261), and since he had some knowledge of the Utilitarian philosophy of law, and was deeply influenced by the journalism of Bentham's followers, it is possible that the postulation of the 'greatest happiness of the greatest number' principle as the foundation of rational legislation influenced this dialectical assessment of eudemonism as the immediate presupposition of the initial level of Objective Spirit (III.451). Once again, the sphere is triadically structured internally, and once again Hegel gives a certain plausibility to this pattern by imposing a teleologically validated syllogistic exposition upon a demonstrably well-founded analysis of the empirical subject-matter (III.233).

The third major sphere of the Psychology, that of Free Spirit (III.265–269), was only worked out in its final form in 1830, and provides us with a good example of the way in which Hegel went about establishing his triads as opposed to his systematic sequences. In 1817 and 1827, what appears in 1830 as § 481, the initiation of this final sphere of the Psychology, was the opening Paragraph of the Philosophy of Objective Spirit. This meant that Psychology, alone of all the major spheres of the Encyclopaedia, was left with a dual rather than a triadic structure. Hegel remedied this typographically, by simply shifting this § into the preceding sphere, but evidently feeling that this was not quite good enough, he proceeded to give some body to the move by inserting the eloquent passage on freedom (§ 482). This did not affect the immediate dialectical progression to any great extent, since in both cases it is "the unity of theoretical and practical spirit" that has to be elicited, but the juggling around did have certain wider implications which are really rather important. If it is to be taken at all seriously philosophically, it seems to suggest, for example, that after 1827 he decided that the immediate unification or sublation of theoretical and practical spirit should be regarded as subjective rather than objective, and that the presentation of freedom as the immediate presupposition of Objective Spirit should, in accordance with the general principles of the dialectic, give added weight to his conception of world history as the realization in a more concrete form of what is already implicit within the general sphere of which it constitutes the conclusion (Enc. §§ 548–552).

There is some evidence that he had originally planned to elicit a triadic structure from the Psychology by dividing Theoretical Spirit (III.103–229) in two. Practical Spirit would then have been the third level, and Thought (III.217–279) the third main moment of the second sphere, corresponding to Actuality in the Logic (Enc. §§ 142–159). He seems to have dropped this idea between 1825 and 1827. Even the most devoted advocate of the absolute validity of the triads can hardly help observing that such an apparently

³³ For the possible background to this, see III.452

casual revision of a major triad, and of the analogies indicated as the result of the parallel structures of the Logic and the Philosophies of Nature and Spirit, contrasts very sharply indeed with the immense care and caution with which he worked out the structural systematization of empiricism expounded within his dialectical progressions. It may therefore be regarded as further evidence, if indeed any more is needed, that the triadicity of his expositions is to be clearly distinguished from the other general principles they involve, if we are attempting to reach a critical appraisal of his work.

Although it is of course essential that we should attempt to come to grips with the general principles of Hegel's system if his philosophical accomplishment is ever to have any worthwhile effect, one has to admit that a mercurial changeableness, a wholly unexpected twist, a completely unforeseen perspective often constitutes the saving grace of the man, the one feature of his expositions which invariably prevents them from settling down into ordinary academic aridity. After labouring at the distinction between theoretical and practical spirit for example, and indicating the progressions in degree of inwardness and objectivity, he reminds us that the whole pattern has a converse side: "Although in what is theoretical the general object certainly becomes subjective in one respect, there is in the first instance another respect in which a certain content of the general object still remains outside the unity with subjectivity. At this juncture therefore, what is subjective only constitutes a form which fails to pervade the object absolutely, so that the object is not that which is thoroughly posited by spirit. In the practical sphere on the contrary, what is subjective, in its immediacy, still has no true objectivity, for as such it is something appertaining to the singularity of the individual, and not the being in and for self of something absolutely universal" (III.103). Carefully considered, this observation can be seen to be the corollary of what has already been established, but it can also take us unawares and force us to rethink our original approach, and it is this process of constant revision and rethinking rather than the precise technicalities of the dialectic, that constitutes the *spirit* of Hegel's rearrangement of the transitions he has formulated.

We should not, perhaps, conclude this section on the Psychology without saying a word or two about the general significance of its finding its end or telos in the law, morality and social ethics. It is most important to remember that like so many of Hegel's major transitions this was a commonplace, and simply part of the general culture of his time. In fact most of the works on Anthropology and Psychology published in Germany between 1800 and 1830 treated the subject-matter of Subjective Spirit as pointing beyond itself to "legal, ethical and religious actuality" (III.454). When one thinks of the emphasis which he laid upon the innate moral sense of those who are hypnotized however (II.317), upon the significance of Pinel's moral treatment of the insane (II.331), upon the ethical nature of freedom (III.231),

Plato's inclusion of "the whole nature of spirit under the justice of spirituality" (III.253) etc., it is difficult to resist the conclusion that this telos was to some extent always in his mind, throughout the whole exposition of Subjective Spirit, and that he was not entirely indifferent to its being regarded as in some sense immanent.[34]

g. SOURCES

Read not to contradict, and confute; nor to believe and take for granted; nor to find talk and discourse; but to weigh and consider.
— SIR FRANCIS BACON.

Since these lectures were designed primarily for undergraduates, it was only natural that Hegel should have given some indication of the sources of his empirical information and of the general ideas he was developing or criticizing. Yet although an educated early nineteenth-century audience may have been familiar with most of his background material, preoccupations and interests have changed, and if a modern reader is to grasp the full implication of his dialectical expositions, he will almost certainly have to make a preliminary study of some of the original sources. The following list, which contains all the works referred to in the lectures on the Philosophy of Subjective Spirit, should facilitate such an investigation. Hegel's library list and his other writings, especially the new Bochum editions of the Jena works and the lectures on the History of Philosophy, have been consulted in order to identify the actual sources of the more general references. In the case of the specialized literature, the first edition is listed if that used is not known, and clearly identifiable works not actually referred to by name are marked with an asterisk. The reader should consult the notes for further particulars concerning the complicated question of possible intermediaries.

It is, perhaps, worth observing, that an informed and critical survey of Hegel's handling of his sources is a necessary complement to any worthwhile treatment of the more formal aspects of his theory of knowledge.

I: Introduction

Aristotle	'De Anima' ('Opera omnia' ed. J. G. Buhle, 5 vols., Strassburg, 1791–1804).
C. Wolff	'Psychologia empirica' (Frankfurt, 1732).
	'Psychologia rationalis' (Frankfurt, 1734).
Anaxagoras	'Fragmenta' (ed. E. Schaubach, Leipzig, 1827).

[34] Cf. Enc. §§503–512. R.-P. Horstmann, 'Subjektiver Geist und Moralität. Zur systematischen Stellung der Philosophie des subjektiven Geistes' (Paper. Hegel Conference, Santa Margherita, May 1973), brings out the complementarity of the two levels.

Cicero	'De Divinatione.'
	'De Natura Deorum.'
R. Descartes	'Oeuvres complètes' (ed. V. Cousin, 11 vols. Paris, 1824/6).
N. Malebranche	'De la recherche de la vérité' (Paris, 1736).
B. Spinoza	'Opera quae supersunt omnia' (ed. H. E. G. Paulus, 2 vols. Jena, 1802/3).
G. W. Leibniz	'Principes de la nature et la grace' (ed. P. Desmaizeaux, in 'Recueil de diverses pièces sur la philosophie' 2 vols. Amsterdam, 1720).
G. Berkeley	'The Works' (2 vols. London and Dublin, 1784).
J. H. Campe	'Kleine Seelenlehre für Kinder' (Hamburg, 1780).
A. C. A. Eschenmayer	'Psychologie in drei Theilen als empirische, reine und angewandte' (2nd ed. Stuttgart and Tübingen, 1822).
H. Steffens	'Anthropologie' (2 vols. Breslau, 1822).

II. Anthropology

Natural Qualities

Hippocrates	'De aere, aquis et locis.'
Cicero	'De Divinatione.'
J. W. Ritter	'Der Siderismus' (Tübingen, 1808).
C.-F. Dupuis	'Origine de tous les Cultes, ou Religion Universelle' (4 vols. Paris, 1795).
C. J. de Ligne	'Fragments de l'Histoire de ma Vie' (written c. 1796; ed. F. Leuridant, 2 vols. Paris, 1928).
P. Pinel	'Traité Médico-Philosophique de l'Aliénation Mentale' (Paris, 1801).
J. C. Reil	'Rhapsodieen über die Anwendung der psychischen Curmethode auf Geisteszerrüttungen' (Halle, 1803).
Xenophon	'Anabasis' ('Opera' ed. H. Stephanus, Geneva, 1581).
G. R. Treviranus	'Biologie, oder Philosophie der lebenden Natur für Naturforscher und Aerzte' (6 vols. Göttingen, 1802/22).
P. Camper	'Dissertation sur les variétés naturelles qui caractérisent la physiognomie des hommes des divers climats et différens ages' (Paris and The Hague, 1791).

J. F. Blumenbach	'De Generis Humani Varietate Nativa' (Göttingen, 1775).
Maximilian of Neuwied	'Reise nach Brasilien' (2 vols. Frankfurt/M., 1820/1).
J. B. Spix and K. F. Martius	'Reise nach Brasilien' (2 vols. Frankfurt/M., 1820).
*H. Koster	'Travels in Brazil' (London, 1816).
I. Kant	'Anthropologie' (Könisgberg, 1798).
*K. Digby	'Second Treatise Declaring the Nature and Operations of Mans Soul' (London, 1669).

Natural Changes:

*J. Evelyn	'Memoirs illustrative of the Life and Writings of John Evelyn' (2 vols. London, 1819).
I. Kant	'Critik der reinen Vernunft' (2nd ed. Riga, 1787).
M. F. X. Bichat	'Traité de la vie et de la mort' (Paris, 1800).

Sensation:

I. Newton	'Optice' (2nd ed. London, 1719).
M. F. X. Bichat	'Traité de la vie et de la mort' (Paris, 1800).
*Plutarch	'Vitae' (Pericles).
J. W. Goethe	'Dichtung und Wahrheit' (Tübingen, 1811/2).

Feeling:

W. Shakespeare	'King Lear' (1608; in 'Wilhelm Shakespears Schauspiele', tr. J. J. Eschenburg, 23 vols. Strassburg and Mannheim, 1778/83, vol. 14).
*J. Brand	'Observations on Popular Antiquities' (2 vols. London, 1813).
J. B. van Helmont	'Ortus medicinae' (Amsterdam, 1648).
*G. T. L. Sachs	'Historia Naturalis duorum Leucaethiopum auctoris ipsius et sororis eius' (Sulzbach, 1812).

Magnetic Somnambulism:

*Plutarch	'Vitae' (Cato the Younger).
Plato	'Timaeus ('Opera omnia', ed. F. C. Exter and J. V. Embser, 11 vols. Strassburg, 1781/7).
J. W. Ritter	'Der Siderismus' (Tübingen, 1808).
K. U. von Salis-Marschlins	'Ueber unterirdische Elektrometrie' (Zürich, 1794).
*J. C. Passavant	'Untersuchungen über den Lebensmagnetismus und das Hellsehen' (Frankfurt/M., 1821).

*J. H. D. Petetin	'Mémoires sur la decouverte des phénomènes que présentent la catalepsie et le somnambulisme' (Lyon, 1787).
C. F. Nicolai	'Beispiele einer Erscheinung mehrerer Phantasmen' ('Berlinische Monatsschrift', May, 1799).
J. G. Scheffner	'Mein Leben, wie ich Johann George Scheffner es selbst beschrieben' (1816/21; Leipzig, 1823).
C. F. L. Schultz	'Ueber physiologe Farbenerscheinungen, insbesondere das phosphorische Augenlicht, als Quelle derselben, betreffend' (1821; Goethe's 'Zur Naturwissenschaft', pt. II, 1823).
J. W. Goethe	'Faust' (Tübingen, 1808).
*J. P. F. Deleuze	'Histoire Critique du Magnétisme Animal' (2 pts. Paris, 1813).
A. M. J. C. de Puységur	'Appel aux savants observateurs du 19 ièm siècle, de la décision portée pars leurs prédécesseurs contre le magnétisme animal et fin du traitement du jeune Hébert' (Paris, 1813).
P. G. van Ghert	'Dagboek eener magnetische Behandeling' (Amsterdam, 1814; Germ. tr. D. G. Kieser 'Archiv für den thierischen Magnetismus', 2, i, 3–188; ii, 3–51, 1817).
D. G. Kieser	'Das zweite Gesicht (second sight) der Einwohner der westlichen Inseln Schottlands, physiologisch gedeutet' ('Archiv für den thierischen Magnetismus', 6, iii, 93–141, 1820).
*J. Brand	'Observations on Popular Antiquities' (2 vols. London, 1813).
*D. von Rosetti	'Winckelmann's letzte Lebenswoche' (Dresden, 1818).
*B. Bendsen	'Beiträge zu den Erscheinungen des zweiten Gesichts' ('Archiv für den thierischen Magnetismus', 8, iii, 125, 1821).
*C. P. Moritz	'Desertion aus einem unbekannten Bewegungsgrunde' ('Magazin für Erfahrungsseelenkunde', II, pp. 16–17, Berlin, 1786).
A. M. J. C. de Puységur	'Mémoires pour servir à l'histoire et l'établissement du Magnétisme Animal' (1786; 3rd edn. Paris, 1820).
*D. G. Kieser	Review of an article by G. Bakker, J. H. G. Walthers and P. Hendriksz (2 pts. Groningen, 1814/18), in 'Archiv für den thierischen Magnetismus', 6, i, 148–160, 1819.

C. A. F. Kluge   'Versuch einer Darstellung des animalischen Magnetismus als Heilmittel' (Berlin, 1811).

K. E. Schelling   'Ideen und Erfahrungen über den thierischen Magnetismus', and 'Weitere Betrachtungen über den thierischen Magnetismus, und die Mittel ihn näher zu erforschen' ('Jahrbücher der Medicin als Wissenschaft', ed. A. F. Marcus and F. W. J. Schelling, vol. II, i, 3–46; ii, 158–190, Tübingen, 1807).

J. B. van Helmont   'Ortus medicinae' (Amsterdam, 1648).

F. K. von Strombeck   'Geschichte eines allein durch die Natur hervorgebrachten animalischen Magnetismus und der durch denselben bewirkten Genesung' (Brunswick, 1813).

P. G. van Ghert   'Mnemosyne' (Amsterdam, 1815; Germ. tr. 1818).

K. J. H. Windischmann   'Ueber Etwas, das der Heilkunst Noth thut. Ein Versuch zur Vereinigung dieser Kunst mit der christlichen Philosophie' (Leipzig, 1824).

Derangement:

P. Pinel   'Traité Médico-Philosophique de l'Aliénation Mentale' (Paris, 1801).

B. Spinoza   'Ethica' in 'Opera' (ed. H. E. G. Paulus, 2 vols. Jena, 1802/3).

J. C. Reil   'Rhapsodieen über... Geisteszerrüttungen' (Halle, 1803).

*Plutarch   'Vitae' (Marcellus).

*J. de la Bruyère   'Les Caractères de Theophraste' (Paris, 1688).

*anon.   'The Morning Chronicle', February 14th, 1823.

J. M. Cox   'Practical Observations on Insanity' (London, 1804; Germ. tr. Halle, 1811).

J. G. Langermann   'Ueber den gegenwärtigen Zustand der psychischen Heilmethode der Geisteskrankheiten und über die erste zu Bayreuth errichtete psychische Heilanstalt' ('Medicinisch- chirurgische Zeitung', ed. J. J. Hartenkeil, Salzburg, 1805, vol. 4, no. 83, pp. 90–93).

A. K. Boerhaave   'Impetum faciens dictum Hippocrati per corpus consentiens' (Leiden, 1745).

*L. Lemnius	'De habitu et constitutione corporis' (Antwerp, 1561).
*W. Pargeter	'Observations on Maniacal Disorders' (Reading, 1792; Germ. tr. Leipzig, 1793).

Gesture :

J. W. Goethe	'Zur Morphologie' (Stuttgart and Tübingen, 1823), vol. II, pt. i.
*G. L. Staunton	'An Authentic Account of an Embassy from the King of Great Britain to the Emperor of China' (2 vols. London, 1797; Germ. tr. 2 pts. Halle, 1798).
*J. J. Engel	'Ideen zu einer Mimik' (2 vols. Berlin, 1785/6).
J. K. Lavater	'Physiognomische Fragmente zur Beförderung der Menschenkenntnis und Menschenliebe' (4 vols. Leipzig and Winterthur, 1775/8).

III: Phenomenology and Psychology

Phenomenology :

I. Kant	'Critik der reinen Vernunft' (2nd ed. Riga, 1787). 'Critik der praktischen Vernunft' (Riga, 1788). 'Critik der Urtheilskraft' (Berlin and Libau, 1790).
K. L. Reinhold	'Versuch einer neuen Theorie des menschlichen Vorstellungsvermögens' (Prague and Jena, 1789).
J. G. Fichte	'Grundlage der gesammten Wissenschaftslehre' (Leipzig, 1794).
B. Spinoza	'Opera' (ed. H. E. G. Paulus, 2 vols. Jena, 1802/3).
*Sextus Empiricus	'Opera' (ed. J. A. Fabricius, Leipzig, 1718).

Psychology :

Dante	'La Divina Commedia.'
É. B. Condillac	'Traité des sensations' (Paris and London, 1754).
I. Kant	'Critik der reinen Vernunft' (2nd ed. Riga, 1787).
Aristotle	'De Anima' ('Opera Omnia', ed. J. G. Buhle, 5 vols. Strassburg, 1791–1804).

J. W. Goethe   'Die Leiden des jungen Werthers' (Leipzig, 1774).

Aristotle   'Metaphysica' ('Opera omnia', ed. J. G. Buhle, 5 vols. Strassburg, 1791–1804).

J. F. Fries   'Neue oder anthropologische Kritik der Vernunft' (3 vols. Heidelberg, 1807).

Signs and Language:

J. L. Lagrange   'Théorie des fonctions analytiques' (Paris, 1797).

W. von Humboldt   'Ueber den Dualis I' ('Abhandlungen der historisch-philologischen Klasse der Königlichen Akademie der Wissenschaften zu Berlin. Aus dem Jahre 1827'. Berlin, 1830, pp. 161–187).

G. L. Staunton   'An Authentic Account of an Embassy from the King of Great Britain to the Emperor of China' (2 vols. London, 1797; Germ. tr. 2 pts. Halle, 1798).

F. H. Jacobi   'Von den göttlichen Dingen und ihrer Offenbarung' (Leipzig, 1811).

Memory:

*G. von Feinaigle   'Mnemonik; oder, Praktische Gedächtnisskunst zum Selbstunterricht' (Frankfurt/M., 1811).

F. A. Mesmer   'Précis Historique des faits relatifs au Magnétisme-Animal jusques en Avril 1781' (London, 1781).

I. Kant   'Critik der reinen Vernunft' (2nd ed. Riga, 1787).

Practical Spirit:

J. Boehme   'Theosophia. Revelata' (2 pts. Amsterdam, 1715).

Plato   'De Republica' ('Opera omnia', ed. F. C. Exter and J. V. Embser, 11 vols. Strassburg, 1781/7).

Aristotle   'Politica' ('Opera omnia', ed. J. G. Buhle, 5 vols. Strassburg, 1791–1804).

Considering the manner in which Hegel's writings are usually interpreted, the most remarkable thing about the way in which he employed his sources in a field such as this, is his relative indifference to the *philosophical works* he seems to have consulted. In fact it would not be very far from the truth to say that Aristotle seems to have been the only philosopher to exercise a positive influence upon this part of the system. There is some evidence that Hegel had second thoughts about certain aspects of his thinking during the 1820's (III.107, 121), but the lectures contain several approving references to him, and numerous details, such as the characterization of sleep, the treatment of sensation and the senses, the description of the human hand, the transition to practicality, are undoubtedly Aristotelian in origin. We are told that the most important lesson to be learnt from 'De Anima' is that: "the Philosophy of Spirit can be neither empirical nor metaphysical, but has to consider the Notion of Spirit in its immanent, necessary development from out of itself into a system of its activity" (I.103). Hegel emphasizes how necessary it is to grasp the significance of the progression in degree of complexity and the distinction between the three main levels of animation central to Aristotle's conception of the soul: "In accordance with his procedure, he begins with what is sensuous and draws a threefold distinction in respect of souls... In man then, one does not only have the presence of a vegetable nature, for he is also, and to the same extent, sentient and thinking; in man these are merely three forms, which, within him, are one and the same" (I.11). This general conception is regarded as providing the broad framework within which a rational and systematic treatment of Subjective Spirit may enter into easy commerce with the necessary empiricism of the various sub-disciplines falling within the general sphere.

Hegel shows himself to be sympathetic to certain aspects of Kant's philosophy. He adopts his broad classification of temperaments, makes use of his doctrine of objective categories in order to distinguish between dreaming and waking, and admits that the philosophical movement he initiated has given rise to a new awareness of the importance of psychology. He notes, however, that his unrealistically exaggerated preoccupation with the subject-object antithesis has set needlessly severe limits to the systematic comprehensiveness of philosophy: "The Kantian philosophy is most accurately assessed in that it is considered as having grasped spirit as consciousness, and as containing throughout not the philosophy of spirit, but merely determinations of its phenomenology" (III.11), and it is this insight which gives rise to his predominantly critical attitude. He takes Kant's thought to be typically north German, implies that it therefore contrasts sharply with his own (II.73), and goes on to point out the limitations of such central Kantian doctrines as the subjectivity of space and time and the distinction between understanding and reason (III.225).

Plato's views on such important topics as the nature of matter, the determination of the absolute, the physiological basis of divination, the political context of ethics, are mentioned with respect. These are quite evidently nothing more than passing references however, and in the final section of the work attention is drawn to the predominant limitation of his thought: "Plato and Aristotle... knew only that man is free through being born a citizen of Athens or Sparta, or by virtue of character, education, philosophy..." (III.268/9).

As is usually the case in his treatment of specialized disciplines such as Anthropology or Psychology, Hegel refers to other philosophical attempts at comprehending them merely in order to pinpoint intellectual aberrations. Descartes, Malebranche, Spinoza, Leibniz and Berkeley are criticized for having fallen into the unjustified dualism of regarding matter and the soul as two subsistent entities mediated by God (§ 389). Spinoza's proposition concerning the correlation between the order and connection of ideas and things is quoted approvingly, but the axiom relating to causality from which it derives is passed over in silence, and Spinoza is subsequently censured for having failed to grasp the significance of free subjectivity's having emerged from substance (III.13). The conceptions of language put forward by Leibniz and Jacobi are criticized, as are Wolff and Campe for having projected Leibnizian dualism into the artificial distinction between rational and empirical psychology. Condillac is censured for failing to distinguish between a natural development and dialectical progression, and Dupuis for the naïve reductionism of his treatment of religion.

Kant's weaknesses are seen as exemplified in the doctrines of his followers. Fichte's philosophy is treated as simply, "a more consistent exposition of the Kantian" (III.281), and the confusions involved in Fries's psychologizing and Reinhold's theory of consciousness as being the natural outcome of their master's teaching. Since putting into practice the Aristotelian approach to these subjects involved the basic analytical procedure of grasping the precise inter-relationship of various spheres and degrees of complexity, Hegel was bound to deplore the way in which Eschenmayer and Steffens were mixing everything up (I.101/3). In the case of Steffens, however, although it was of course quite fantastic that a subject such as geology should have been treated as directly relevant to Anthropology, there is a clear similarity between his broad approach and Hegel's (I.158).

This sweeping dismissal of the various philosophical attempts at grasping the significance of Subjective Spirit contrasts sharply with the close and respectful attention paid to the relevant *non-philosophical literature* of the time. What Hegel evidently wanted from his sources was not a supply of half-baked philosophical generalities, but empirical information based upon immediate experience, direct observation and professional insight: "If the Germans frequently make merry over the defective theories of the French, it can be affirmed... that there is much more to be said for the naïve metaphysics the French employ... than there is for the inveterate dreaming of German savants, whose theorizing is not infrequently as warped as it is lame" (II.303).

Failure to realize this can easily give rise to disastrous misinterpretations of what he is attempting to do in the Philosophy of Subjective Spirit. Although it is its subject-matter rather than its supposed position in the history of philosophy that has to be considered first if we are attempting to interpret it, even specialists will tend to regard its empirical aspect as original or unusual if they are unfamiliar with its cultural background. It should not be over-looked that the subject-matter of the work simply reflects the general interests of the period. In fact the great bulk of its basic material is so much a part of the general professional knowledge of the time, that it is practically impossible to indicate Hegel's sources with any certainty. One might easily list a dozen contemporary works on Anthropology which begin by quoting Thales' injunction for example (I.3), and treatments of national character-istics (II.67) or attempts to characterize the various stages of life (II.103) were no less common in the anthropological literature of the time. Hegel's many attempts at ranging the senses in a rational sequence have numerous contemporary counterparts (II.167), and although his comments upon the connection between disposition and good-heartedness (II.225) might strike us as being no less odd than the classification of nostalgia (II.247), in both instances he is simply retailing commonplaces. Seemingly out of the way information concerning the Camisards (II.259), the effect of music on sleep-walkers (II.265), cretins (II.357), the English malady (II.365), turns out to be perfectly ordinary once its sources and background have been investi-gated. Even the much discussed treatment of language and the transition from Subjective to Objective Spirit (III.269) are part and parcel of the general attitudes and preoccupations of the time.

When Hegel refers to the representation of the brothers death and sleep (II.135), the fatal combat between father and son (II.239), to the Orang-Utang's use of a stick (II.415), Dante's conception of truth (III.83), Mesmer's experiment with words and thought (III.205), he is drawing upon sources not usually associated with the subjects he is discussing. The reference to la Bruyère's 'Characters' (II.359) may be an example of his borrowing second-hand from a treatise on mental derangement, but the various instances drawn from Plutarch (II.195, 247, 359), and the invoking of Boehme (III.243), are clearly examples of his citing works in no way directly related to the subject-matter being dealt with in this part of the Encyclo-paedia.

His comments on Windischmann (II.321) and Lavater (II.423) indicate a sharp eye for irrelevant theorizing and the exaggeration of the significance of a subject. When illustrating a general point he will often refer to writers who provide him with unadorned and unpretentious accounts of immediate experiences, — Sachs (II.237), Nicolai (II.269), Scheffner (II.271), Helmont (II.307) etc. We know from the general body of his unpublished manuscript notes (II.477), as well as his lectures (II.63) and printed works (III.183), that he took a great interest in travel books and often made use of what he had extracted from them. He takes an ordinary, commonsense view of the status of such knowledge: "The sensuous presence of things has so light a hold upon objective consciousness, that I can also know of what is not sensuously present to me. I can, for example, be familiar with a distant country merely through what has been written about it." (II.205). Many of the writers mentioned in the lectures, — Ritter, Schultz, Puységur, van Ghert, Karl Schelling, Strombeck, Cox, Langermann etc., had actually investigated problems, performed experiments or been professionally involved with the subject-matter Hegel is surveying. When consulting general accounts of wider fields of research, he takes care to confine himself to works by such recognized authorities as Camper (II.51), Blumenbach (II.51), Bichat (II.135), Kieser (II.285), Humboldt (III.183), the accuracy and reliability of which could not reasonably be regarded as a matter for debate. He was not averse to making use of a text-book the presuppositions of which he found unsatisfactory (II.303, 19), as long as it provided him with straightforward factual material. He evidently consulted Reil for facts rather than ideas for example, whereas Pinel (II.331), his other major source in the Anthropology, seems to have helped him in both respects.

The literary origin and reliability of his empirical information are sometimes open to conjecture on account of our only knowing a source which he is unlikely to have consulted (II.93, 12; 265, 3; 379, 29), or possible sources which he may or may not have been acquainted with (II.287, 37; 291, 19; 357, 27; 367, 15), or, occasionally, because the ultimate origin of his data is now untraceable (II.291, 14; 381, 28). In certain instances, we know where he may have obtained his information despite his not mentioning his sources (II.275, 24; 277, 8; 279, 31; 297, 15; 307, 9; 385, 12; 419, 22. III.141, 33; 205, 2; 305, 36), although we cannot be certain that he is not making use of an unknown intermediary.

Some of his expositions can be convincingly related to his own experience. His quoting of Kent's remark to Lear may well be the direct outcome of his admiration for a teacher at the Stuttgart Grammar School for example (II.497), and his views on youth and education (II.115), as well as his accounts of Jean Paul's way of lulling children to sleep (II.145), pendulation (II.263) and family feeling (II.273) can also be related to what we know of his personal contacts and activities. He often refers to the Bible as a matter of course, and without much trace of self-consciousness.

Of recent years, it has become increasingly apparent that he derived a great deal of his general information from newspapers, periodicals and reviews. The Prince de Ligne's witticism concerning revolutions (II.37) and the account of Madam Westermann (II.317) probably derive from such a source. What is more, it looks as though the much quoted passage in his Berlin inaugural may well have been influenced by an article he read while preparing his lectures on Animal Magnetism (II.323, 27). It is indeed remarkable how often English newspapers and periodicals provided him with material. Even an apparently chance remark on the nature of language turns out, on investigation, to have been lifted from 'The Morning Chronicle' (II.425, 3), and his information concerning the Portsmouth Case (II.365), the conditions in Bedlam (II.373), the plight of George III (II.377, 383) was almost certainly derived from the same source. There are several instances of his having extracted material from 'The Quarterly Review' (II.65, 18; 101, 22. III.187, 4), and at least one of his having made use of 'The Edinburgh' (III.183, 35).

The reference to Treviranus's 'Biologie' (II.45) is something of a curiosity, since although he used this work extensively for his lectures on the Philosophy of Nature, and continued to include this reference to it in the later editions of the Encyclopaedia, he seems never to have enlarged upon this particular subject in the lecture-room. Finally, it should perhaps not be overlooked that there are instances of his having apparently confused or misrepresented his sources (II.267, 29; 373, 19; 385, 12; 385, 37; 415, 35).

The reader is referred to the notes for more detailed information concerning the general background to these lectures. The following abbreviations have been used for the other Hegelian works referred to throughout the commentary.

Jub.	'Sämtliche Werke. Jubiläumsausgabe' (ed. H. Glockner, 26 vols. 4th ed. Stuttgart-Bad Cannstatt, 1965).
Hoff. Dok.	'Dokumente zu Hegels Entwicklung' (ed. J. Hoffmeister, 2nd ed. Stuttgart, 1973).
Nohl	'Hegels theologische Jugendschriften' (ed. H. Nohl, Tübingen, 1907).
Zur Psych.	'Zur Psychologie und Transzendentalphilosophie. Ein Manuskript. 1794.' (To be published in 'Gesammelte Werke', I.167–192, Felix Meiner Verlag, Hamburg.)
Jen. Real.	'Jenaer Realphilosophie 1805/6' (ed. J. Hoffmeister, Berlin, 1969).
Phen.	'The Phenomenology of Mind' (tr. J. B. Baillie, 2nd ed. London and New York, 1949).
Phil. Prop.	'Philosophische Propädeutik' (ed. K. Rosenkranz, Jub. vol. 3, 4th ed. Stuttgart-Bad Canstatt, 1965).
L. Logic	'Hegel's Science of Logic' (tr. A. V. Miller, London and New York, 1969).
Logic	'Hegel's Logic' (tr. W. Wallace, 3rd ed. Oxford, 1975).
Phil. Nat.	'Hegel's Philosophy of Nature' (tr. M. J. Petry, 3 vols. London and New York, 1970).

Phil. Sub. Sp.   'Hegel's Philosophy of Subjective Spirit' (tr. M. J. Petry, 3 vols. Dordrecht, 1977).

Vera   'Philosophie de l'esprit de Hégel' (tr. A. Vera, 2 vols. Paris, 1867/9).

Phil. Right   'Hegel's Philosophy of Right' (tr. T. M. Knox, Oxford, 1962).

World Hist.   'Lectures on the Philosophy of World History. Introduction: Reason in History' (tr. H. B. Nisbet, Cambridge, 1975).

Phil. Hist.   'The Philosophy of History' (tr. J. Sibree, New York, 1956).

Aesthetics   'Aesthetics. Lectures on Fine Art' (tr. T. M. Knox, 2 vols. Oxford, 1975).

Phil. Rel.   'Lectures on the Philosophy of Religion' (tr. E. B. Speirs and J. B. Sanderson, 3 vols. London, 1962).

Hist. Phil.   'Lectures on the History of Philosophy' (tr. E. S. Haldane and F. H. Simson, 3 vols. London and New York, 1963).

h. THE TEXT

Il n'y aurait point d'erreurs qui ne périssent d'elles-mêmes, rendues clairement. — VAUVENARGUES.

The present German text has various sources, the central part of it being the photographic reproduction of what Hegel published in the third edition of the *Encyclopaedia* (1830). This basic material consists of a hundred and six numbered Paragraphs (§§ 377–482), forty-six of which include supplementary passages or Remarks, and two of which are also supplied with footnotes (II.248; III.82). Although this edition was the last to be seen through the press by Hegel himself, and since the manuscript has disappeared has to be regarded as definitive, various faults have had to be corrected. For the most part they are of minor importance, being either fairly obvious omissions (III.8, 14; 102, 26) and errors (I.24, 11; II.160, 19; 162, 3; 322, 30; III.82, 21; 182, 7; 222, 15), or straightforward inconsistencies (II.242, 29; III.76, 25; 104, 9; 120, 28; 154, 22), though they do involve a few debatable emendations (I.14, 32; II.2, 24; 242, 30; 410, 14; III.70, 10; 92, 30; 108, 1; 166, 15; 230, 20).

When Boumann republished this material from the Encyclopaedia in 1845,[1] it was not without making some acceptable corrections (III.62, 13; 92, 30; 108, 1; 166, 15), but although it is possible that he had access to

[1] 'Werke' (18 vols. Berlin, 1832–1845), vol. VII, pt. 2.

relevant manuscript or lecture material which is now no longer available, the peculiarities of his text are such that they cannot be taken very seriously. There would certainly have been no point in reprinting it as it stands, since he reproduces errors (121, 13; 355, 16), introduces misprints (104, 13; 302, 30; 309, 29; 346, 28), omits and inserts words (105, 4; 240, 24; 333, 23; 348, 28) and phrases (29, 18; 154, 15; 294, 9), substitutes erroneously (274, 1), changes constructions and meanings to little apparent purpose (164, 20; 232, 7; 240, 19; 271, 20; 291, 4; 311, 2), and even violates basic grammar in pursuit of an interpretation (280, 6).

The best text of the Encyclopaedia to date is that published by F. Nicolin and O. Pöggeler in 1959.[2] It is not entirely free of misprints (320, 26; 367, 11), and its modernization of Hegel's punctuation is not an unqualified success (see 311, 7; 341, 39; 355, 34 etc.), but it does incorporate all but two (II.242, 29; III.92, 30) of the emendations suggested in this edition.

As in the corresponding translation of the Philosophy of Nature (Enc. §§ 245–376), material which appeared in the first edition of the Encyclopaedia (1817) is distinguished from that which first appeared in the second (1827) and third editions as follows:

i. passages dating from 1817, *reproduced* in 1830.

ii. **passages first printed in either 1827 or 1830.**[3]

Analysing the text in this way is a necessary procedure if any attempt is to be made to trace the development of Hegel's ideas, since it brings out the sections of the Encyclopaedia which remained unrevised (§§ 384, 393, 444), as well as those on which he lavished the most attention (§§ 377–380; 396–410; 457–471), and highlights the introduction of new subjects in already established contexts (§§ 394, 395, 415), as well as providing evidence of the way in which he tinkered with his triads (§§ 385, 449, 482). There would have been little point in following Hoffmeister and attempting to note all the differences between the second and third editions,[4] but the dropping of significant passages from the 1817 text has been indicated (III.107; 121), as has any extensive variation in the 1827 text (§§ 409, 442, 476), and any important addition made in 1830 (§§ 413, 426, 436). In respect of the development of the printed text, it is interesting to note, for example, that all the Paragraphs of such a reputedly neglected section as the Phenomenology were revised in either the second or the third edition, and that new material was added to well over a third of them in 1830.

When Boumann was preparing the Additions to this published text from

---

[2] 'Enzyklopädie' (Hamburg, 1959).
[3] 'Hegel's Philosophy of Nature' (3 vols. London, 1970), I.123.
[4] J. Hoffmeister, 'Encyclopädie' (5th ed. Leipzig, 1949).

the records of Hegel's lectures, he had seven manuscript sources at his disposal all but one of which have since been lost.[5] Since most of the lecture courses were based upon the first edition of the Encyclopaedia, present editorial standards would require that such records should be published separately in so far as they illustrate the development of Hegel's ideas.[6] In accordance with the principles of the first edition of the collected works however, Boumann conflated everything into these Additions to the 1830 text, and since it is no longer possible to analyse his editorial work at all exhaustively, it has to be accepted much as it stands. In the present edition, misprints have been noted (II.122, 35; 190, 35; 208, 7; 232, 9; 286, 2; III.20, 15; 32, 10; 102, 5; 130, 31; 132, 4; 234, 7), punctuation (II.48, 11; 212, 3) and grammar (II.14, 34; 52, 17) corrected, and a couple of sentences improved (II.370, 20; III.72, 15), but apart from this the original text has not been tampered with. Like the Paragraphs, it has been photographically reproduced, with the page-numbering of the 1845 edition indicated in the left-hand margin. Since Glockner also reproduced Boumann's text photographically, it is rather curious that his edition should contain a number of fresh misprints (163, 24; 220, 1; 247, 30; 310, 25; 373, 1).[7]

The merits and demerits of Boumann's work in assembling these Additions, in so far as they can be ascertained from the material now available, have been discussed in detail in the commentary. Unlike some of the other early editors, he seems to have realized the importance of reproducing Hegel's terminology accurately (III.49, 27), although there is some evidence that he may occasionally have distorted meanings by 'normalizing' phrases (III.151, 28). He provides us with the only connected account we have of certain material (III.339, 34), and with a convincing but hitherto un-corroborated lay-out for one of the most original and extensively erudite expositions of the whole work (II.303, 20), but where his reproduction of the dialectical pattern can be checked, it is not always found to be completely reliable (III.163, 15). Unlike Michelet, the editor of the Philosophy of Nature, he evidently had little interest in Hegel's references to his sources, and in this respect it has therefore been necessary to supplement his work extensively from the hitherto unpublished lecture-notes. Since by and large he has a good eye for material of more purely philosophical significance, it is all the more surprising that at several crucial junctures he should display such complete lack of judgement in this respect. Had he edited several important Paragraphs more circumspectly (§§ 377, 378, 381, 389, 438, 439), and given a better account of the Phenomenology, the general reception of

---

[5] I.cliii.

[6] K.-H. Ilting, 'Vorlesungen über Rechtsphilosophie' (4 vols. Stuttgart-Bad Canstatt, 1973/4).

[7] 'Jubiläumsausgabe' (ed. H. Glockner, 4th ed. Stuttgart-Bad Canstatt, 1965), vol. 10.

the Philosophy of Subjective Spirit over the last century and a quarter would certainly have been more enlightened.

Hegel *lectured* regularly on this section of the Encyclopaedia throughout the whole of the Heidelberg and Berlin period, treating it in general terms while dealing with his system as a whole (Winter 1816/17, Summer 1818, Winter 1818/19, Winter 1826/7), and devoting separate courses to expounding it in detail (Summer 1817, Summer 1820, Summer 1822, Summer 1825, Winter 1827/8, Winter 1829/30). Since the average attendance at these lectures was somewhat higher than that at the corresponding courses on the Philosophy of Right, it is rather disappointing that only three sets of student lecture-notes are known to have survived, — Hotho's from the summer of 1822, and Griesheim's[8] and Kehler's from the summer of 1825.[9] The Hotho manuscript is shorter than the others, consisting of two hundred and four untidily written pages, and since we have other material from Hegel's own hand illustrating the state and development of his ideas during the early 1820's, it has not been brought into consideration in editing this text. Both the Griesheim and the Kehler manuscripts provide us with valuable and comprehensive records of an important series of lectures however, and I am most grateful to the Staatsbibliothek Preussischer Kulturþesitz and the Jena University Library for allowing me to make such extensive use of them in supplementing Boumann's Additions.

Griesheim's account runs to three hundred and eighty-three quarto pages. Like the rest of his Hegel manuscripts, it is written in such a beautifully clear and consistently neat hand, that he must have used shorthand during the lectures and then copied out longhand in his own time, and he did this with such care and accuracy, that it has been possible to reproduce his text much as it stands. The revision necessary has involved no more than inserting a heading (III.270, 1), placing the page numbers in brackets, correcting the punctuation in the translation, putting right obvious slips (II.108, 35; III.270, 2; 272, 22; 328, 31) and a couple of misspellings (II.414, 32; III.344, 35), and bringing a few names (II.70, 33; 178, 34; 291, 33; 302, 31), words (II.316, 31; III.196, 4; 326, 18) and inflexions (II.324, 10; III.192, 1; 292, 35; 292, 36; 294, 5; 296, 1; 306, 11; 340, 4; 342, 32; 352, 28) into line with normal usage.

The Kehler manuscript consists of two hundred and forty-three pages, very closely written in a semi-shorthand, evidently while Hegel was actually lecturing. The two respects in which it is clearly superior to Griesheim's are

---

[8] H. Hotho, *Philosophie des Geistes. Nach dem Vortrage des Herrn Professor Hegel. Im Sommer 1822.* Ms. germ. quarto 1298. K. G. J. von Griesheim, *Philosophie des Geistes, vorgelesen von Professor Hegel. Sommer 1825.* Ms. germ. quarto 544. Staatsbibliothek Preussischer Kulturbesitz, Handschriftenabteilung, Archivstrasse 12–14, 1 Berlin 33.
[9] H. von Kehler, *Philosophie des Geistes nach Hegel. Sommer 1825.* Ms. Chron. 1906.6. Universitätsbibliothek, Goetheallee 6, 69 Jena.

its dating of the individual lectures, and its inclusion of a quotation from Aristotle which Hegel evidently read out in Greek. Frequently recurring words such as *allgemeine, auch, durch, nicht, Thätigkeit, zusammen* are written in shorthand, and nearly all inflexions are omitted, so that much of the consistency and purity of the printed text is simply due to its having been deciphered. The Greek (I.10) and the spelling (II.262, 38; 302, 33) have been standardized however, one inflexion has been altered (II.270, 17), and a couple of names have been corrected (II.35, 36; 306, 31).

In general, material from these manuscripts has been added to Boumann's text when it illustrates the development of Hegel's ideas, when it makes his general philosophical position clearer, and when it indicates what his sources were. Griesheim's Phenomenology is published in full (III.270–357), not only because of the exceptional importance of this section to Hegelianism in general, but also on account of the very unsatisfactory nature of Boumann's version, and in order to give an idea of what the lectures were actually like. Having two sets of notes from the same course makes it possible to analyse the merits of both quite closely, and in order to facilitate this the corresponding pages in the two manuscripts are given at the head of each extract. Any significant differences already noted have been either bracketed and incorporated in the text, or discussed in the commentary. Related contextual problems arising from differences between the first and third edition of the Encyclopaedia have also been dealt with in the notes (II.321, 39; III.257, 14 etc.).

During the Summer Term of 1818, Hegel *dictated* a number of passages, evidently in order to bring out leading ideas, indicate the general lay-out of his expositions, and clarify the significance of transitions from one sphere to another more effectively than he had in the printed text of the Encyclopaedia. An unknown student's notes of two of these dictated passages have been included in the present text, the only alterations being the standardization of the spelling of a couple of words (I.64, 39: II.424, 36). Notes made by Hegel himself, evidently as a basis for dictation during the 1820 and 1822 courses, have recently been published, and extracts from them have been included at the relevant junctures. In order to make this excessively disjointed material more intelligible, it has been partly repunctuated, mainly by the addition of full-stops, capitals and commas. I should like to acknowledge my indebtedness to the original editors of these dictation notes for permission to publish them in this edition.[10]

The *notes* on the first edition of the Encyclopaedia which Hegel had before him while lecturing on Subjective Spirit have also been published recently, and since they throw a great deal of light upon the development of his ideas

[10] F. Nicolin, 'Unveröffentlichte Diktate aus einer Enzyklopädie-Vorlesung Hegels' ('Hegel-Studien', vol. 5, pp. 9–29, 1969); H. Schneider, 'Unveröffentlichte Vorlesungs-Manuskripte Hegels' ('Hegel-Studien', vol. 7, pp. 9–59, 1972).

between 1817 and 1822, especially in respect of Psychology, they have been referred to extensively in the commentary to volume three.[11] There is direct (pp. 35, 50, 76) as well as indirect (III.229, 27; 241, 5; 305, 36) evidence that Hegel used them for his lectures on Phenomenology and Psychology as late as the Summer Term of 1825, and some reason at least for attempting to relate them to an even later stage in his development (III.93, 21), but the introduction (pp. 18–19) clearly pre-dates the fragment on the Philosophy of Spirit (1822), many characteristic features of the more detailed expositions can hardly be dated later than this (III.35, 10; 55, 33; 89, 31; 145, 24; 235, 9; 257, 15; 263, 10; 321, 44; 329, 15), and there is no trace of even some of the more important material dealt with in (III.179, 9) and after (III.143, 8; 247, 9; 253, 23) 1825.

The fragment on the Philosophy of Spirit (I.90–139) just referred to seems to date in the main from the summer of 1822, and was the outcome of Hegel's long-standing intention to publish a separate work on Subjective Spirit which would evidently have been comparable in detail and scope to the Philosophy of Right (I.91, 1). In fact he got no further than sketching the introduction and parts of the Anthropology, but the text as we have it, though fragmentary, is so well formulated and of such intrinsic interest on account of the light it throws upon the development of certain central concepts, that it was thought worthwhile to present the main body of it in full. I am most grateful to Professor Nicolin for permission to make use of the version he published in Hegel-Studien some years ago.[12] The marginalia and critical apparatus have been omitted, what he presents as an appendix has been included in the main text as a footnote (I.92), and the position of the Umlaut in such words as *äussere, häufig, räumlich* has been normalized, but apart from this the two versions are identical.

i. LANGUAGE

I am not yet so lost in lexicography, as to forget that words are the daughters of earth, and that things are the sons of heaven. Language is only the instrument of science, and words are but the signs of ideas. — DR. JOHNSON.

Since Hegel took an intelligent if not specialized interest in the various empirical procedures of *comparative philology*, there is an encouraging con-

[11] F. Nicolin and H. Schneider, 'Hegels Vorlesungsnotizen zum Subjektiven Geist' ('Hegel-Studien', vol. 10, pp. 11–77, 1975).
[12] 'Ein Hegelsches Fragment zur Philosophie des Geistes' ('Hegel-Studien', vol. I, pp. 10–48, 1961). This manuscript is also in the possession of the Staatsbibliothek Preussischer Kulturbesitz, — Hegel Nachlaß, vol. 2, acc. ms. 1889, 252.

creteness and versatility about his philosophic treatment of language (§ 459), and if we have some knowledge of the attitudes that have become entrenched in this field over the last two hundred years, we shall certainly be fascinated by the way in which he approaches some of the main issues of his day. His accommodation of what are often regarded as mutually exclusive attitudes is surely one of the most striking characteristics of his exposition. It is certainly intriguing, for example, to find that although he falls in with the reductionists in calling attention to the significance of the physical sensations so basic to verbal expression, he also criticizes those who regard the proliferation of this sensuous material as the highest accomplishment of a language. When we find him involving himself in the bewildering complexity of the anthropological, historical and cultural factors out of which language arises, we hardly expect him to go on to generalize, as he does, and in such a lucid and incisive manner, about the superiority of an alphabetic over a hieroglyphic script. And how are we to square his questioning of the value of Leibniz's attempt to mathematicize concepts, with his insistence upon the prime importance of the universal logical determinations in language? The commentary to volume three of the present edition should help to call attention to the hitherto neglected point that a full-scale investigation of his treatment of such matters by someone fully qualified in the history of comparative philology would be an extremely worthwhile undertaking. The stimulating analysis of the wide range of ordinary academic issues included in this treatment should not, however, lead us to overlook the fact that it is the *dialectical* structuring and placing of language that has to be regarded as the most significant feature of Hegel's linguistics.

He was considering language as essentially an aspect of psychology as early as 1794, and this basic conception never changed. The precise nature of his mature view of it as *phantasy*, presupposing intuition, recollection and imagination, and being the immediate presupposition of memory and thought, has been dealt with in some detail in the commentary, and need not be reconsidered here. Its most important features are probably the relegation of subject-object problems to the subordinate sphere of phenomenology, and the close identification of language with thought. It is worth noting, however, that such a neat and illuminating dialectical progression as he provides, could only have been worked out by first analyzing the inherent and proximate levels of complexity in the general sphere of subject-matter under consideration, and then synthesizing in the light of his knowledge of the further levels of complexity still to be dealt with. For Hegel, therefore, language is the subject of philosophical enquiry in a very specifically differentiated sense, it being the task of philosophy to sort out the various factors or levels it involves, and then give them their systematic placing in relation to all other levels of complexity. Curiously enough, although in § 459 he refers back to the physiological foundations of language, he makes very little reference to subsequent dialectical developments. One might have expected him to observe that since there could be no law, no legislative assemblies, no poetry, no religion, no philosophy without language, it might very well be regarded as *the* presupposition of the subsequent Philosophy of Spirit. It should be remembered, however, that he sees it as simply *one* presupposition *in context*. For him, it is language that is the subject of philosophical enquiry, not philosophy that is the subject of linguistic enquiry.

Anthropologically primitive, geographically and historically contingent, culturally fluid though it is, language for Hegel is also an adequate means for philosophizing effectively, for analyzing and establishing the relationships between categories, conveying a rational conception of the natural sciences, exploring the whole complexity of the philosophy of spirit. He puts a high value upon the human experience embodied in verbal utterance, the circumstantial rationality and pragmatic effectiveness of the words we use, and his dialectical assessment of linguistics was evidently the outcome rather than the justification of this basic attitude. When discussing the physiological effects of certain sensations for example, he observes that, "Every one of us is already familiar with the main phenomenon of this embodiment on account of language, which contains much that is relevant and cannot very well be explained away as an age-old error." (II.187, 24). When dealing with the nature of memory, he insists that it is absurd, "to regard thought as defective and handicapped on account of its being bound to the word, for although it is usually precisely the *inexpressible* that is regarded as most excellent, this is a vain and unfounded opinion, for the truth is that the inexpressible is merely a turbid fermentation, which only becomes clear when it is capable of verbalization. It is therefore the word which endows thoughts with their worthiest and truest determinate being. It is of course also possible to fling words about without dealing with the matter. The fault here derives not from the word however, but from the deficiency, indeterminateness and incapacity of the thinking" (III.205, 20).

An interesting outcome of this faith in the inherent wisdom of ordinary language is the way in which he occasionally appeals to it in order to back up what he thinks may be an unacceptable or difficult point. While dealing with the complicated and elusive connection between our inner sensations and our bodily organs for example, and noting that there are various exceptions to the general rule he is attempting to establish, he observes that, "even language has enough understanding to equate *heart* with courage, *head* with intelligence, and not equate heart and intelligence for example" (II.189, 30). The important transition from intuition to presentation, which he was not able to formulate satisfactorily until the very end of his career, is eventually clarified by the curious procedure of dwelling upon the literal significance of the perfect tense: "If I say, 'I have seen this', when speaking of an intuition sublated into a presentation, language therefore has everything to be said for it. What is expressed here is not merely past but also presence, the past here being merely *relative*, since it occurs only in the *comparing* of the *immediate* intuition with what we now have within presentation. The wholly peculiar significance of the word '*have*' used in the perfect tense is however that of presence, for what I have seen is not merely something I had, but something I still have, and which is therefore still present within me." (III.143, 22). He is quite fond of punning as a means of establishing a point, and when dealing with the associative imagination evidently goes out of his way to justify the propensity: "the deepest passion can indulge in such play, for a great mind knows how to set everything which confronts it in relation to its passion, even in the sorriest of circumstances.' (III.165, 21). The use he makes of shades of meaning common in his native Swabian but unusual in standard German was probably very largely unconscious, although he does occasionally show that he was not entirely unaware of it, as for example when he observes that his fellow countrymen do not distinguish between smell and taste: "A flower is said to have a '*nice taste*' instead of a '*nice smell*', so that in so far as we also smell with our tongue, the nose tends to be superfluous." (Phil. Nat. II.161, 20).

This respect for the inherent wisdom of ordinary language does not however, hinder him from enlisting the words he has at his disposal in the *service* of his thought, *extending* their meaning in order to express what he has to say. He may occasionally have appealed to the popular wisdom embodied within them for support in establishing a philosophical point, but this should not lead us to overlook the fact that his predominant and overriding procedure is that of remoulding, clarifying and giving a new precision to language by submitting it to philosophical analysis. From the list of definitions provided at the end of this chapter, one might pick out Accidenz, Erkenntniß, Erscheinung, Inhalt, Sache, Schein, Urteil, as examples of words the precise meanings of which depend upon the analytic and synthetic work on the relationships between *universal categories* carried out in the 'Logic', the first part of the 'Encyclopaedia'. An enquiry into the background, the history, the normal usage of such words as Abbild, Empfindsamkeit, Fähigkeit, Gegenstand, Leidenschaft, soon brings to light the extent to which he changes and develops meanings in order to establish hitherto unfamiliar connections, draw what he considers to be essential distinctions, or reassess the subject-matter being dealt with. Although he was probably unaware of the Swabian background determining his use of such words as auffassen, befangen, Eigenheit, Sache, Zerstreuung, it is tempting to suppose that this is not the case with Gedächtnis, and that he must have been aware of what §§ 464–465, the transition from memory to thought, owed to the language of his native province. The most characteristic procedure in this general tendency to develop and transform ordinary language is that of dwelling upon the literal meanings of words and phrases in order to extract the required significance from them.

Begreifen, Einbildung, Empfindung, Erinnerung, Urteil, Verrücktheit are examples of this, and many more may be found in the main body of these lectures. His use of the unusual word 'convolution' in connection with the various objects of the intuited world for example, was evidently determined by his knowledge of its literal meaning (II.209, 12), and the establishment of the precise nature of habit, which he confesses to be one of the most difficult tasks facing the philosophy of anthropology, is taken to involve the literal interpretation of a common phrase: "Habit has quite rightly been said to be second nature, for it is *nature* in that it is an immediate being of the soul, and a second nature in that the soul *posits* it as an immediacy, in that it consists of an inner formulation and transforming of corporeity pertaining to both the determinations of feeling as such and to embodied presentations and volitions" (II.391, 31). There are many other instances of this basic procedure, — perception finds its dialectical placing within the Phenomenology on account of the literal meaning of Wahrnehmen (note III.27, 3), the etymological connection between 'whole' and 'holy' is noticed in order to get subject-object problems into perspective (note III.85, 29), a pleonastic language-game clinches the dialectical definition of 'interest' (note III.253, 33) etc. It is worth noticing, perhaps, that although this use of literal meanings, real or imagined, constitutes Hegel's main method of transforming ordinary language, it is also the outcome of his belief in the inherent rationality of the popular medium with which he is working. As he seems to have conceived of the matter, to extend and clarify thought by searching out the basic constituents of the language in which it is expressed, is to strip away the superficialities of thoughtless convention in order to reveal the wealth of human experience and intelligence embodied in our linguistic inheritance.

To some extent his use of such words as Beisichsein, herabsetzen, übergreifen, umfassen, Vereinzelung, Wahrheit, can be understood from the ordinary language of his day, but since he also uses them in connection with one of the central principles of the dialectic, some idea of the more purely philosophical aspect of his work is also necessary if their full significance is to be grasped. As has been pointed out in the definitions, they all derive their special meaning from the general conception of what might be thought of as the logical connection subsisting between the terms of a syllogism, or, if we are also taking the subject-matter of the expositions into account, the asymmetrical relationship between various degrees of complexity. The procedure of distinguishing levels of complexity, which is involved in working out the relationship between this general conception and the subject-matter of Subjective Spirit, has given rise to the dialectical determining of a further group of words, — beseelt, Empfinden, Gefühl, Gemüt, Kenntnis, Naturell, Neigung, schauen, Selbstgefühl, Trieb, verleiblichen, Vorstellung, Wissen. In cases such as these, as has been pointed out, it is not only the formal, but also the real, the historical, the contingent, the empirical aspect of the expositions that has to be taken into consideration if the full meaning of the text is to be grasped.

It is to be hoped that the present commentary to this section of the 'Encyclopaedia' will go some way toward opening it up to the ordinary procedures of historical and empirical research, for without the firm foundation of substantially uncontroversial knowledge to be provided by such an investigation, there is not much chance of expounding its wider philosophical significance in any very worthwhile manner. Some ordinary empirical definitions, Besonnenheit, Disposition, Gefühlsleben, Objekt, Stoff, Vermögen for example, quite a few of which involve reference to Hegel's academic preoccupations rather than his use of the everyday language, have been listed at the end of this chapter. Most of them have, however, been provided in the commentary. The history of the various reactions to these lectures shows only too clearly that the precise significance of the broad distinctions between the soul, consciousness and spirit should be firmly grasped before any attempt is made to interpret the various sub-divisions of the three major disciplines of Subjective Spirit (notes II.3, 1; III.79, 1), and with regard to the details of the text, the problem of insufficient basic information can be even more pressing. Words such as Idiosynkrasie (note II.93, 14), Herz (notes II.155, 34; 225, 33), Ton (note II.411, 4), Beiwesen (note III.139, 35) are easily misinterpreted for example, and if discussed without reference to the contexts from which they derive their significance, soon give rise to a distortion of Hegel's meaning. The most difficult section of the work in respect of the technical terminology employed is that devoted to mental derangement (§ 408). In order to avoid basic misunderstandings, concepts such as Delirium, Faselei, Narrheit, Verrücktheit, Wahnsinn, Zerstreutheit, have had to be discussed and defined not simply with regard to the general background literature of the time, but also with reference to the works Hegel is known to have drawn upon. In this particular case, it has also been necessary to pay close attention to the methods of classification then employed in pathology, to the whole field of early nineteenth-century nosology. — In the spheres of botany and zoology, the artificial classificatory system of Linnaeus had tended to stimulate progressive and fruitful research, but in that of pathology, it merely encouraged attempts to simplify and impose apparent order upon diseases which were very imperfectly understood, even at a predominantly physical level. In the case of mental derangement, as in the case of physical disease, imperfect diagnosis, carried out in the light of erroneous, over-simplified and arbitrarily systematized principles, gave rise to much practical inefficiency and a terminological chaos. The truth of the matter was, that the medical know-how of the day, unlike the botanical and zoological knowledge, was incapable of providing a satisfactory empirical foundation for the sort of elaborate classificatory system being formulated.

Since most of the. English works on the subject soon appeared in German versions, translating Hegel's treatment of the psychiatry of his day presents us with some extremely complicated problems. It is apparent from a contemporary survey of the work of such influential psychiatrists as Haslam, Pinel, Cox and Arnold ('Quarterly Review', II.155–80, 1809), that Hegel's English contemporaries were well aware of the importance and difficulty of defining basic terms, even in their own language. A lexicographical analysis of the Anglo-German psychiatric literature of the period, especially the translations, would, therefore, be an extremely valuable undertaking.

It may not be out of place to conclude this general survey of Hegel's language by making a few observations on the business of translating it. Two contemporary German dictionaries, both compiled on Johnsonian principles, are essential, — J. C. Adelung's 'Grammatisch-kritisches Wörterbuch der Hochdeutschen Mundart' (4 pts. Vienna, 1807–8) and J. H. Campe's 'Wörterbuch der Deutschen Sprache' (5 pts. Brunswick, 1807–11). They should be supplemented by reference to J. and W. Grimm, 'Deutsches Wörterbuch' (16 vols. Leipzig, 1854–1954), and A. Götze, 'Trübner's Deutsches Wörterbuch' (ed. W. Mitzka, 8 vols. Berlin, 1939–56), both of which are firmly based on historical principles, and give copious references to their sources. Hegel's Swabian inheritance is dealt with by R. Schneider in his 'Schellings und Hegels schwäbische Geistesahnen' (Würzburg, 1938), and the linguistic aspect of it may be explored in exhaustive detail in the monumental 'Schwäbisches Wörterbuch' (6 vols. Tübingen, 1904–36), by H. Fischer and W. Pfleiderer. W. T. Krug's 'Allgemeines Handwörterbuch der philosophischen Wissenschaften' (1827–34; 2nd ed. 5 vols. Leipzig, 1832–8) is most useful as a comprehensive survey of the philosophical terminology of the period, although like J. Ritter's 'Historisches Wörterbuch der Philosophie' (Darmstadt, 1971ff.), it has to be used with caution, since accepted philosophical usage was by no means always identical with Hegel's, and evidence drawn from the 'Phenomenology' of 1807 or H. Glockner's 'Hegel-Lexikon' (2 vols. Stuttgart, 1957) is not necessarily a reliable guide in a detailed analysis of the Berlin lectures. As will be apparent from the commentary, the only way to make certain that one is doing justice to Hegel's treatment of the specialized disciplines surveyed in 'Subjective Spirit', is to get to grips with the Anglo-German literature of the time, especially the translations. There is a good contemporary 'Dictionary of the... German and English Languages' (2 vols. Karlsruhe, 1828) by J. L. Hilpert however, and an apparently little known but extremely valuable 'Dictionary of the... German and English Languages adapted to the present state of literature, science, commerce and arts' (4 pts. Bremen and London, 1854–8), by N. I. Lucas.

In 1894 William Wallace published a translation of §§ 377–577 of the Encyclopaedia, which has recently been re-issued together with an English version of Boumann's Additions (1971). The work shows that apart from the lexical matters already discussed, the principal elementary factors to which English translators should pay particular attention when dealing with such a text are the *emphasized words*, especially those in the §§ actually published by Hegel, the *difference between singulars and plurals* (I.14, 3; 82, 31; II.10, 31; 26, 21; 28, 28; 30, 18; 78, 8; 84, 9; 84, 13; 124, 14; 166, 32; 172, 30; 180, 26), *tenses* (I.16, 8; 18, 15; 64, 14; II.2, 1; 30,6; 252, 35), and the *philosophical implications of active and passive verbs* (II.14,10; 14, 34; 80, 31; 98, 4; 120, 19; 184, 22–26; 210, 9; 264, 8–15; 294, 25; 370, 4). Our general understanding of most of Hegel's works has improved considerably since late Victorian times, and in some cases has been revolutionized by recent research. When compared with the sober precision of the German original, the exuberance of the associative picture-thinking which led Wallace into producing such phrases as 'man's genuine reality' (for I.2, 7), 'a shadow cast by the mind's own light' (I.70, 8), 'the superstructure of specialized and instrumental consciousness' (II.216, 30), 'the more public consciousness is so liberal' (II.224, 26), 'the self is to be stamped upon and made appear' (II.386, 11), 'this spontaneity of self-centred thought' (II.396, 10) etc., though certainly interesting, can hardly fail to strike the modern Hegel scholar as misleadingly imprecise and impressionistic. The re-examination of the original manuscripts carried out over the last seventy years has made it quite clear that we cannot afford to overlook the care and consistency with which Hegel chose his words, and that it is absolutely essential to the effective communication of his thought that concepts as basic to it as implicitness (I.44, 20), relationship (I.30, 8; II.34, 5; 118, 25) or sublation (I.28, 7; 52, 26; 76, 15; II.148, 31–3; 254, 33), should not be lost sight of in translation. When dealing with a text as painstakingly prepared as Boumann's, we should take care not to confuse coming with showing (I.44, 26), ideality with being of an ideal nature (II.106, 14; 166, 21; 202, 26; 270, 11), shape with figure (II.408, 29), and make certain that the distinctions drawn in the 'Logic' between things and objects (I.74, 1), subsistence and existence (I. 28, 26), existence and determinate being (I.44, 21), determinations and categories (II.10, 29), as well as those established elsewhere between empiricism and induction (I.10, 1), sensation and feeling (I.32, 33; II.24, 14–19; 184, 11), peculiarity and idiosyncrasy (I.8, 2; II.92, 8), feeling and consciousness (II.204, 2), insanity and derangement (II.326, 22) etc., are not obliterated. With regard to the subject-matter, the actual material dealt with in the work, one might well ask what point there can be in encouraging anyone to think that Hegel ascribed urges to plants (I.30, 17), that he thought light expands (II.168, 13) and that writing is therapeutic (II.200, 14), or that he was unable to distinguish between comparison and non-entity (II.230, 4), presence and existence (II.268, 9), a terrain and a continent (II.54, 8), an impulse and an instinct (II.58, 30), a sword and a dagger (II.92, 13), the

Far East and India (II.58, 22), peoples and races (II.66, 9), Indians and Hindus (II.56, 12), Africans and savages (II.74, 31), Englishmen and their rights (II.80, 2).

In the English version of the text published here, the attempt has been made to reproduce the element of consistency in Hegel's terminology. A preliminary survey was carried out, equivalents for recurrent terms not already defined in the corresponding edition of the Philosophy of Nature were decided upon, and as long as the outcome was reasonably compatible with normal language, they were then employed throughout. The policy has obvious advantages in respect of indexing, tracing developments and examining the interaction between vocabulary and context, and there turned out to be surprisingly little need to circumvent awkward, inconsistent or unintelligible phrases by modifying it.

Definitions

Most of the words defined here present difficulties on account of problems arising out of the subject-matter dealt with in the 'Philosophy of Subjective Spirit'. The reader is referred to the corresponding edition of the 'Philosophy of Nature' (I.146–177) for definitions of recurrent terms more closely related to the 'Logic' or the general principles of Hegelianism.

Abbild: likeness. Since the word is often used of an artist's *immediate portrayal* of something, it differs from *picture* or *image* in necessarily implying liveliness, a direct relation to reality. It first came into use in the seventeenth century, but since it was still a rarity at the end of the eighteenth (Adelung), Hegel's use of it in relating such extremely generalized entities as *spirit, Idea, soul, Notion, understanding* etc., though certainly idiosyncratic, cannot be regarded as simply an extension of the everyday language. There is no evidence in these lectures that he associated it with the imagist theory (Abbildtheorie) of perception put forward by Leucippus and Democritus.

    I.3, 31 'The consideration of spirit... comprehends spirit as a likeness of the eternal Idea.'

    II.23, 9 'The natural determinateness of the soul is to be grasped... as being a likeness of the Notion.'

    II.429, 18 'The human soul... has transformed its body into the likeness of its ideality and freedom.'

    III.31, 25 'This simple difference is the realm of *the laws* of their appearance, their quiescent and universal likeness.'

Accidenz: accidence. A precise definition is provided in the treatment of the sub-categories of actuality and essence (Logic §§ 150/2); cf. Aristotle, 'Posterior Analytics', 73a35 – 73b24; Kant, 'Critique of Pure Reason', A 186,

B 229. Hegel does not often use the word, and the unusual English equivalent enables one to distinguish it from the much commoner *accident* (Zufall). Cf. John Pinkerton (1758–1826), 'Petralogy. A treatise on rocks' (London, 1811), Introd. p. 4, "Petralogy... is divided into twelve domains... six being distinguished by circumstances or accidences of various kinds."

II.207, 26 'This content still relates itself to the feeling soul as accidences do to substance.'

II.405, 24 'The soul's... body becoming an accidence brought into harmony with its substance, which is freedom.'

Affection: affection. The German word, which entered the language from Latin (afficere) during the sixteenth century, involves the concept of *being affected* by something. Hegel, like many of his contemporaries, regarded affections as being closely associated with sensations (§ 401), and as therefore differing from *passions* such as ambition, covetousness and revenge (§§ 473/4), in that they are intense and ephemeral, and not the outcome of a slowly developing or persistent desire. If an affection's effect upon a person's disposition remained confined to the soul, it was a feeling; if it influenced his body, it was an 'Affekt'. Cf. Kant, 'Anthropologie' (1798), §§ 73–88; J. G. E. Maaß, 'Versuch über die Gefühle, besonders über die Affecten' (2 pts. Halle, 1811/12); J. F. Pierer, 'Medizinisches Realwörterbuch', I.116 (1816); notes II.163, 27; 367, 10. III.129, 15.

I.129, 11 'The determinateness of sensation is still determined as an immediate affection within the soul.'

II.163, 27 'The most interesting aspect of a psychic physiology would be... the *embodiment* assumed by spiritual determinations, especially as affections.' (Affecte.)

II.175, 3 'Even in the affection of heat, feeling may be said to be involved to some extent with *solid* corporeality.'

III.119, 24 'The *form* of feeling is certainly a *determinate* affection.'

III.243, 14 'Practical feeling certainly knows itself... as subject to affections.'

Anlage: endowment. Although Paracelsus first introduced the word into German as the equivalent of the Latin *dispositio*, that is to say, as a term for the whole somato-psychic potentiality of a person, it did not come into common use until the middle of the eighteenth century. Hegel uses it, as did his contemporaries, in order to call attention to the natural, physical, inherited foundation of what is psychic. Cf. Goethe, 'Maxims and Reflections', 423: "Anlagen entwickeln sich zwar auch naturgemäß, müssen aber erst durch den Willen geübt und nach und nach gesteigert werden."

II.85, 25 'What is *natural* is understood to consist of natural endowments... These endowments include *talent* and *genius*.'

II.223, 29 'The child's... further endowments in respect of bodily shape, temper, character, talent, idiosyncrasies etc.'

II.397, 32 'The religious or moral etc. content... is in him neither merely *implicitly* as an endowment, nor as a transient sensation.'

auffassen: take up. In earlier usage, the word was associated with the rapid movement of seizing a sword, — which might give us 'quick on the uptake' as a suitable English equivalent. Since it was also used for arresting in the name of the law, 'apprehend' might also be considered. Originally, however, it referred to the procedure of keeping and preserving liquids such as water, milk, oil, wine by 'taking them up' into *vats* (Fässer), and since this meaning had been preserved in Swabian (Fischer I.376), it almost certainly constituted an aspect of the associative thinking involved in Hegel's use of the word.

I.79, 3 'Cognition is taken up not simply as a determinateness of the logical aspect of the Idea.'

II.117, 4 'The youth... no longer recognizes his ideal in the person of a man, but takes it up as a universal.'

II.129, 10 '*The being-for-self* of the waking soul, taken up *concretely*, is *consciousness* and *understanding*.'

III.125, 26 '*Attention*, without which the taking up of the object... is impossible.'

Auffassung, the corresponding noun, has been rendered by *conception*, hence:

I.17, 2 'All those finite conceptions of spirit... have been called in question.'

II.109, 16 'It is however speech which enables man to conceive of things as being universal.'

II.363, 20 '*Together with* their distorted view of one point, fools also have... a correct conception of things.'

III.309, 33 'In so far as we conceive of necessity as inner connectedness, we are omitting the form of immediacy.'

Band: bond. The original meaning of the German word, as of its English equivalent, was simply that with which a *bundle* of things, such as sticks, is *bound* together. Cf. notes III.139, 28; 165, 1.

III.139, 27 'The spiritual bond holding together all the singularities.'

III.165, 1 'Intelligence gives the images a *subjective* bond, replacing their *objective* one.'

III.209, 4 'The ego... is at the same time the power over the different names, the empty bond which fixes series of them within itself.'

befangen: constrained. Just as the *boundary* of a plot of land can be *fixed*, so a spiritual capacity can be *cramped* or *restricted*. Although this analogical transference from the material to the spiritual had already taken place in Middle High German, and survived in Swabian (Fischer I.752), it fell out of use in standard German, and was not revived until the end of the eighteenth century: see Goethe, 'Faust', I.3818, 2824; note II.303, 7.

II.79, 28 'The originality of the English... is not unconstrained and natural.'

II.227, 25 'The two sides... in derangement stand in a relation to one another which is as yet unconstrained.'

II.235, 20 'The individual... in *dreaming* is constrained within a simple and immediate *self-relation*.'

II.403, 22 'Even the unconstrainedly religious person will not readily entertain this vain opinion.'

begreifen: comprehend. The fourteenth century mystics transferred the original meaning of grasping manually to the spiritual sphere, and so developed the double meaning of grasping intellectually and assimilating inwardly: P. Strauch, 'Margaretha Ebner und Heinrich von Nördlingen' (Freiburg/Br. and Tübingen, 1882), p. 43; Fischer, I.765. Christian Wolff (1679-1754) first brought the corresponding noun, *Begriff* (Notion), into general philosophical use: Paul Piur, 'Studien zur sprachlichen Würdigung Christian Wolffs' (Halle, 1903), p. 85, but at the end of the eighteenth century it was still regarded as too academic for an acceptable sermon, see P. H. Schuler (d. 1814), 'Geschichte der Veränderungen des Geschmacks im Predigen, insonderheit unter den Protestanten' (3 pts. Halle, 1792/4), VIII.

I.15, 22 'The need to comprehend is still more pressing on account of the oppositions which immediately present themselves.'

II.243, 30 'In this field, even believing what one sees, let alone comprehending it, has freedom from the categories of the understanding as its basic condition.'

III.193, 31 'In German for example, the Notion, spiritual comprehension, corresponds to the natural procedure of grasping with the hand.'

III.283, 25 'Fichte wanted to comprehend consciousness itself.'

Beisichsein: being with self; self-communion. Ordinary German usage indicates that Hegel must have associated this word with living alone, keeping something to oneself, thinking to oneself etc. Judging from the use he makes of it, one might feel justified in also applying it to a unity which is able to maintain sub-units within itself, a whole which pervades its parts, a syllogism including its premisses and conclusion, a healthy organism etc.

I.39, 24 'Spirit is therefore far from being torn out of its simplicity and being with itself.'

II.165, 30 'In contrast to natural things, I am the universal which is with itself in determinateness.'

II.347, 3 'This consciousness is therefore not truly with itself, but remains engrossed in the negation of the ego.'

In order to distinguish its positive character from its mere pervasiveness, *self-communion* has also been used as its English equivalent:

II.245, 10 'The form of feeling considered here constitutes the surrender of the individual's existence as a self-communing spirituality.'

II.399, 13 'The undisturbed self-communion of the soul in all the particularity of its content, is entirely a matter of necessity.'

II.401, 1 'It is this self-communion that we call *habit*.'

beseelt: animated. Literally, 'besouled'. The word first appeared in print in the works of the Stuttgart poet G. R. Weckherlin (1584–1653), and implies, perhaps, rather more spirituality than its English equivalent. Cf. F. W. Gotter (1746–1797), 'Gedichte' (Gotha, 1787), p. 158: "Für alles, was sie sonst beseelte, Ist sie nun kalt." Hegel seems to have employed it with the distinction between living being and the *soul* in mind (note II.3, 1):

II.43, 32 'It is in (animals) rather than man that the sympathetic moods of animated life become most evident.'

II.165, 25 'The universal nature of the animated individual also displays itself.'

Unfortunately, it has not been possible to reproduce this distinction:

I.33, 32 'The sex-relationship is therefore the culminating point of animate (*lebenden*) nature.'

I.125, 2 'In sensation therefore, the soul is free animation (*Lebendigkeit*).'

III.315, 7 'Consciousness of a certain general object is present, animation (*Lebendigkeit*), I relate myself to a living being.'

Besonnenheit: self-possession. The root of the noun is the verb 'besinnen', i.e. to consider, deliberate, reflect upon. Hegel's use of it is almost certainly influenced, directly or indirectly, by the definition of the cardinal virtue of soundness of soul in Plato's 'Charmides'. Cf. H. North, 'Sophrosyne, self-knowledge and self-restraint in Greek literature' ('Cornell Studies in Classical Philology', vol. 35, 1966); note II.243, 4.

II.143, 35 'The self-possessed person also dreams, but he is none the wiser when he awakes.'

II.243, 4 'As form, as a state of the self-conscious, cultured, self-possessed person, the life of feeling is a disease.'

III.247, 1 'A self-possessed person... can encounter something conforming to his will without breaking out into the feeling of joy.'

Betrachtung: consideration. Although its basic Latin element (tractare, to handle) came into German very early on, and the word was used in its modern sense in Luther's translation of the Bible, it has never become truly popular. In the everyday language it simply means *consideration, weighing, reflection*, but the late mediaeval mystics connected it with losing oneself in God, and in certain contexts it is still associated with the 'vita contemplativa'.

I.3, 28 'The consideration of spirit is only truly philosophical when it recognizes the living development and actualization of the Notion of spirit.'

II.141, 21 'The ego... renders the general objects free of its subjectivity, while also considering them as totalities.'

III.245, 32 'I am also perfectly capable of contenting myself with a calm consideration of the matter.'

Bildung: instruction, culture, cultivation, training. The contexts in which Hegel uses the word make it impossible for the translator to confine himself to a single English equivalent. On its complicated general history, which is not, however, directly relevant to an understanding of the present text, see I. Schaarschmidt, 'Der Bedeutungswandel der Worte bilden und Bildung' (Dissertation, Königsberg, 1931).

I.81, 17 'That which constitutes instruction... is to be... excluded from the progression to be considered here.'

I.91, 17 'The specific standpoint of Greek culture.'

II.195, 28 'The various kinds of laughter are therefore indicative... of an individual's level of culture.'

II.419, 17 'It is not only cultivation and lack of cultivation which are apparent in the particular manner of walking.'

II.407, 3 '*Training* is required if this pervasion is to become *determinate*.'

Darstellung: representation. Cf. the definition of *Vorstellung*. The basic and original meaning, which is simply that of placing something somewhere, see Luther's translation of Exodus XL.4, shades naturally into the secondary meaning of *exhibiting*, — Luther's translation of I Samuel XVII.6. Hegel sometimes has in mind the pictorial representation or portrayal which gave rise to 'Darsteller' for *actor* during the latter part of the eighteenth century: H. P. Sturz (1736–1779), 'Schriften' (Leipzig, 1779), II.283, "Wenn England seine Darsteller neben Königen begräbt".

I.29, 21 'External nature resembles spirit in that it also is rational, divine, a representation of the Idea.'

II.135, 28 'In antiquity we find sleep and death represented as brothers.'

II.167, 13 'The rational necessity of this... is demonstrated in that we grasp the senses as representing moments of the Notion.'

III.175, 1 'The strength of Jupiter, for example, is represented by the eagle.'

I.57, 19 'Spirit exhibits the unity of form and content.'

II.197, 25 'The expression in the eye exhibiting the changing portrait of the soul, the canvas onto which it is breathed as it were.'

Disposition: disposition. Hegel's use of the word is in harmony with Aristotle's definition of διάθεσις as 'the order of that which has parts' (Met. 1022 b 1); cf. Aquinas' definition of *dispositio* (Met. Arist. Comm. no. 1058).

II.43, 35 'He felt justified in concluding, from the sound dispositions he had divined in animals, that he might expect something similar.'

II.265, 20 'In a person disposed in a specific manner it can be brought about by purely external things.'

II.293, 1 'By virtue of a certain particular disposition.'

The English word has also been used for Hegel's 'Gemüt'.

Eigenheit: characteristic property. The German word often has the slightly negative connotation we associate with *peculiarity, singularity* or *oddity:* Goethe, "Eigenheiten, die werden schon haften, Kultiviere deine Eigenschaften", see P. Fischer, 'Goethe-Wortschatz' (Leipzig, 1929), p. 170, but this seems to be absent from Hegel's use of it, as it was from the Swabian of his day (Fischer II.571).

II. 153, 16 'Its *most particularized* and characteristic *natural* property.'

II.153, 38 'In sensation... content... is therefore posited as my *most characteristic property.*'

II.155, 32 'While man's *thinking* is the *most characteristic property* distinguishing him from beasts, he has sensation in common with them.'

Einbildung: imagination, formulation. From an analysis of the relationship between actuality and what we believe about it, Aristotle concluded that "imaginations persist in us and resemble sensations" ('De Anima', 429 a 5). The object-subject relationship implied by the word was exploited by the late mediaeval mystics, and influenced Paracelsus' use of its German equivalent ('Werke', ed. K. Sudhoff, I/9, 251ff.). This relationship is quite evidently in Hegel's mind when he uses closely related forms to convey the idea of formulating one thing within another, — a highly idiosyncratic use of the word which seems not to have been Swabian in origin (Fischer, II.591).

II.249, 11 'Clairvoyance is *exposed* to all the *contingency* incident to feeling, imagining etc.'

II.367, 29 'Madness... need not stem from a *vacant imagination.*'

III.157, 27 'The intelligence active within this possession is the *reproductive imagination.*'

II.389, 38 'The formulation of presentative determinations within corporeity' (Einbildung).

II.391, 13 'This formulation... of feeling within the *being* of the soul' (Sicheinbilden).

II.427, 2 'The soul as formative within its corporeity' (Hineinbildung).

III.51, 6 'The relationship of desire to the general object is not *formative*' (des *Bildens*).

Empfinden: sentience. Hegel uses the word to emphasize sensation's involvement in objectivity. In his fully mature terminology therefore (notes II.215, 18; III.119, 22), sentience falls below sensation and feeling in degree of inner coherence.

II.161, 5 'One sphere of sentience distinguishes itself principally as the determination of corporeity (of the eye etc., of the parts of the body in general), and becomes sensation in that it is *recollected*, internalized within the being-for-self of the soul.'

II.161, 23 'Sentience is the individual spirit living in healthy partnership with its corporeity.'

II.425, 12 'The actual soul, in its *habitual* sentience and *concrete* self-awareness.'

Empfindsamkeit: sensitivity. The direct, basic, almost physical aspect of sensation. This is a curious, and perhaps significant alteration of the fashionable late eighteenth century meaning of the word (I.5, 37). Lessing suggested to J. J. C. Bode (1730–1793) that he might use it in translating Sterne's 'Sentimental Journey', see 'Empfindsame Reise' (Hamburg, 1768/9), preface, and it then swept Germany. Goethe subsequently spoke of the 'Empfindsamkeitskrankheit' from which he had freed himself by writing 'Werther' ('Werke', 43, p. 318), and after a couple of decades the Stuttgart journalist C. F. D. Schubart (1739–1791) was condemning the word as ridiculous on account of its having been so abused by 'idiotic scribblers': 'Vaterlandschronik' (Stuttgart, 1787/91), 1789, p. 785.

II.175, 27 'A certain determinateness of the subject's sensitivity, a reaction of subjectivity to externality.'

II.177, 2 'Through this peculiar measure of sensitivity... exterior sensation becomes something which is peculiarly anthropological'.

II.203, 15 'Sensation involves sensitivity, and there is reason for maintaining therefore, that... sensation puts more emphasis upon the passive aspect of feeling.'

Hegel uses *Sensibilität* (sensibility) for the immediate organic foundation of sensitivity, i.e. the capability a nerve possesses of conveying the sensation produced by the contact of another body along with it (Phil. Nat. III.302).

II.135, 24 'Animal life however, which *Bichat* takes to include the system of sensibility and irritability, the activity of the nerves and muscles.'

II.305, 23 'Whereas *outward-going* sensibility contradicts this *simple natur-ality.*'

Empfindung: sensation. Making use of the *literal* meaning of the word (note II.147, 9), Kant had *reversed* general usage and maintained that we *sense* heat in so far as we judge there to be a physical basis for the sensation in the hot body, and *feel* it in so far as it is pleasant or unpleasant. Hegel took some time about accepting the implications of this linguistic revolution, but he seems to have done so by 1822 (note II.215, 18), and in the 1827 and 1830 editions of the Encyclopaedia, §§ 399–402, gives his reasons (cf. note III.119, 22). Lack of consistency in translating this word can therefore give rise to a total miscomprehension of his thought and development in respect of this part of the 'Anthropology'.

II.151, 15 'If one and the same water were simultaneously both water in general and coloured, this differentiated determinateness would be for the water itself, and the water would therefore have sensation; for anything which maintains itself in its determinateness as a generality or universal possesses sensation.'

II.203, 14 'Sensation involves sensitivity, and there is reason for maintaining therefore, that while sensation puts more emphasis upon the passive aspect of feeling, upon *finding*, i.e. upon the immediacy of feeling's determinateness, feeling refers more to the selfhood involved here.'

III.121, 6 'The determinateness of sensation is derived from an *independent general object* which is either external or internal.'

erfassen: apprehend. Cf. the definition of *auffassen* for the origin of the word. Although apprehending, understanding, the assimilating of knowledge is its usual meaning, it can also be used of a wolf's *seizing* a sheep, or of a person's being *possessed* by an idea. In Swabian, as in English, its usual meaning is not part of the man in the street's everyday language (Fischer, II.792).

II.209, 3 'The simple unity of the soul... does not yet apprehend itself as distinct from what is external.'

III.7, 33 'It is only in that I come to apprehend myself as ego that the other becomes objective to me.'

III.355, 21 'I... relate myself wholly objectively... apprehending general objects as they are.'

Erinnerung: recollection. Quite distinct from memory (§§ 461–464), which presupposes imagination and is closely associated with thought. The original meaning of the German word, — to *inwardize* something in someone else, to ram it home, is immediately apparent, and played an important part in determining Hegel's dialectical definition (notes III.17, 11; 125, 6). It should, perhaps, be observed, that in Luther's translation of John XIV.26,

Christ tells his disciples that the Holy Ghost will 'inwardize' whatever He had said to them. For Hegel, however, recollection is a rather basic level of Psychology (§§ 452–454), concerned primarily with the inwardizing or subjectivizing of sensation (note III.149, 25).

I.133, 4 'This is not yet the place for an affection's passing into recollection proper.'

II.161, 8 'One sphere of sentience... becomes sensation in that it is *recollected, internalized* within the... soul.'

III.125, 4 'Although this is active *recollection*, the moment of *appropriation*, it is so as the still *formal* self-determination of intelligence.'

Erkenntniß : cognition. Cf. the definition of *Kenntnis* and *Wissen.* The word only occurs in Swabian when the language is being used for admonitory, hortatory or theological purposes (Fischer, II.809), hence Hegel's Logic §§ 223–232. Information (*Kenntniß*) is the knowledge involved in psychology, cognition is the systematic knowledge worked out in the philosophies of logic, nature and spirit (note III.125, 29).

I.79, 1 'In developing itself in its ideality, spirit has being as being *cognitive.*'

I.107, 10 'The only possible cognition of its true relationship is a speculative one.'

II.245, 17 'Scientific cognitions or philosophic notions and general truths require another foundation.'

II.249, 18 'A condition of heightened truth capable of conveying cognition of *universal* validity.'

III.107, 27 'What intelligence does as theoretical spirit has been given the name of *cognition.*'

III.355, 16 'There is the belief and trust that I am able to cognize things.'

Erscheinung : appearance, phenomenon. Cf. the definition of *Schein*, and the Logic §§ 131–141, where Hegel deals with sub-categories such as content and form, relationship, force and its expression etc. *Appearance* has been used when reference is being made to this general sphere of categories.

I.97, 14 'The cognition of the *appearances* is not only of the utmost importance, but completely indispensable.'

II.193, 16 'The connection between these physiological appearances and the motions of the soul.'

III.227, 4 'The circumstances conditioning an appearance still pass for independent existences.'

Hegel does occasionally use the word *Phänomen* (II.284, 33; note III.3, 1), the seventeenth century German debut of which, see Christian Weise (1642–1708), 'Die drei ärgsten Erznarren in der ganzen Welt' (1673; ed. W. Braune, Halle, 1878), p. 164, seems to have set the scene for its extraordinary career in modern philosophy. He means by it something which simply appears, is there, and when he uses *Erscheinung* in this sense, *phenomenon* has been introduced as its English equivalent.

II.237, 8 'The phenomena that go by the name of *birthmarks*.'

II.271, 4 'The miraculous in phenomena of this kind.'

II.417, 13 'We shall... simply make mention of the commonest of these phenomena.'

Fähigkeit : aptitude. Strictly speaking, the German word can only be used of a *potential* ability, its original meaning being *capable of grasping*: Fischer, II.917; Goethe, 'Wahlverwandtschaften', I.5, "Fähigkeiten werden vorausgesetzt, sie sollen zu Fertigkeiten werden." Hegel, however, also uses it to mean a positive capacity.

II.223, 7 'Genius is the entire totality of determinate being, life and character, not merely as a possibility, aptitude or implicitness, but as effectiveness.'

II.315, 1 'Clairvoyants often acquire the aptitude for knowing something which is not a matter of their own... vision.'

III.327, 2 'The aptitude for satisfying the drive being this identity.'

fassen : grasp. Cf. the definition of *auffassen*. The original meaning of filling a vat with something was still common in the Swabian of Hegel's day (Fischer, II.966/7).

II.133, 23 'Spirit, which is to be grasped in its truth as pure activity.'

II.347, 29 'Derangement has to be grasped as essentially an illness of *both the spirit and the body*.'

III.165, 26 'The association of presentations is therefore also to be grasped as a subsumption of singulars under a *universal*.'

Gedächtnis : memory. Quite distinct from recollection (§§ 452–454), which presupposes intuition and is closely associated with imagination. The original meaning of the German word, — devotion, concentrating one's thoughts upon a person, a cause, God, had by no means disappeared from the Swabian of Hegel's day, in which a person's *general ability to think* could be referred to as his *Gedächtnis*: see Fischer, III.141, "Er hat ein gutes Gedächtnis, bloss behalten kann er die Sache net so gut" (note III.211, 21). For Hegel, memory (§§ 461–464) has language as its immediate presupposition, and is itself the immediate presupposition of thought.

II.277, 3 'A boy... gradually lost his memory to such an extent that he no longer knew what he had done an hour before.'

II.391, 23 'Like memory, habit is a difficult point in the organization of spirit.'

III.199, 18 'The internalizing of this externality in recollection constitutes *memory*.'

III.211, 19 'Even our language gives memory the high assessment of immediate affinity with thought.'

Gefühl: feeling. Cf. the definition of *Empfindung*. In the anthropological sphere, the elaborate section on the feeling soul (§§ 403–10) was extensively revised during the 1820's in order to distinguish more effectively between sensation and feeling. In the light of Hegel's final analysis of this context, sensation is more directly involved, by means of the senses, in what is physical, while feeling is the foundation of inwardness, the ego (note II.215, 18). It is important to note that he also drew a clear distinction in the psychological sphere between the basic feeling involved in intuition (§ 447; note III.119, 22), and feelings such as pleasure and pain, joy and repentance, which are treated as a sub-sphere of practical psychology (§§ 471–472).

I.129, 13 'Even in the feeling of freedom, the soul is not free.'

II.215, 19 'The feeling individual is *simple ideality*, subjectivity of sensation.'

III.121, 21 'However, the form possessed by spirit in feeling is the lowest and the worst, that of selfhood and singularity, within which spirit lacks the freedom of infinite universality.'

III.239, 25 'Only feelings that are self-seeking, bad and evil come under consideration in the treatment of practical feelings and dispositions, for the singularity which maintains itself in the face of what is universal possesses no other.'

Gefühlsleben: life of feeling. Since Hegel uses the word when referring to the *primitive* foundation of more advanced psychic states, one cannot assume that it embodies his mature distinction between sensation and feeling. He almost certainly derived the meaning he attached to the word from an article on the second sight by D. G. Kieser (1820), "Consequently, when man's *life of feeling* sees at a distance in time and space, this is parallel to the *life of reason*'s doing so. Both procedures are therefore properties of the human soul, the former being a *feeling at a distance*, the latter a *cognition at a distance*" (note II.243, 4). F. X. Baader (1765–1841), 'Ueber die Abbreviatur' (1822;) 'Gesammelte Schriften', 1853, pp. 110/12) criticized Kieser and Hegel severely for drawing this distinction.

II.225, 2 'In the immediate existence of this life of feeling... (the self) is also present as another distinct individual.'

II.251, 1 'This life of feeling lacks the personality of the understanding and the will.'

II.387, 13 'The self is to be posited within this life of feeling.'

Gegenstand: general object; subject-matter. Cf. the definition of *Objekt*. The word first appeared in German as a translation of the Latin *objectum* in 1625, but was not widely used in philosophical contexts until the eighteenth century. In High German it was still too academic for an acceptable sermon while Hegel was studying at the Tübingen Seminary: P. H. Schuler, loc. cit., see the definition of *begreifen*. In Swabian, however, evidently on account of its being associated with the rather more everyday Gegenüberstehende, it was quite a normal synonym for *impediment, opposition, obstacle, resistance* (Fischer, III.180). In these lectures at least (cf. Phil. Nat. I.163/4), Hegel's language is consistent in that he uses *Gegenstand* for what is general, unspecified undifferentiated, and *Objekt* for what is particular, specific, distinct (note III.7, 21).

III.7, 21 'The ego allows its other to unfold itself into a *totality* equal to its own, and precisely by this means, by becoming something *confronting* the soul *independently* rather than a corporeal being *belonging* to it, to become a *general object* in the strict sense of the word.'

III.111, 28 'Intelligence does not confine itself to this taking up of the immediately presented content of objects, but purifies the general object of that within it which shows itself to be purely external.'

III.287, 8 'The general object is an object, out there for itself, posited immediately.'

He also uses the word when the subject-object antithesis is not under consideration, and in such cases *subject-matter* has been regarded as a more suitable English equivalent.

II.185, 38 'Our subject-matter at this juncture is simply the *embodying* of inner sensations.'

II.203, 26 'Through the immanent progression in the development of our subject-matter.'

Gehalt: capacity. Cf. the definition of *Inhalt*. Hegel's use of the word in these lectures indicates that he took it to refer to the *container together with its content*. In § 409 for example, the containing (*gehaltvolle*) truth of the specific sensations and desires is their conscious incorporation into the *psychology* of individuals, as, for example in the pursuit of happiness (note II.387, 18). It originally meant 'what is held', in custody for example. Since the fifteenth century it has also meant the *standard* of a precious metal, and this meaning gave rise to Luther's using it in order to refer to a person's *intrinsic worth* or *merit*. Towards the close of the eighteenth century it was introduced into aesthetics in order to distinguish between the *original living experience* basic to art, and the various artistic forms in which this experience finds expression.

I.11, 1 'The metaphysical factor failed to acquire any inwardly concrete determination and capacity.'

II.387, 16 'This is not the containing truth of the specific sensations.'

III.167, 2 'Intelligence is not only a general form, for its inwardness is *internally determined concrete subjectivity*, with a capacity of its own deriving.'

Gemüt: disposition. Cf. the definition of *Disposition*. Hegel confines the use of the Latin word to the more general meaning, and uses *Gemüt* for what is psychological. The looseness of the related English and German terminology makes it difficult to define the word satisfactorily, but it is perhaps useful to bear in mind that Hegel took it to be synonymous with 'heart', to be closely associated with feeling, and to constitute a basic or primitive factor in awareness (note II.225, 33).

II.225, 32 'This concentrated individuality also reveals itself as what is called the *heart* or *disposition*.'

II.393, 25 'The disposition becomes hardened to misfortune.'

III.91, 37 'Having a purpose through some kind of interest arising out of intelligence or disposition.'

III.165, 13 'Various dispositional determinations such as gaiety or gloominess.'

Gemüthsbewegung: dispositional disturbance. The word occurs in Dutch (*gemoedsbeweging*), and seems to have been introduced into German by J. J. Bodmer (1698–1783) in order to define the psychological cause of certain facial expressions and gestures: 'Critische Betrachtungen über die poetischen Gemählde der Dichter' (Zürich and Leipzig, 1741), p. 290. The basic conception it embodies is that of a person's composure of disposition (*Ruhe des Gemüths*) being disturbed by a pleasant or unpleasant presentation (*Vorstellung*). If the disturbance remains confined to the soul, it is a feeling (Gefühl), if it influences the body, it becomes an affection (*Affekt*): note II.163, 27.

II.223, 23 'The violent dispositional disturbances and injuries etc. experienced by the mother.'

II.239, 12 'Unpleasant dispositional disturbances will spoil the mother's milk.'

III.303, 31 'Regardless of the content of internal general objects, be they dispositional disturbances, spirit, God etc. they may also be known to us immediately.'

Gesinnung: conviction. The sense in which Hegel uses the word dates only from the middle of the eighteenth century. Although it was known in Swabia, it was usually confined to the written language (Fischer, III.530). He almost certainly associated it with the political disturbances of the revolutionary period, when the kings and princes of Germany appealed to their 'upright and well-intentioned' subjects (*Rechtschaffenen und Wohlgesinnten*) to stand firm against the disruptive tendencies of the time. By 1848 the left was reacting to the phrase with scorn and derision. The wider significance assumed by the word during this period is illustrated by A. W. Schlegel's letter of July 7th, 1807 to K. von Hardenberg: "Mir erscheint die Poesie in diesen verworrenen Zeiten besonders als Zeugniß der Gesinnung": see Josef Koerner, 'Briefe von und an A. W. Schlegel' (2 pts. Zürich, 1930) I.205; note III.269, 21.

II.43, 20 'In the case of a battle, it is not only ethical conviction, but also the level of morale, the feeling of physical strength that matters.'

II.83, 5 'The Germans... could... not often be brought to give proof of this their substantial conviction.'

III.269, 21 'Such existence so infuses the individual with the conviction of what is ethical.'

Cf. II.369, 18; III.277, 27; III.305, 29.

herabsetzen: reduce. Since one of the fundamental principles of a dialectical progression is the asymmetrical relationship between various degrees of complexity, it is only natural that Hegel should have used this word to mean the *regression* to what is less complex, more basic. There seems to be nothing to choose between his use of the German word and its Latin equivalent (reducirt). A previous English translator of these lectures (1971; 9, 5; 24, 29; 106, 23) actually used the word for aufheben (Phil. Nat. I.172–3): see Phil. Sub. Sp. I.28, 7; 76, 15. II.254, 33.

II.133, 25 'Spirit... reduces itself to the form of a natural or immediate being.'

II.197, 30 'A person... is reduced to a sufferer, the ideality or *light* of his soul is dimmed.'

II.393, 26 'The reduction of the sensation... to an externality, an immediacy.'

II.167, 15 'The quintuplicity of the senses reduces (*reducirt*) itself quite naturally to three classes.'

Inhalt: content. Cf. the definition of *Gehalt.* Hegel's use of the word is a clear instance of his being influenced by logical and philosophical considerations rather than the everyday language. Unlike Gehalt, the word refers not to the form but to what it contains. The most common use of it in the Swabian of Hegel's day was in connection with letters, articles, laws, sermons etc. (Fischer, IV.37).

I.55, 34 'The true content contains the form within itself therefore, and the true form is its own content.'

II.167, 2 'It is on account of its content's being of a sensuous nature that the sensation of what is external generally falls apart.'

III.149, 16 'Intelligence posits the *content* of *feeling* within its inwardness.'

Kenntnis: acquaintance, information. Cf. the definitions of *Erkenntnis* and *Wissen.* Its distinctive characteristic, according to Hegel, is that it is philosophically unpretentious, unsystematic (note III.125, 29).

II.95, 29 'A more concrete determination... would anticipate acquaintance with mature spirit.'

III.125, 29 'It is through attention that spirit first becomes present in the matter, acquires it, by gaining *information* about it.'

Leidenschaft: passion. Throughout the eighteenth century, the attention that had been paid to the passions by Descartes, Hobbes, Spinoza and Locke, led German psychologists to classify them as somewhat vaguely related to feeling and the baser appetitive faculty: M. Dessoir, 'Geschichte der... Psychologie' (2nd ed. Berlin, 1902), 439–445. Kant, 'Anthropologie', §§ 73–88 perpetuated this general attitude, and in such a run of the mill work as J. G. C. Kiesewetter's 'Erfahrungsseelenlehre' (2 pts. Vienna, 1817), the distinction is drawn between an irrational *passion* such as that of motherhood, giving rise to doting and molly-coddling, and natural *inclination* such as sexuality, giving rise to reasonable and healthy love between parents and children (note II.367, 10). Consequently, when Hegel (§ 474) determines passion as presupposing the will and as the immediate presupposition of free spirit, by the standards of his time he is giving it an extraordinarily high assessment.

I.95, 30 'Their passions, can easily be mistaken for *that which constitutes their* substantial character and will.'

II.79, 8 'In the midst of the storm of revolutionary passion their understanding displayed itself in... resolution.'

III.247, 25 'In so far as the totality of practical spirit interferes with any one of the many limited determinations posited within the general opposition, it is *passion.*'

III.251, 10 'Nothing great has been nor can it be accomplished without passion.'

nachdenken: think over. Hegel deals with the general philosophical significance of the procedure in the *Encyclopaedia* §§ 2–9 and 21–26. The fundamental everyday meaning is that of *directing* one's thoughts to something in order to come to grips with it, understand it better. To meditate, reflect, ponder, ruminate, consider may also be regarded as English equivalents. Cf. II.163, 30; 357, 4.

III.141, 30 'During the execution of his work a genuine poet has to *consider* it and *think it over.*'

III.355, 18 'I determine (things) through the spiritual activity of my thinking them over.'

Naturell: natural disposition. The word came into German from French during the second half of the seventeenth century: 'Zeitschrift für deutsche Wortforschung' (Strassburg, 1901–1914), vol. VIII, p. 81; see Johann Hübner (1668–1731), 'Natur... Lexicon' (Leipzig, 1717), "Naturel heißet eines Menschen Geburts-Art, Sinn, humeur, inclination." Hegel defines the word dialectically by reproducing the major transition from nature to the soul in the minor transition from natural disposition to temperament and character (note II.85, 27).

II.67, 30 'The distinct variety of natural disposition in the continent displays itself everywhere as bound to the soil.'

II.85, 25 'What is natural is understood to consist of natural endowments as distinct from whatever the person has become by means of his own activity.'

Neigung: inclination. Cf. the definition of *Leidenschaft.* There is no evidence in these lectures that Hegel associated the word with Kant's distinction (note II.367, 10), or that he had considered it in connection with Schiller's famous criticism of Kantian rigorism in 'Ueber Anmut und Würde' (1793). He regarded inclinations as quite distinct from natural dispositions in that they presuppose the *will* and are closely associated with passions (§ 473).

I.9, 23 'The supposedly petty intentions, inclinations and passions of these heroes.'

II.225, 26 'The totality of the individual in this compact condition is not the same as the unfolding of its consciousness into... inclinations etc.'

III.247, 22 'The will is at first *natural* and immediately identical with its determinateness, — *impulse* and *inclination.*'

III.347, 13 'The forms, which are those of feeling, inclination, benevolence, love friendship, do not concern us.'

Objekt: object. Cf. the definition of *Gegenstand,* which Hegel takes to be general, unspecified, undifferentiated (note III.7, 21). He uses *Objekt* in order to refer to what is particular, specific, distinct, and often for the counterpart of the ego. It might be worth investigating whether this usage was derived from the works of Kant, Fichte and Schelling, or whether Hegel was the first to draw the distinction.

II.123, 23 'The subject's interest in an object disappears at the same time as the opposition between them.'

III.11, 16 'The progressive logical determination of the object is what is *identical* in *subject* and *object.*'

III.97, 1 'This standpoint differs from that of consciousness in that the object is no longer determined as the *negative* of the ego.'

III.351, 6 'At this juncture, reality is self-consciousness, ego in relationship with an object presented to it externally.'

Sache: matter. The English word has only been indexed when it occurs as the equivalent of *Materie*. In the Logic (§§ 147–154), Hegel treats *Sache* as a sub-category of actuality, as presupposing conditions, and being the immediate presupposition of activity and cause (*Ursache*). The definition of the logical category is not always directly relevant to an understanding of his contextual use of the word however, which often conveys shades of meaning derived from everyday German, and, indeed, from his native Swabian: see R. Schneider, 'Schellings und Hegels schwäbische Geistesahnen' (Würzburg, 1938), p. 55. The word originally meant *contention* or *strife*, and in both German and English (sake) subsequently assumed the meaning of *cause, action* or *lawsuit*. This legal significance was primary in the Swabian of Hegel's day, and had given rise to four further meanings: 1) the object of a lawsuit, 2) a concern or affair, 3) a fact or circumstance, 4) a tangible object or thing capable of being a human possession (Fischer, V.513–515). Hegel's use of the word in these lectures can be categorized in the light of this linguistic background:

2) I.71, 25 'It is... regarded not only as a matter of understanding but also of moral and religious concern that the *supremacy* of the *standpoint* of finitude should be firmly maintained.'
III.171, 24 'As reason,... the action of intelligence is its making a *being* or *matter* of itself.'
III.205, 30 'It is of course also possible to fling words about without dealing with the matter.'

3) II.159, 6 'Whoever does so withdraws from what is common to all, the field of reasons, thought and the matter in hand, into his singular subjectivity.'
III.105, 31 'It is not for philosophy to take such imperfections of determinate being and presentation for truth, to regard what is bad as the nature of the matter.'

4) I.69, 20 'In property the matter is posited as what it is, i.e. as something which lacks independence.'
III.159, 20 'The same matter is subject to so many laws, that what occurs tends to be quite the opposite of a law.'

schauen: envision. Hegel takes intuition (*Anschauung*) to involve apprehension of, 'the genuine substance of the general object' (III.139, 2), and seems to have fixed the meaning he attaches to envision partly by drawing upon everyday usage and partly by interpreting the word literally (note III.145, 9). He evidently regarded the dropping of the *an* in *anschauen* (§ 449) as implying a much less certain involvement in objectivity, and uses *schauen* mainly in connection with animal magnetism (§ 406).
II.249, 26 'The revelations of somnambulistic vision.'

II.279, 14 'Examples of this kind of envisioning are much commoner from earlier times.'

II.279, 37 'The envisioning soul raises itself above the determination of time.'

II.307, 11 'The magnetic state displays itself... in the *spiritual* form of *vision.*'

Schein: apparency, show. Cf. the definition of *Erscheinung*, and I.109, 25; III.89, 7; 93, 12. In the Logic (§§ 112–14), Hegel deals with apparency as presupposing the categories of being, and as constituting the *basic* characteristic of the explanatory categories assessed within the general sphere of essence (§§ 112–159). The predominant meaning is closely related to the literal one of *shining forth* (Fischer, V. 743–4), as a content does in its form, a whole in its parts, an actuality in its contingency, a cause in its effect etc. *Apparency* and *show* have been treated as interchangeable in the translation.

I.111, 22 'The Notion is the eternal positing of this apparency to itself.'

II.21, 27 'The mediation whereby spirit has being is such as to sublate itself, to prove itself only a show.'

II.151, 4 'The immediacy of the soul is however reduced... to a *show.*'

III.111, 7 'Even *spirit*... still has the form of immediacy, and consequently the *apparency* of being *self-external.*'

III.345, 38 'I am, and I appear as such, and this apparency has being within the other.'

Selbstgefühl: self-awareness. Cf. the definition of Gefühl. In the Organics (§§ 337–376) see Phil. Nat. III.402, Hegel uses the word to mean *sentience*, in the Phil. Hist. (241–3), to mean self-confidence, which is its more usual meaning. In these lectures (§§ 403–10), it has the precise and literal meaning of the *feeling* involved in being aware of oneself. Such feeling adumbrates the ego of consciousness (§ 412), but since the subject is not yet able, 'to work up the particularity of its self-awareness into ideality and so overcome it' (II.327, 9), it is not yet consciousness proper. See H. B. Weber 'Vom Selbstgefühle und Mitgefühle' (Heidelberg, 1807) p. 36: "The former refers to the particular consciousness of our *subject* as an effective force, the second generally to the universal consciousness of our self-subsistent ego." Cf. note II.323, 28.

II.387, 6 'Self-awareness, in that it is immersed in the particularity of such feelings... is not distinguished from them.'

III.55, 25 'The immediacy... has its own self-awareness, as well as its being *for others.*'

III.245, 8 'When I am frightened I have a *sudden* sense of the discord between something external and my positive self-awareness.'

Stoff: material. The German word is ultimately related to *stopfen*, to plug or stop up, but re-entered the language about 1660 from Dutch as a term for cloth, stuff, material. As in English, the original verb probably helped to establish the secondary meaning of material or matter in general. It is perhaps of interest to note that the word was not popular in Swabian (Fischer, V.791), and that at the end of the eighteenth century the sense in which Hegel uses it was regarded as too *academic* for an acceptable sermon: P. H. Schuler, loc. cit., see the definition of *begreifen*.

I.17, 20 'It is not the random retailing of the material to hand... which constitutes the only scientific method.'

II.207, 3 'Consciousness effects its independence of the material of sensation.'

III.91, 27 'That whereby this material is spiritualized and its sensuousness sublated, being misconstrued and overlooked.'

Täuschung: delusion, deception. Cf. the definition of *Wahn*. From the original meaning of *untruthful talk*, the word came to be associated with deception, disappointment, illusion. Hegel uses it to imply a greater degree of self-awareness than *Wahn*.

II.243, 25 'They simplify their consideration of the matter by dismissing accounts of it... as delusion and imposture.'

II.269, 18 'He... was aware that this was simply a delusion.'

II.283, 30 'In this development there is every possibility of deception.'

Trennung: division. The word was rare in Swabian (Fischer, II.367), and is evidently used by Hegel to convey the concept of an *unnatural* separation, or at least of a cleavage capable of being overcome or healed.

II.15, 16 'The standpoint of this division is not, therefore, to be regarded as finally and absolutely true.'

II.255, 19 'Such obstruction and division sometimes progresses so far, that the particular activity of a system establishes itself as a centre.'

II.295, 3 'The essence of disease *in general* has to be posited as residing in the dividing off of a particular system.'

Trieb: drive, impulse. A complete survey of Hegel's use of this word (note III.249, 20) seems to indicate that when translating it into English his meaning is most accurately captured if drive is employed in metaphysical, physical (Phil. Nat. III.350), and, generally, more basic (§ 427) contexts, and impulse in connection with what is more distinctly psychological (§ 473). To imply, as did a former English translation of these lectures (1971), that Hegel regarded plants as having *urges* (9, 45), was quite evidently misleading.

I.31, 17 'In the plant there is already... something to which we ascribe a drive.'

II.117, 10 'In this subjectivity... lies not only its opposition to the extant world, but also the drive to sublate this opposition.'

III.259, 36 'It is interest, therefore, which expresses the determination distinguishing our having impulses, from self-consciousness having drives.'

I.67, 14 'The child is still immersed in naturality, having only natural impulses.'

II.365, 20 'An uncontrollable impulse to suicide.'

III.257, 20 'It is in this way that such a particularity of impulse is no longer immediate.'

übergreifen: include. Cf. the definitions of *begreifen* and of *sublation* (Phil. Nat. I.172). The original meaning of the word was the legal one of *encroaching*, on someone's land, rent, services, power etc. It subsequently assumed the more general meaning of *overlapping*, in the way that a wide-eaved roof over-laps the walls of a house for example, and it was evidently with this second meaning in mind that Hegel re-interpreted its literal significance. From the contexts in which he uses it, it seems reasonable to assume that he had in mind a relationship in which what might be regarded as an *active* unity includes its sub-units.

II.15, 13 'The true universal, in its inclusion of particularity, relates itself to the particular.'

II.271, 11 'The soul... *includes the other* rather than *standing onesidedly* opposed *to it.*'

III.77, 16 'The pure *ego*, the pure form which includes and encompasses the *object* within itself.'

III.219, 1 'The true universal, which is the inclusive unity of itself with its other.'

umfassen: embrace, contain. Cf. the definition of *auffassen*. The everyday meaning of the word is simply the physical act of embracing. Hegel's general usage indicates that he took it to be synonymous with *übergreifen*, (II.395, 31), apart from its implying a more passive relationship between the unity and its sub-units.

I.87, 21 'The universality which embraces both sides.'

II.273, 20 'Each particular point... is embraced within the feeling of totality incident to my actuality.'

III.249, 15 '*Impulse* embraces a *series* of satisfactions, is something of a *whole*, a *universal.*'

III.117, 9 '*Presentation*... also contains three stages.'

III.163, 13 'This third stage contains the *phantasy* which symbolizes.'

Urteil: judgement; basic division; primary component. The original meaning of the word was the legal one of *administering* or *dispensing* justice (erteilen).

The use of the corresponding noun (Urteil) was confined to legal contexts until the end of the seventeenth century, when Leibniz introduced it into logic: see Hegel's Logic, §§ 166–180. Hegel's use of the word's literal meaning (basic division, primary component), which was almost certainly influenced by developments in eighteenth century logic (notes II.151, 26; III.121, 8; 241, 26), was not entirely original: J. H. Campe (1746–1818), 'Wörterbuch zur Erklärung und Verdeutschung der unserer Sprache aufgedrungenen fremden Ausdrücke' (Brunswick, 1813), 282 b.

II.423, 18 'Assessment based on physiognomic expression is therefore only of value as an *immediate* judgement.'

III.121, 3 'What is original is not the simplicity of feeling but *judgement* in general, the division of consciousness into a subject and object.'

II.219, 28 'In this basic division the soul is subject in general.'

II.325, 1 'The essence of the feeling totality is to divide itself internally, and to awaken *to the basic internal division.*'

II.127, 2 'Individuality, distinguished from the mere *being* of its immediate and *primary component.*'

II.151, 19 'Awakening may be called a *primary component* of the individual soul... on account of this state's eliciting a *division* of the soul.'

Vereinzelung: singularization. Cf. the definition of *singularity* in Phil. Nat. I.172. Since Hegel makes the attempt to present the most complex levels of each sphere as singularities, he is perpetually attempting to show that singularity as such *includes* the less complex levels of universality and particularity: see Logic § 163.

II.23, 18 'The transition to the second section consists of these *universal* particularizations or varieties being taken back into the unity of the soul, or, and it is the same thing, being led on into singularization.'

II.83, 21 'The soul is singularized into the individual subject.'

III.135, 2 'This universality derives from spirit, and has still not developed any *actual* singularization.'

verleiblichen: embody. Unlike *corporality* and *corporeality* (Phil. Nat. I.151), embodiment presupposes and is an outcome of sentience (§ 401) and feeling (§ 411).

II.163, 13 'It would be worth treating and developing the *system* of inner sentience in its self-embodying *particularization* as a distinct science, as a *psychic physiology.*'

II.165, 7 'The intestines and organs... constitute the systematic embodying of what is spiritual.'

II.407, 20 'The soul finds... an ever-increasing capacity for immediately embodying its inner determinations.'

Vermögen: faculty. Hegel objected to the 'empirical' psychology of his day which attempted to expound the subject in terms of aggregates of fixed faculties instead of analysing it into revisable levels of complexity (I.13, 97; II.107). He is, however, not averse to using this word in a neutral, general and undogmatic sense when referring to such fields of enquiry as divination (II.249, 28) and sight (II.397, 1). For a survey of the background to his use of it, see note III.107, 8.

I.13, 19 *'Empirical psychology* which starts by observing and describing particular spiritual faculties, finds its place between observation which is concerned with the contingent singularity of spirit, and pneumatology, dealing as this does solely with the unmanifest essence of spirit.'

I.97, 6 'Although *psychology* is also fundamentally empirical, it sorts out the appearances into general classes, describes them under the headings of *psychic powers, faculties* etc., and considers spirit in accordance with the particularities into which this procedure dissects it.'

III.107, 6 'Like power, faculty is represented as being the *fixed determinateness* of a *content*, as intro-reflection.'

Verrücktheit: derangement. Hegel's definition of the word (II.327, 20) derives from its *literal* meaning, which is closely related to the popular concept of the mind as a piece of clockwork ('there is a screw loose'). J. C. Hoffbauer's 'Psychologische Untersuchungen' (Halle, 1807), p. 373: "One is justified in calling every derangement an aberration, since the essence of derangement consists precisely in its involving a disturbance of the proper relationship between various faculties of the soul." In early nineteenth century German, the broadly generic term for mental disturbances was Wahnsinn (insanity); see note II.329, 27. On the narrow distinction between derangement and the most basic kind of conscious belief, see III.275, 33.

I.85, 21 'It is here that the conditions of derangement and somnambulism belong.'

II.327, 24 'The subject therefore finds itself involved in a contradiction between the totality systematized in its consciousness and the particular determinateness which is not fluidified and given its place and rank within it. This is *derangement.*'

II.335, 20 'In so far as it is deranged, it cleaves to a *merely* subjective identity of the subjective and the objective *rather than* to an *objective* unity of these two sides.'

III.273, 15 'Although we have... certainly also spoken of consciousness in respect of derangement, we have done so by anticipation.'

Vorstellung: presentation. Cf. the definition of *Darstellung.* It was quite common to dwell upon the literal meaning of the German word when attempting to establish its precise significance, see G. E. Schulze (1761–1833),

'Psychische Anthropologie' (3rd ed. Göttingen, 1826), p. 134, "The primary point in the formation of the German word *Vorstellen*, is that by means of the similarity between their content and what is perceived, presentations facilitate the formation of cognition adequate to the properties of what is perceived. The basic point of the Latin repræsentare, is that by means of presentations, cognition of what has been sensed in the past is renewed. On account of its derivation from νοῦς however, the Greek ἐννοέω indicates that presentations are determinations of our ego, and so constitute something merely subjective." Hegel also dwells upon the literal significance of the word in his dialectical definition of it (§ 451) as presupposing intuition and being the immediate presupposition of thought: note III. 145, 9. *Representation* has only been employed in translating this word when the sense requires it: see III.107, 6. Cf. William Thomson (1819–1890), who made a close study of C. Wolff's 'Philosophia rationalis' (1740) and D. A. Wyttenbach's 'Praecepta philosophiae logicae' (Amsterdam, 1782), 'Outline of the laws of Thought' (London, 1842), § 46, "The impression which any object makes upon the mind may be called a Presentation."

I.91, 7 'Man... lives, senses, looks about, presents, thinks, wills and accomplishes.'

II.343, 23 'A presentation may be said to be *deranged* when the deranged person regards an *empty abstraction* and a *mere possibility* as something concrete.'

III.227, 36 'This mutual self-penetration of thinking subjectivity and objective reason is the final result of the development of theoretical spirit through the stages of intuition and presentation which precede pure thought.'

Wahn: delusion. Cf. the definition of *Täuschung*, which implies a greater degree of self-awareness. As the first element in *Wahnsinn* or insanity (note II.329, 27) the word derives from a root signifying 'wanting', so that it might be regarded as almost synonymous with *vacuity* (note II.355, 12). Hegel, however, seems to use it with its more normal signification in mind. This derived from the earlier meaning of 'hope' or 'expectation', which had completely passed out of use by the end of the eighteenth century (Adelung).

II.355, 12 '*That which confronts* the delusion of the soul at the *same time* has being *for it.*'

II.379, 1 'Now although his delusion was not called in question, he was forbidden to rave.'

II.383, 12 'A doctor cured him of this delusion.'

Wahrheit: truth. Cf. the definitions of the related words *Beisichsein, herabsetzen, übergreifen, Vereinzelung, sublation* (Phil. Nat. I.172), and the distinction between truth and certainty (§ 416 Addition). Hegel takes the truth of anything to be the unity within which it is sublated or included. In the case of the syllogism for example, the conclusion is the truth of its premises. It is therefore relative to the particular context under consideration, in which case Hegel speaks of a *proximate* truth, and absolute in the sense that spirit includes logic and nature within itself. The analysis of complexity relationships is therefore an essential preliminary to the recognition of truth. Hegel also uses the word to mean simply the opposite of falsity (I.101, 12; III.105, 30).

I. 25, 18 '*For us*, spirit has *nature* as its *presupposition*. It is the *truth* of nature, and, therefore, its *absolute prius*. Nature has vanished in this truth.'

I.41, 33 'Christian theology also conceives of God or truth as spirit.'

II.25, 21 'The *universal soul*, as *world soul*, must not as it were be fixed as a subject, since it simply constitutes that universal *substance* which has its actual truth only as *singularity*, subjectivity.'

II.219, 2 'The real extrinsicality of corporeity has no truth for the feeling soul.'

III.25, 28 'The *proximate* truth of the *immediately singular* is therefore its *being related* to another.'

III.237, 25 'That which is *rational* and has the shape of rationality in that it is thought, constitutes the same content as that possessed by *good* practical feeling, but it constitutes it in its universality and necessity, its objectivity and truth.'

Wissen: knowledge, knowing. Cf. the definitions of *Erkenntnis* and *Kenntnis*. *Wissen* comes between the two in degree of certainty. Despite its constituting an element in the German word for *science* (*Wissenschaft*), Hegel uses it with reference to rather basic or primitive kinds of knowledge.

II.247, 36 'Characteristic of this knowledge (clairvoyance) is therefore that consciousness in this immanence should have *immediate* knowledge.'

II.357, 30 '*Absent-mindedness* consists of *not knowing* what is in the immediate vicinity.'

III.79, 3 '*Spirit* has determined itself into the truth of the simple immediate totality of the soul and of consciousness. The latter is knowledge, which is now infinite form.'

Zerrissenheit: disruption. The German word suggests tearing asunder, dilaceration, mangling, dismemberment more forcibly than does its English equivalent. Hegel uses it mainly in connection with mental derangement.

II.333, 1 'Our interpretation of derangement... is not to be taken to imply that *every* spirit, *every* soul, must pass through this stage of extreme disruption.'

II.367, 19 'Only we know of this, the fool *himself* being untroubled by any feeling of inner disruption.'

II.403, 1 'By overcoming the disruption of its inner contradiction... the soul has become a self-relating identity.'

Zerstreuung: diversion. Although the predominant meaning Hegel associated with this word is clear enough from the instances of his use of it given below, its complicated involvement in the dialectical definition of desipience and absent-mindedness should not be overlooked (note II.357, 31). When considering its involvement in Hegel's general doctrine of inwardness, it is certainly worth remembering that in Swabian (Fischer VI.1153) the word also meant *opposition*. In a pre-Lutheran Swabian translation of the Bible for example, "He that is not with me is against me" (Matthew XII.30), is rendered as, "Der nit sament mit mir, der zerstreuet."

II.127, 17 'Sleep invigorates this activity... as withdrawal from the world of *determinateness*, from the diversion of becoming fixed in singularities.'

II.295, 12 'More than is commonly the case amid the diversions of the waking soul, (the human soul) attains to a profound and powerful feeling of the *entirety* of its *individual* nature.'

III.259, 13 "Will is initially the *unending process* of diversion, of sublating one inclination or enjoyment by means of another.'

v

VORWORT

Aus Mangel an Zeit sieht sich der unterzeichnete Herausgeber dieser Ausgabe des dritten Theiles der Hegelschen Encyklopädie genöthigt, auf eine wissenschaftliche Betrachtung dieses Theiles hier zu verzichten und sich auf die Angabe der bei Bearbeitung desselben von ihm gebrauchten Materialien, so wie auf eine kurze Bezeichnung der Grundsätze zu beschränken, nach welchen er jenen Stoff in den zu machenden Zusätzen benutzen zu müssen geglaubt hat.

Was den ersten Punkt betrifft, so hat der Unterzeichnete zuvörderst zu bemerken, daß er nur den die Lehre vom subjectiven Geist enthaltenden Abschnitt der Encyklopädie mit Zusätzen zu versehen beauftragt worden ist, da die den objectiven und den absoluten Geist behandelnden Abschnitte jenes Werkes schon — theils in der neuen Ausgabe der Rechtsphilosophie, — theils in den veröffentlichten Vorlesungen Hegels über die Philosophie der Geschichte, über die Aesthetik, die Religionsphilosophie und die Geschichte der Philosophie — ihre genügende Erläuterung erhalten haben. Die sich auf die Lehre vom subjectiven Geist beziehenden Paragraphen des fraglichen Buches konnten aber aus den nothwendigerweise sehr kurz gefaßten Vorlesungen Hegels über die gesammte Encyklopädie keine wesentliche Erläuterung bekommen. Dagegen boten die

vi

FOREWORD

Lack of time has forced the undersigned editor of this
edition of the third part of the *Hegel*ian Encyclopaedia to
drop the idea of providing a scientific consideration of
the work, and to confine himself to giving an account of
the materials used in compiling it, and a brief outline of the   5
procedure it was thought advisable to adopt in making use
of these sources in the preparation of the Additions.

He should state at the outset with regard to the first point,
that he has only been commissioned to furnish the doctrine
of *Subjective Spirit* with Additions, the sections of the work   10
devoted to objective and absolute spirit having already been
sufficiently elucidated by means of the new edition of the
Philosophy of Right and the publication of *Hegel*'s lectures
on the Philosophy of History, Aesthetics, the Philosophy of
Religion and the History of Philosophy. It was out of the   15   +
question that the Paragraphs relating to the doctrine of
Subjective Spirit should have received satisfactory treatment
in the necessarily extremely condensed lectures *Hegel*
delivered on the Encyclopaedia as a whole. The bulk of the   +
material for the task in hand was therefore provided by the   20

ausschließlich über jenen einzelnen Zweig der Philosophie
von Hegel gehaltenen Vorlesungen einen reichen Stoff zur
Benutzung für den in Rede stehenden Zweck dar. — Dem
Herausgeber lagen zuvörderst die eigenen Collegien-Hefte
Hegels vor. Das eine derselben, mit dem Datum „Mai
1817" versehen, zeigt in seinen einzelnen Theilen eine
große Ungleichheit der Ausführlichkeit; — das andere, in
Berlin verfaßt und zum ersten Male im Sommersemester
des Jahres 1820 gebraucht, ist etwas gleichmäßiger aus-
geführt. Beide Hefte enthalten jedoch nicht eine zu fest-
verbundenen Sätzen ausgearbeitete Entwicklung des Gegenstandes, sondern meistentheils nur allgemeine Umrisse
und abgerissene Worte. — Zu diesen Urquellen kamen die
nachgeschriebenen Hefte, in der Zahl von fünfen. Außer
seinem eigenen Hefte benutzte der Herausgeber erstlich zwei
Hefte, die Hegel selber sich hat abschreiben lassen und deren Eines von ihm bei seinen Vorlesungen in den Jahren 1828 und 1830 zu Grunde gelegt und zu diesem
Behufe mit vielfachen eigenhändigen Randbemerkungen aus-
gestattet worden ist. Nicht geringere Ausbeute gewährte
endlich das sehr fleißig ausgearbeitete, umfangsreiche Heft
des Herrn Majors v. Griesheim aus dem Jahre 1825,
und das etwas gedrängtere, in guter Ordnung befindliche
des Herrn Dr. Mullach aus dem Jahre 1828.

In Bezug aber auf den zweiten hier zu besprechenden
Punkt, — die Weise der Benutzung jenes Materials, —
ist der Unterzeichnete von der Ansicht ausgegangen, daß
ihm die unerläßliche Pflicht oblag, den vergleichungsweise
rohen Stoff gedachter Vorlesungen in diejenige künstlerische
Form zu bringen, die auch von einem wissenschaftlichen Werke
mit Recht gefordert wird. Ohne eine solche Umgestaltung

lectures dealing exclusively with this particular branch of philosophy. — *Hegel*'s own note-books constituted the editor's primary sources. The first of these, dated May 1817, displays great unevenness in the working out of its various sections. The second, which was prepared in Berlin and first used   5 during the summer term of 1820, is written up somewhat more uniformly. Both are devoid of a treatment of the subject matter in rounded sentences however, and consist for the most part of nothing but general reminders and isolated words. —- Five sets of lecture notes were also available.   10   + Apart from his own, the editor made use of two which *Hegel* himself had had prepared. The lectures given by *Hegel* in 1828 and 1830 were based on one of these, with the result that the manuscript is profusely annotated with his own marginal notes. Major von *Griesheim*'s very carefully pre-   15 pared and comprehensive record of the lectures given in 1825, and the somewhat more compact but well-arranged notes taken by Dr. *Mullach* in 1828, were also found to be extremely useful.   +

With regard to the second point mentioned above, the   20 procedure adopted in making use of this material, the editor has assumed that he was necessarily obliged to give the comparatively raw material of these lectures the artistic form justifiably required of a scientific work. Had the sources not been recast in this manner, there would in this case have   25

würde in dem vorliegenden Falle eine widerwärtige Disharmonie zwischen dem zu erläuternden Buche und den zu
demselben gemachten Zusätzen entstanden seyn. Zur Be-
seitigung dieser Disharmonie war aber große Anstrengung
nöthig. Denn Hegels Vorlesungen hatten zwar, weil er
mit vieler Freiheit vorzutragen pflegte, alle Frische und
allen Zauber einer erst im Augenblick geschaffenen Ge-
dankenwelt; doch führte dies mehr oder weniger vollstän-
dige Improvisiren nicht selten auch den Uebelstand unbe-
wußter Wiederholungen, der Unbestimmtheit, des Zuweit-
gehens und des Springens mit sich. Diese Mängel muß-
ten in der Bearbeitung sorgfältig vermieden werden. Die
nöthig werdenden Veränderungen sind aber nur in dem
eigenen, unzweifelhaften Sinne Hegels unternommen worden. Diesem Sinne glaubt der Unterzeichnete auch dadurch
entsprochen zu haben, daß er aus seinen Zusätzen Dasjenige nicht weggelassen hat, was die Seele der Hegelschen
Vorlesungen ausmacht, — nämlich die dialektische Entwicklung, welche Hegel in den Vorlesungen meistens mit grö-
ßerer Ausführlichkeit und zum Theil in tieferer Weise,
als in dem gedruckten Text zu geben für nothwendig er-
achtete, weil dieselbe dort, wegen der außerordentlichen
Gedrängtheit der Darstellung, zuweilen den Schein der
Aeußerlichkeit und der bloßen Versicherung bekommt. Obgleich daher in den Zusätzen zuweilen ein schon in dem
darüber stehenden Paragraphen behandelter Punkt hat be-
sprochen werden müssen; so ist Dies doch hoffentlich niemals in der Form einer überflüssigen Wiederholung, sondern immer auf die Art geschehen, daß solcher Punkt dadurch eine vollständigere Entwicklung und Faßlichkeit gewonnen hat.

viii

been a disagreeable lack of harmony between the book to be elucidated and the Additions prepared for this purpose. It was however by no means easy to overcome this discordance. Since he was in the habit of delivering them freely, *Hegel*'s lectures had all the freshness and fascination of a world of 5 thought in the process of being created. As they were more or less entirely a matter of improvisation however, they were not infrequently marred by inadvertent repetitions, want of precision, disconnectedness and lack of proportion. These flaws had to be painstakingly removed in the working over 10 of the sources. Although alterations were necessary however, they were only made in accordance with what was indubitably *Hegel*'s own procedure. The editor believes that he has also kept to the sense of this procedure by not omitting from these Additions the dialectical development which con- 15 stitutes the soul of *Hegel*ian lecturing. Since this development, as rendered in the extraordinarily condensed exposition of the printed text, often assumes the appearance of being an imposition and a mere assertion, *Hegel* usually considered it necessary to expound it more fully in the 20 *lectures*, and to some extent in greater depth. It is often the case therefore, that a point made in the initial Paragraph has also had to be dealt with in the subsequent Addition. It is to be hoped however, that this has been *so* carried out that it never constitutes a superfluous repetition, but always 25 establishes a more complete development and clarification of the point at issue.

So bleibt dem Herausgeber nichts mehr übrig, als der Wunsch, daß diese Ausgabe des dritten Theiles der Hegelschen Encyklopädie, neben den auf denselben Gegenstand sich beziehenden, sehr verdienstvollen Arbeiten von Michelet, Rosenkranz und Daub, einen ehrenvollen Platz behaupten möge.

Berlin den 12. April 1845.

Boumann.

Nothing remains for the editor but to hope that this edition
of the third part of the *Hegel*ian Encyclopaedia may prove a   +
worthy companion to the highly distinguished works relating
to the same subject already published by *Michelet, Rosenkranz*
and *Daub*.   5   +

Berlin, 12 April 1845.

.

                         Boumann.   +

Die Philosophie des Geistes

The Philosophy of Spirit

DIE PHILOSOPHIE DES GEISTES

Der dritte Teil der Enzyklopädie der philosophischen Wissenschaften im Grundrisse

3

Einleitung

§. 377.

Die Erkenntniß des Geistes ist die concreteste, darum höchste und schwerste. Erkenne dich selbst, diß absolute Gebot hat weder an sich, noch da wo es geschichtlich als ausgesprochen vorkommt, die Bedeutung, nur einer Selbsterkenntniß nach den particulären Fähigkeiten, Charakter, Neigungen und Schwächen des Individuums, sondern die Bedeutung der Erkenntniß des Wahrhaften des Menschen, wie des Wahrhaften an und für sich, — des Wesens selbst als Geistes. Eben so wenig hat die Philosophie des Geistes die Bedeutung der sogenannten Menschenkenntniß, welche von andern Menschen gleichfalls die Besonderheiten, Leidenschaften, Schwächen, diese sogenannten Fälten des menschlichen Herzens zu erforschen bemüht ist, — eine Kenntniß, die theils nur unter Voraussetzung der Erkenntniß des Allgemeinen, des Menschen und damit wesentlich des Geistes Sinn hat, theils sich mit den zufälligen, unbedeutenden, unwahren Existenzen des Geistigen beschäftigt, aber zum Substantiellen, dem Geiste selbst, nicht bringt.

Zusatz. Die Schwierigkeit der philosophischen Erkenntniß des Geistes besteht darin, daß wir es dabei nicht mehr mit der vergleichungsweise abstracten, einfachen logischen Idee, sondern mit der concretesten, entwickeltsten Form zu thun haben, zu welcher die Idee in der Verwirklichung ihrer selbst gelangt. Auch der endliche oder subjective Geist, — nicht bloß der absolute — muß als eine Verwirklichung der Idee gefaßt werden. Die Betrachtung des Geistes ist nur dann in Wahrheit philosophisch, wenn

4

sie den Begriff desselben in seiner lebendigen Entwicklung und Verwirklichung erkennt, d. h. eben, wenn sie den Geist als ein

THE PHILOSOPHY OF SPIRIT

The third part of the Encyclopaedia of
the Philosophical Sciences in outline

Introduction

§ 377

**Knowledge of spirit is knowledge of the most concrete
and consequently of the sublimest and most difficult
kind. Know thyself, this absolute commandment,
is not concerned with a mere self-knowledge, with
the particular abilities, character, inclinations and   5
foibles of the individual, but in its intrinsic import, as
in the historical contexts in which it has been formulated, it is concerned with cognition of human truth,
with that which is true in and for itself, — with
essence itself as spirit. Concern with what is called a   10
cognition of human nature, involving the attempt
to investigate the peculiarities, passions and foibles
of other people, the so-called recesses of the human
heart, is equally alien to the philosophy of spirit.
Cognition of this kind is of significance only if it   15
presupposes cognition of that which is universal,
of man, and hence, essentially of spirit. Since it is
concerned with nothing but the contingent, insignificant and untrue existences of spiritual being, it
necessarily fails to penetrate to the substantial   20
being of spirit itself.**

Addition. The philosophical cognition of spirit is difficult
since it involves our being concerned with the most concrete
and developed form attained in the self-actualization of the
Idea, and no longer with the comparatively abstract   25
simplicity of the logical Idea. Not only absolute but also
finite or subjective spirit has to be grasped as an actualization
of the Idea. The consideration of spirit is only truly philosophical when it recognizes the living development and
actualization of the Notion of spirit, that is to say when it   30
comprehends spirit as a likeness of the eternal Idea. It is
however in the nature of spirit to cognize its Notion. The

Abbild der ewigen Idee begreift. Seinen Begriff zu erkennen gehört aber zur Natur des Geistes. Die vom delphischen Apollo
an die Griechen ergangene Aufforderung zur Selbsterkenntniß hat
daher nicht den Sinn eines von einer fremden Macht äußerlich
an den menschlichen Geist gerichteten Gebots; der zur Selbsterkenntniß treibende Gott ist vielmehr nichts Andres, als das eigene
absolute Gesetz des Geistes. Alles Thun des Geistes ist deßhalb
nur ein Erfassen seiner selbst, und der Zweck aller wahrhaften
Wissenschaft ist nur der, daß der Geist in Allem, was im Himmel und auf Erden ist, sich selbst erkenne. Ein durchaus Andres
ist für den Geist gar nicht vorhanden. Selbst der Orientale verliert sich nicht gänzlich in dem Gegenstande seiner Anbetung; die
Griechen aber haben zuerst Das, was sie sich als das Göttliche
gegenüberstellten, ausdrücklich als Geist gefaßt; doch sind auch sie
weder in der Philosophie noch in der Religion zur Erkenntniß
der absoluten Unendlichkeit des Geistes gelangt; das Verhältniß
des menschlichen Geistes zum Göttlichen ist daher bei den Griechen noch kein absolut freies; erst das Christenthum hat durch
die Lehre von der Menschwerdung Gottes und von der Gegenwart des heiligen Geistes in der gläubigen Gemeine dem menschlichen Bewußtsein eine vollkommen freie Beziehung zum Unendlichen gegeben, und dadurch die begreifende Erkenntniß des Geistes
* in seiner absoluten Unendlichkeit möglich gemacht.

---

* *Kehler Ms.* SS. 2–3; vgl. *Griesheim Ms.* SS. 4–6:... dem allgemeinen Geist
der Welt, und was so das innerste, Bestimmung der Religion ist, muß sich
die Philosophie auch zu ihrem Gegenstand machen. Durch diesen höheren
Standpunkt gestaltet sich die Aufgabe die zunächst dieselbe Begrenzung
hätte, wie bei den Griechen, ganz anders. Durch das Bewußtsein vom
unendlichen Geist, wird der Kreis des Geistes, der für die Griechen war,
herabgedrückt zu einem nur endlichen Geist; so verändert sich sogleich die
Stellung desselben Kreises. Auf der anderen Seite ist der menschliche Geist,
den wir endlich zu nennen pflegen, zugleich in Beziehung auf den Geist, den
wir den unendlichen Geist nennen. Durch diese Beziehung ist er denn der
Geist, den wir den endlichen nennen, selbst, der in sich, für sich den höheren
Standpunkt, freien Boden für sich gewinnt, und damit in ein anderes
Verhältnis zu der Natur tritt, in ein Verhältnis der Unabhängigkeit, daß
dieselbe ihm ein äußerliches ist. Oft was man in ästhetischen Erzählungen
beklagen hört, daß wir nicht in der schönen Einheit mit der Natur stehen
wie die Griechen, das kann die Empfindsamkeit, aber nicht die Vernunft

self-knowledge demanded of the Greeks by the Delphic
Apollo is therefore not to be regarded as a commandment
of an alien power imposed upon the human spirit from
without; for the deity which induces self-knowledge, is
indeed nothing other than spirit's own absolute law. Conse-   5
quently, everything done by spirit is merely an apprehension
of itself, and the purpose of all true science is simply that in
all that is, in heaven and on earth, spirit should recognize
itself. For spirit, there is a complete absence of any thorough-
going alterity. Although even the oriental does not wholly   10
lose himself in the general object of his worship, the Greeks
were the first to grasp expressly that the Divinity with which
they had confronted themselves was spirit. Neither in   +
philosophy nor in religion did they succeed in cognizing the
absolute infinity of spirit however, so that among them, the   15
relationship of the human spirit to the Divine is still not
absolutely free. It was Christianity, through the doctrine of
the Incarnation of God and of the presence of the Holy Spirit
in the communion of believers, which first gave human
consciousness a completely free relation to the infinite, and   20
so made possible the Notional cognition of spirit in its
absolute infinity.*   +

---

* *Kehler* Ms. pp. 2–3; cf. *Griesheim* Ms. pp. 4–6: Philosophy also has to make
the universal spirit of the world, and what is most internal, the determination
of religion, into its general object. The task would initially have had the   +
same scope as it did with the Greeks, but on account of this higher standpoint, it presents itself in a wholly different manner. Through consciousness
of infinite spirit, the cycle of spirit as it was for the Greeks is reduced to a
spirit which is merely finite, the status of this cycle undergoing an immediate
alteration, while human spirit, which is usually said to be finite, is at the
same time in relation with what we call infinite spirit. It is, then, through
this relation that the spirit said to be finite wins for itself within itself the
being-for-self of the higher standpoint, the free foundation from which it
enters into another relationship with nature, one of independence, one in
which nature is external to it. One frequently hears it deplored, in writings   +
on aesthetics, that we should differ from the Greeks in lacking the beauty of   +
being at one with nature. It is for sensitivity to regret this however, not for
reason, for it is of the essence of spirit to be free, and so to be free for itself,

Nur eine solche Erkenntniß verdient fortan den Namen einer philosophischen Betrachtung. Die Selbsterkenntniß in dem gewöhnlichen trivialen Sinn einer Erforschung der eigenen Schwächen und Fehler des Individuums hat nur für den Einzelnen, — nicht für die Philosophie — Interesse und Wichtigkeit, selbst aber in Bezug auf den Einzelnen um so geringeren Werth, je weniger sie sich auf die Erkenntniß der allgemeinen intellectuellen und moralischen Natur des Menschen einläßt, und je mehr sie, von den Pflichten, dem wahrhaften Inhalt des Willens absehend, in ein

5

bedauern, das Wesen des Geistes ist, daß er frei sei, dazu gehört, daß er frei für sich sei, nicht in dem unmittelbaren natürlichen stehen bleibe. Auf unsrer Stellung, nach der wir auffassen, was wir menschlichen Geist nennen, haben wir den Geist in einem Verhältniß, das ein Verhältniß der Mitte ist zwischen zwei Extremen: die Natur und Gott; der Anfangspunkt für den Menschen, der andere der absolute Endzweck, das absolute Ziel. Betrachten wir den Geist nach dieser Stellung, die in unserem Bewußtsein ganz geläufig ist, fragen wir überhaupt, was der Geist ist, so ergibt sich aus der Stellung, die der Geist zwischen zwei Extremen hat, daß diese Frage noch in sich schließt: wo der Geist herkommt und wo der Geist hingeht. Diese beiden Fragen nach seinem Anfang und Ziel scheinen weitere Fragen zu sein, als die, was der Geist ist, aber es wird sich bald zeigen, daß beide Fragen und ihre Beantwortung es in der That wahrhaftig ist, wodurch das erkannt wird, was der Geist ist. Der Geist fängt von der Natur überhaupt an, man muß nicht bloß an die äußerliche Natur denken, dahin gehört ebenso die sinnliche Natur des Menschen selbst, sein sinnliches, leibliches Sein, Em (3) pfinden, Verhalten zu anderen Gegenständen, bloßes Empfinden kommt bloß den Thieren zu; das Extrem, wo der Geist hingeht, ist seine Freiheit, Unendlichkeit, an und für sich sein. Das sind die zwei Seiten, aber fragen wir, was der Geist ist, so ist die unmittelbare Antwort diese, der Geist ist diese Bewegung, dieser Prozeß, diese Thätigkeit, von der Natur auszugehen, sich von der Natur zu befreien; es ist dieß das Sein des Geistes selbst, seine Substanz. Nach der gewöhnlichen Weise pflegt man zu sprechen, der Geist ist Subjekt, thut dieß, und außer seiner That, dieser Bewegung, Prozeß, ist er noch besonders, seine Thätigkeit so mehr oder weniger zufällig; die Natur des Geistes ist diese absolute Lebendigkeit, dieser Prozeß selbst zu sein, von der Natürlichkeit Unmittelbarkeit zugehen, seine Natürlichkeit aufzuheben zu verlaßen, und zu sich selbst zu kommen, und sich zu befreien, das ist er, nur als zu sich gekommen, als ein solches Produkt seiner selbst ist er; seine Wirklichkeit ist nur, daß er sich zu dem gemacht hat was er ist. Das erste Sein ist nur Begriff (und der Geist ist nicht in seinem Begriff vollendet, sondern wesentlich Idee,) und dieser Begriff ist nur sein natürliches Sein; er ist dies erste Sein im Begriff, wodurch der Geist nicht erschöpft ist; er ist als Geist nur als Resultat seiner selbst.

Henceforth it is only cognition of this kind that may be referred to as a philosophical consideration. *Self-knowledge*, in the usual trivial sense of an investigation of the foibles and faults peculiar to the individual, is of interest and importance only to the individual, not to philosophy. Even with regard 5 to the individual however, its value diminishes to the extent that it fails to concern itself with knowledge of the general intellectual and moral nature of man, and, disregarding the duties which constitute the true content of the will, degenerates into the individual's being self-complacently absorbed 10

---

not to remain within the immediacy of what is natural. On account of the position from which we are assessing what we call human spirit, we have spirit within a relationship as the middle between two extremes: nature and God; the one being, for man, the point of departure, the other being the absolute and ultimate end, the absolute goal. Within our consciousness this position is a wholly familiar one, and if we consider spirit from it, if we raise the general question of what spirit is, it becomes apparent from its position between the two extremes that the question implies the further question of where it comes from and whither it tends. Although the questions concerning its beginning and goal appear to be distinct from enquiry into what spirit is, the literal truth that cognition of this is to be gained through the asking and answering of them will soon become apparent. Spirit has its beginning in nature in general. One must not think merely of external nature, but also of the sensuous nature of man himself, his sensuous, bodily being, (3) sensing, being in relation with other general objects; mere sensing is confined solely to animals. The extreme to which spirit tends is its freedom, its infinity, its being in and for itself. These are the two aspects, but if we ask what spirit is, the immediate answer is that it is this motion, this process of proceeding forth from, of freeing itself from nature; this is the being, the substance of spirit itself. Spirit is usually spoken of as subject, as doing something, and apart from what it does, as this motion, this process, as still particularized, its activity being more or less contingent; it is of the very nature of spirit to be this absolute liveliness, this process, to proceed forth from naturality, immediacy, to sublate, to quit its naturality, and to come to itself, and to free itself, it being itself only as it comes to itself as such a product of itself; its actuality being merely that it has made itself into what it is. The initial being is merely its Notion (and spirit within its Notion is not completed, but essentially Idea); this Notion being only its natural being. Within the Notion it is this initial being, but this does not exhaust it, for it is only as the result of itself that it has being as spirit.

selbstgefälliges Sicherumwenden des Individuums in seinen ihm
theuren Absonderlichkeiten ausartet. — Dasselbe gilt von der gleich=
falls auf die Eigenthümlichkeiten einzelner Geister gerichteten soge=
nannten **Menschenkenntniß.** Für das Leben ist diese Kenntniß
allerdings nützlich und nöthig, besonders in schlechten politischen
Zuständen, wo nicht das Recht und die Sittlichkeit, sondern der
Eigensinn, die Laune und Willkür der Individuen herrschen, —
im Felde der Intriguen, wo die Charactere nicht auf die Natur
der Sache sich stützen, vielmehr durch die pfiffig benutzte Eigen=
thümlichkeit Andrer sich halten und durch dieselben ihre zufälligen
Zwecke erreichen wollen. Für die Philosophie aber bleibt diese
Menschenkenntniß in eben dem Grade gleichgültig, wie dieselbe
sich nicht von der Betrachtung zufälliger Einzelnheiten zur Auf=
fassung großer menschlicher Charactere zu erheben vermag, durch
welche die wahrhafte Natur des Menschen in unverkümmerter Rein=
heit zur Anschauung gebracht wird. — Sogar nachtheilig für die
Wissenschaft wird jene Menschenkenntniß aber dann, wenn sie, —
wie in der sogenannten pragmatischen Behandlung der Geschichte
geschehen — den substantiellen Character weltgeschichtlicher In=
dividuen verkennend, und daß Großes nur durch große Charactere
vollbracht werden kann, nicht einsehend, aus der zufälligen Ei=
genthümlichkeit jener Heroen, aus deren vermeintlichen kleinen Ab=
sichten, Neigungen und Leidenschaften die größten Ereignisse der
Geschichte abzuleiten den geistreich sein sollenden Versuch macht;
ein Verfahren, bei welchem die von der göttlichen Vorsehung
beherrschte Geschichte zu einem Spiele gehaltloser Thätigkeit und
zufälliger Begebenheiten herabsinkt.

§. 378.

Der **Pneumatologie** oder sogenannten **rationel=
len Psychologie,** als abstracter Verstandesmetaphysik,
ist bereits in der Einleitung Erwähnung geschehen. Die
empirische Psychologie hat den concreten Geist
zu ihrem Gegenstande, und wurde, seitdem nach dem Wie=
deraufleben der Wissenschaften die Beobachtung und Er=
fahrung zur vornehmlichen Grundlage der Erkenntniß des
Concreten geworden, auf dieselbe Weise getrieben, so daß

in his own precious peculiarities. — The same is true of the
so-called *knowledge of human nature*, which is also concerned
with the peculiarities of particular minds. It must be admitted   +
that this knowledge is useful and necessary in life, particularly
in bad political situations, dominated not by law and ethics   5
but by the capriciousness, whims and wilfulness of individuals, and in the field of intrigue, where rather than comply-
ing with the nature of the matter, characters survive and
attempt to gain their fortuitous ends by artfully exploiting
the peculiarity of others. — It is however precisely to the   10   +
extent that this knowledge of human nature is unable to
raise itself from the consideration of contingent particularities to the comprehension of the great human personalities
through which the true nature of man is made apparent with
untroubled clarity, that it remains a matter of indifference   15
to philosophy. What is more, this knowledge of human
nature becomes a hindrance to science when, as in the so-
called pragmatic treatment of history, it misconstrues the
substantial character of world-historical individuals, fails to
realize that only greatness of character can accomplish   20
anything great, and makes the supposedly ingenious attempt
to derive the greatest historical events from the fortuitous
peculiarity, the supposedly petty intentions, inclinations
and passions of these heroes; a procedure which degrades
the Divine Providence of history to an interplay of futile   25
activity and adventitious occurrences.   +

§ 378

**Pneumatology, or what is called rational psy-
chology, has already been mentioned in the Introduction as an abstract metaphysics of the understanding. Empirical psychology has concrete   30   +
spirit as its general object, and after the revival of the
sciences, when observation and experience became
the principal foundation for the cognition of what is
concrete, it was also cultivated in this way, so that**

theils jenes Metaphysische außerhalb dieser empirischen Wissenschaft gehalten wurde und zu keiner concreten Bestimmung und Gehalt in sich kam, theils die empirische Wissenschaft sich an die gewöhnliche Verstandesmetaphysik von
Kräften, verschiedenen Thätigkeiten u. s. f. hielt und die
speculative Betrachtung daraus verbannte. — Die Bücher
des **Aristoteles über die Seele** mit seinen Abhandlungen über besondere Seiten und Zustände derselben sind
deswegen noch immer das vorzüglichste oder einzige Werk
von speculativem Interessen über diesen Gegenstand. Der
wesentliche Zweck einer Philosophie des Geistes kann nur
der seyn, den Begriff in die Erkenntniß des Geistes wieder einzuführen, damit auch den Sinn jener aristotelischen
* Bücher wieder aufzuschließen.

Zusatz. Ebenso, wie die im vorigen §. besprochene, auf
die unwesentlichen, einzelnen, empirischen Erscheinungen des Geistes gerichtete Betrachtungsweise, ist auch die im geraden Gegentheil nur mit abstract allgemeinen Bestimmungen, mit dem ver-
meintlich erscheinungslosen Wesen, dem Ansich des Geistes sich
beschäftigende sogenannte **rationelle Psychologie** oder Pneumatologie von der echt speculativen Philosophie ausgeschlossen, da
diese die Gegenstände weder aus der Vorstellung als gegebene aufnehmen, noch dieselben durch bloße Verstandeskategorien bestimmen

---

* *Kehler Ms.* S. 56; vgl. *Griesheim Ms.* S. 74: Das Beste was vom Geist gesagt
ist, ist von Aristoteles, wenn man den Geist speculativ kennen lernen will, so
hat man sich nur an ihn zu wenden, er geht nach seiner Weise vom sinnlichen aus, unterscheidet dreierlei Seelen: eine vegetabilische Seele, eine
animalische Lebendigkeit und eine geistige Lebendigkeit, diese drei erscheinen als Pflanzenreich, Thierreich und Menschenreich:
1) νοητικόν, 2) αἰσθητικόν, 3) θρεπτικόν, deren *ἀντικείμενα* sind: *νοητόν,
αἰσθητόν, τροφή. 3) ἡ γὰρ θρεπτικὴ ψυχὴ καὶ τοῖς ἄλλοις ὑπάρχει, καὶ πρώτη
καὶ κοινοτάτη δύναμίς ἐστι ψυχῆς, καθ' ἣν ὑπάρχει τὸ ζῆν ἅπασιν. ἧς ἐστιν ἔργα
γεννῆσαι καὶ τροφῇ χρῆσθαι· 2) ἡ δ' αἴσθησις ἐν τῷ κινεῖσθαί τε καὶ πάσχειν
συμβαίνει ... 1) τὸ μόριον τῆς ψυχῆς ᾧ γινώσκει τε ἡ ψυχὴ καὶ φρονεῖ, ...
ἀπαθὲς ἄρα δεῖ εἶναι, ... ἀνάγκη ἄρα, ἐπεὶ πάντα νοεῖ, ἀμιγῆ εἶναι, ὥσπερ φησὶν
᾿Αναξαγόρας, ἵνα κρατῇ, τοῦτο δ' ἐστὶν ἵνα γνωρίξῃ· παρεμφαινόμενον γὰρ κωλύει
... καὶ ἀντιφράττει* Ar. *περί ψυχῆς*, lib. 2 & 3.
Im Menschen aber ist dies vorhanden, daß er vegetabilischer Natur ist,
ebenso aber auch empfindend und denkend; im Menschen sind dies nur drei
Formen an ihm dieser Einen und desselben.

while the metaphysical factor failed to acquire any inwardly concrete determination and capacity on account of its being excluded from the empirical science, the empirical science kept to the conventional metaphysics of the understanding, to powers, various activities, etc., and drove out speculative consideration. — It is on account of this that Aristotle's books on the soul, as well as his dissertations on its special aspects and conditions, are still by far the best or even the sole work of speculative interest on this general topic. The essential purpose of a philosophy of spirit can be none other than re-introducing the Notion into the cognition of spirit, and so reinterpreting the meaning of these Aristotelian books.*

Addition. The manner of consideration discussed in the preceding Paragraph, occupied as it is with the inessential singular, empirical manifestations of spirit, like its direct opposite, what is called *rational psychology* or pneumatology, which concerns itself solely with the abstract and general determinations of what is supposed to be the unmanifest essence or implicit being of spirit, is alien to truly speculative philosophy, which may neither take up general objects as given in presentation, nor, as did the psychology which

---

* *Kehler* Ms. p. 56; cf. *Griesheim* Ms. p. 74: The best that has been said of spirit has been said by Aristotle, and if one wants to know spirit speculatively one has only to consult him. In accordance with his procedure, he begins with what is sensuous and draws a threefold distinction in respect of souls, distinguishing a vegetable soul, animal and spiritual animation, the three of which appear as the plant, animal and human kingdoms: 1) thought, 2) sense, 3) nutrition, the objects of which are: the object of thought, the object of sense, and nutriment. 3) "The nutritive soul belongs to other living beings as well as man, being the first and most widely distributed faculty, in virtue of which all things possess life. Its functions are reproduction and assimilation of nutriment. 2) Sensation consists in being moved and acted upon... 1) The part of the soul with which it knows and understands,... must be impassive... Spirit, then, since it thinks all things, must needs, in the words of Anaxagoras, be unmixed with any, if it is to rule, that is, to know. For by intruding its own form it hinders and obstructs..." Ar. 'De Anima' bks. 2 & 3.

In man then, one does not only have the presence of a vegetable nature, for he is also, and to the same extent, sentient and thinking; in man these are merely three forms, which, within him, are one and the same.

darf, wie jene Psychologie that, indem sie die Frage aufwarf, ob der Geist oder die Seele einfach, immateriell, Substanz sei. Bei diesen Fragen wurde der Geist als ein Ding betrachtet; denn jene Kategorien wurden dabei, nach der allgemeinen Weise des Verstandes, als ruhende, feste angesehen; so sind sie unfähig, die Natur des Geistes auszudrücken; der Geist ist nicht ein Ruhendes, sondern vielmehr das absolut Unruhige, die reine Thätigkeit, das Negiren oder die Idealität aller festen Verstandesbestimmungen, — nicht abstract einfach, sondern in seiner Einfachheit zugleich ein Sich = von = sich = selbst = unterscheiden, — nicht ein vor seinem Erscheinen schon fertiges, mit sich selber hinter dem Berge der Erscheinungen haltendes Wesen, sondern nur durch die bestimmten Formen seines nothwendigen Sichoffenbarens in Wahrheit wirklich, — und nicht (wie jene Psychologie meinte) ein nur in äußerlicher Beziehung zum Körper stehendes Seelending, sondern mit dem Körper durch die Einheit des Begriffs innerlich verbunden.

In der Mitte zwischen der auf die zufällige Einzelnheit des Geistes gerichteten Beobachtung und der sich nur mit dem erscheinungslosen Wesen desselben befassenden Pneumatologie steht die auf das Beobachten und Beschreiben der besonderen Geistesvermögen ausgehende empirische Psychologie. Aber auch diese bringt es nicht zur wahrhaften Vereinigung des Einzelnen und Allgemeinen, zur Erkenntniß der concret allgemeinen Natur oder des Begriffs des Geistes, und hat daher gleichfalls keinen Anspruch auf den Namen echtspeculativer Philosophie. Wie den Geist überhaupt, so nimmt die empirische Psychologie auch die besonderen Vermögen, in welche sie denselben zerlegt, als gegebene aus der Vorstellung auf, ohne durch Ableitung dieser Besonderheiten aus dem Begriff des Geistes den Beweis der Nothwendigkeit zu liefern, daß im Geiste gerade diese und keine anderen Vermögen sind. — Mit diesem Mangel der Form hängt nothwendig die Entgeistigung des Inhalts zusammen. Wenn in den bereits geschilderten beiden Betrachtungsweisen einerseits das Einzelne, andrerseits das Allgemeine als etwas für sich Festes genommen wurde, so gelten der empirischen Psychologie auch die Besonde-

gave rise to the question of the spirit or soul being simple,
immaterial or a substance, determine them by means of the
mere categories of the understanding. These questions, true
to the general character of the understanding, in that they
took spirit to be a thing, assumed these categories to be static 5
and fixed. As such the categories are incapable of expressing +
the nature of spirit however, for far from being anything
static, spirit is absolute unrest, pure activity, the negating or
ideality of all the fixed determinations of the understanding.
It is not abstractly simple, for it differentiates itself from 10
itself in its simplicity, nor is it already complete prior to its
being manifest, an essence maintaining itself behind the
range of its manifestations, for it is only truly actual through
the determinate forms of its necessary self-revelation. This
psychology imagined it to be a thing, a soul standing in a 15
merely external relation to the body, but it is inwardly
bound to the body through the unity of the Notion. +

Empirical psychology which starts by observing and describing particular spiritual faculties, finds its place between
observation which is concerned with the contingent singular- 20
ity of spirit, and pneumatology, dealing as this does solely
with the unmanifest essence of spirit. However, since it too
fails to further the true union of the singular with the universal, cognition of the concrete universal nature or Notion of
spirit, it is also unworthy of being regarded as truly specula- 25
tive philosophy. Empirical psychology takes up spirit in
general, as well as the particular faculties into which it
divides it, as these are given in presentation. It fails to derive
these particularities from the Notion of spirit and so provide
proof of the necessity of spirit's containing precisely these and 30
no other faculties. — The despiritualization of the content is +
necessarily bound up with this lack of form. In the two
modes of consideration already described, on the one hand
it was that which is singular, and on the other that which is
universal, which was regarded as separate and fixed. In 35
empirical psychology, it is the particularizations into which

rungen, in welche ihr der Geist zerfällt, als in ihrer Beschränkt=
heit starre; so daß der Geist zu einem bloßen Aggregat von selbst=
ständigen Kräften wird, deren jede mit der anderen nur in Wech=
selwirkung, somit in äußerlicher Beziehung steht. Denn, obgleich
8   diese Psychologie auch die Forderung eines zwischen den verschie=
denen Geisteskräften hervorzubringenden harmonischen Zusammen=
hangs macht — ein bei diesem Gegenstande häufig vorkommendes,
aber ebenso unbestimmtes Schlagwort, wie sonst die Vollkommen=
heit war — so ist damit nur eine sein sollende, nicht die ur=
sprüngliche Einheit des Geistes ausgesprochen, noch weniger aber
die Besonderung, zu welcher der Begriff des Geistes, seine an
sich seiende Einheit, fortgeht, als eine nothwendige und vernünf=
tige erkannt; jener harmonische Zusammenhang bleibt daher eine
sich in nichtssagenden Redensarten breit machende leere Vorstel=
lung, welche gegen die als selbstständig vorausgesetzten Geistes=
kräfte zu keiner Macht gelangt.

§. 379.

Das Selbstgefühl von der lebendigen Einheit des
Geistes setzt sich von selbst gegen die Zersplitterung dessel=
ben in die verschiedenen, gegeneinander selbstständig vorge=
stellten Vermögen, Kräfte oder was auf dasselbe hin=
auskommt, ebenso vorgestellten Thätigkeiten. Noch
mehr aber führen die Gegensätze, die sich sogleich darbie=
ten, von der Freiheit des Geistes und von dem Deter=
minirtwerden desselben, ferner von der freien Wirksam=
keit der Seele im Unterschiede von der ihr äußerlichen
Leiblichkeit, und wieder der innige Zusammenhang beyder,
auf das Bedürfniß hier zu begreifen. Insbesondere ha=
ben die Erscheinungen des animalischen Magnetis=
mus in neuern Zeiten auch in der Erfahrung die sub=
stantielle Einheit der Seele und die Macht ihrer
Idealität zur Anschauung gebracht, wodurch alle die festen
+   Verstandesunterschiede in Verwirrung gesetzt, und eine
speculative Betrachtung für die Auflösung der Widersprüche
unmittelbarer als nothwendig gezeigt wird.

spirit is divided which are regarded as being rigidly distinct, so that spirit is treated as a mere aggregate of independent powers, each of which stands only in reciprocal and therefore external relation to the other. It is true that this psychology also demands the establishment of a harmonious integration 5 between the various powers — and in this context reiterates and fails to define this commonplace as persistently as it once reiterated and failed to define perfection — but this is merely an unsubstantiated acknowledgement of the original unity of spirit, it is not the expression of it. Even less is it the 10 cognition of the necessity and rationality of the particularization to which the Notion or implicit unity of spirit progresses. This harmonious integration is therefore no more than an empty presentation, masquerading in meaningless locutions, and remaining ineffective in the face of what are 15 presumed to be independent spiritual powers.

§ 379

The self, in that it is aware of the living unity of spirit, is itself opposed to its being split up into what are presented as different and mutually independent faculties, powers, or, what amounts to the same 20 thing, activities. In this context however, the need to comprehend is still more pressing on account of the oppositions which immediately present themselves between the freedom of spirit and its being determined, and not only between the distinctness 25 and free activity of the soul and the corporeity external to it, but between this and their intimate interdependence. In more recent times, the substantial unity of the soul and the power of its ideality have even become apparent as a matter of experience, 30 particularly in the phenomena of animal magnetism. This has discredited all the rigid distinctions drawn by the understanding, and it has become immediately obvious that if contradictions are to be resolved, a speculative consideration is a necessity. 35

Zusatz. Alle jene, in den beiden vorhergehenden Paragraphen geschilderten endlichen Auffassungen des Geistes sind theils
durch die ungeheure Umgestaltung, welche die Philosophie überhaupt in neurer Zeit erfahren hat, theils, von der empirischen
Seite selbst her, durch die das endliche Denken vor den Kopf
stoßenden Erscheinungen des animalischen Magnetismus verdrängt
worden. — Was das Erstere betrifft, so hat sich die Philosophie
über die seit Wolf allgemein gewordene endliche Betrachtungsweise
des nur reflectirenden Denkens — auch über das Fichtesche Ste-
henbleiben bei den sogenannten Thatsachen des Bewußtseyns —
zur Auffassung des Geistes als der sichselbstwissenden wirklichen
Idee, zum Begriff des sich auf nothwendige Weise insichselbst-
unterscheidenden und aus seinen Unterschieden zur Einheit mit sich
zurückkehrenden lebendigen Geistes erhoben, damit aber nicht bloß
die in jenen endlichen Auffassungen des Geistes herrschenden Abstractionen des nur Einzelnen, nur Besonderen und nur Allgemeinen überwunden und zu Momenten des Begriffs, der ihre
Wahrheit ist, herabgesetzt, sondern auch, statt des äußerlichen Be-
schreibens vorgefundenen Stoffes, die strenge Form des sich selbst
mit Nothwendigkeit entwickelnden Inhalts als die allein wissenschaftliche Methode geltend gemacht. Wenn in den empirischen
Wissenschaften der Stoff als ein durch die Erfahrung gegebener
von außen aufgenommen und nach einer bereits feststehenden allgemeinen Regel geordnet und in äußerlichen Zusammenhang gebracht wird; so hat dagegen das speculative Denken jeden seiner
Gegenstände und die Entwicklung derselben in ihrer absoluten Nothwendigkeit aufzuzeigen. Dieß geschieht, indem jeder besondere Begriff aus dem sich selbst hervorbringenden und verwirklichenden allgemeinen Begriff oder der logischen Idee abgeleitet wird. Die
Philosophie muß daher den Geist als eine nothwendige Entwicklung der ewigen Idee begreifen, und Dasjenige, was die besonderen Theile der Wissenschaft vom Geiste ausmacht, rein aus dem
Begriffe desselben sich entfalten lassen. Wie bei dem Lebendigen
überhaupt, auf ideelle Weise, Alles schon im Keime enthalten ist,
und von diesem selbst, nicht von einer fremden Macht hervorgebracht wird; so müssen auch alle besonderen Formen des lebendi-

Addition. All those finite conceptions of spirit outlined in the two preceding Paragraphs have been called in question, not only by the prodigious transformation undergone by philosophy in general in recent times, but even empirically, through the phenomena of animal magnetism having proved   5 intractable to finite thinking. In the former case, the finite manner of consideration employed by simply reflective thought, which has become general since Wolf, and the   + Fichtean preoccupation with the so-called facts of consciousness, have both been superseded, and philosophy has   10   + raised itself to the conception of spirit as the Idea in the actuality of its self-awareness, the Notion of living spirit differentiating itself within itself in a necessary manner and returning from its differences into unity with itself. In doing so, philosophy has not only overcome the abstractions that   15 dominate these finite conceptions of spirit — the simply individual, the simply particular and the simply universal — and reduced these to moments of their truth, the Notion, but has also established that it is not the random retailing of the material to hand, but the rigorous form in which the content   20 necessarily develops itself, which constitutes the only scientific method. The material of the empirical sciences is   + taken up from without as it is provided by experience, and brought into external connection by being ordered in accordance with the precept of a general rule. Speculative   25 thought, however, has to exhibit each of its general objects and their development, in their absolute necessity. This is brought about in that each particular Notion is derived from the self-producing and self-actualizing universal Notion, from the logical Idea. Philosophy has therefore to grasp   30   + spirit as a necessary development of the eternal Idea, and to allow that which constitutes the particular parts of the science of spirit to unfold itself purely from the Notion of spirit. Just as everything in living being generally, is brought forth not by an alien power, but by the germ in which it is   35 already contained in an ideal (ideell) manner, so all

10

gen Geistes aus seinem Begriffe als ihrem Keime sich hervortrei-
ben. Unser vom Begriff bewegtes Denken bleibt dabei dem eben-
falls vom Begriff bewegten Gegenstande durchaus immanent; wir
sehen der eigenen Entwicklung des Gegenstandes gleichsam nur
zu, verändern dieselbe nicht durch Einmischung unserer subjectiven
Vorstellungen und Einfälle. Der Begriff bedarf zu seiner Verwirklichung keines äußeren Antriebs; seine eigene, den Widerspruch der Einfachheit und des Unterschieds in sich schließende
und deßwegen unruhige Natur treibt ihn, sich zu verwirklichen,
den in ihm selbst nur auf ideelle Weise, das heißt, in der widersprechenden Form der Unterschiedslosigkeit vorhandenen Unterschied
zu einem wirklichen zu entfalten, und sich durch diese Aufhebung
seiner Einfachheit als eines Mangels, einer Einseitigkeit, wirklich
zu dem Ganzen zu machen, von welchem er zunächst nur die
Möglichkeit enthält.

Nicht minder aber, wie im Beginn und Fortgang seiner Entwicklung, ist der Begriff im Abschließen derselben von unserer Willkür unabhängig. Bei bloß räsonnirender Betrachtungsweise erscheint der Abschluß allerdings mehr oder weniger willkürlich;
in der philosophischen Wissenschaft dagegen setzt der Begriff selber
seinem Sichentwickeln dadurch eine Grenze, daß er sich eine ihm
völlig entsprechende Wirklichkeit giebt. Schon am Lebendigen sehen
wir diese Selbstbegrenzung des Begriffs. Der Keim der Pflanze —
dieser sinnlich vorhandene Begriff — schließt seine Entfaltung mit
einer ihm gleichen Wirklichkeit, mit Hervorbringung des Samens.
Dasselbe gilt vom Geiste; auch seine Entwicklung hat ihr Ziel
erreicht, wenn der Begriff desselben sich vollkommen verwirklicht
hat — oder was dasselbe ist — wenn der Geist zum vollkommenen Bewußtseyn seines Begriffs gelangt ist. Dieß Sich-in-Eins-
Zusammenziehen des Anfangs mit dem Ende, — dieß in seiner
Verwirklichung zu sich selber Kommen des Begriffs erscheint aber
am Geiste in einer noch vollendeteren Gestalt als am bloß Lebendigen; denn während bei diesem der hervorgebrachte Samen nicht
derselbe ist mit dem, von welchem er hervorgebracht worden, ist

11

in dem sich selbst erkennenden Geiste das Hervorgebrachte Eins
und Dasselbe mit dem Hervorbringenden.

particular forms of living spirit have to drive themselves
forth from its Notion, which is their germ. In that our   +
thought is motivated by the Notion, it therefore remains
entirely immanent within the similarly motivated general
object. We, as it were, simply observe the general object's   5
own development, without changing it through the introduction of our subjective presentations and fancies. The
Notion needs no external impulsion for its actualization. Its
own nature, restless in that it includes within itself the contradiction of simplicity and difference, drives it to actualize 10
itself by unfolding into actuality the difference which is only
present within it in an ideal (ideell) manner as the contradictory form of lack of difference. Through this sublation of
its simplicity as of a deficiency, a onesidedness, it is driven to
actually constitute the whole, which in the first instance it   15
contains only as possibility.

The Notion is, moreover, no less independent of our
wilfulness in the conclusion, than it is in the beginning and
progression of its development. Reached merely by means of
superficial rationalization, the conclusion will of course   20
appear to be more or less arbitrary. In philosophic science
however, the Notion itself sets a limit to its self-development
by giving itself an actuality which is entirely adequate to it.
This limiting of itself may already be seen in living being,
where a sensuous representation of the Notion, the germ of   25
the plant, completes its development by bringing forth an
actuality like itself, the seed. Spirit is the same in this respect,   +
for its development has also reached its goal when its Notion
has completely actualized itself, that is to say, when spirit
has attained complete consciousness of its Notion. Neverthe-   30
less, the beginning and the end drawing themselves together
into one, as the Notion comes to itself in its actualization,
appears in a more complete shape in spirit than it does in
mere living being. In the latter, the seed which is brought
forth is not identical with that from which it is engendered,   35
but in self-knowing spirit, that which is brought forth and
that which brings it are one and the same.   +

Nur wenn wir den Geist in dem geschilderten Proceß der Selbstverwirklichung seines Begriffs betrachten, erkennen wir ihn in seiner Wahrheit; (denn Wahrheit heißt eben Uebereinstimmung des Begriffs mit seiner Wirklichkeit). In seiner Unmittelbarkeit ist der Geist noch nicht wahr, hat seinen Begriff noch nicht sich gegenständlich gemacht, das in ihm auf unmittelbare Weise Vorhandene noch nicht zu einem von ihm Gesetzten umgestaltet, seine Wirklichkeit noch nicht zu einer seinem Begriff gemäßen umgebildet. Die ganze Entwicklung des Geistes ist nichts Anderes als sein Sichselbsterheben zu seiner Wahrheit, und die sogenannten Seelenkräfte haben keinen anderen Sinn als den, die Stufen dieser Erhebung zu sein. Durch diese Selbstunterscheidung, durch dieß Sichumgestalten und durch die Zurückführung seiner Unterschiede zur Einheit seines Begriffs ist der Geist, wie ein Wahres, so ein Lebendiges, Organisches, Systematisches, und nur durch das Erkennen dieser seiner Natur ist die Wissenschaft vom Geiste gleichfalls wahr, lebendig, organisch, systematisch; — Prädicate, die weder der rationellen noch der empirischen Psychologie ertheilt werden können, da jene den Geist zu einem von seiner Verwirklichung abgeschiedenen, todten Wesen macht, diese aber den lebendigen Geist dadurch abtödtet, daß sie denselben in eine vom Begriff nicht hervorgebrachte und zusammengehaltene Mannigfaltigkeit selbstständiger Kräfte auseinanderreißt.

Wie schon bemerkt, hat der thierische Magnetismus dazu beigetragen, die unwahre, endliche, bloß verständige Auffassung des Geistes zu verdrängen. Diese Wirkung hat jener wunderbare Zustand besonders in Bezug auf die Betrachtnng der Naturseite des Geistes gehabt. Wenn die sonstigen Zustände und natürlichen Bestimmungen des Geistes sowie die bewußten Thätigleiten desselben wenigstens äußerlich vom Verstande aufgefaßt werden können, und dieser den in ihm selbst wie in den endlichen Dingen herrschenden äußeren Zusammenhang von Ursach und Wirkung, — den sogenannten natürlichen Gang der Dinge — zu fassen vermag; so zeigt sich der Verstand dagegen unfähig, an die Erscheinungen des thierischen Magnetismus auch nur zu glauben, weil in denselben das nach der Meinung des Verstandes durchaus feste Ge-

12

Since it is precisely the agreeing of the Notion with its
actuality which constitutes truth, it is only when we regard
spirit in the delineated process of the self-actualization of its
Notion, that we know the truth of it. Spirit in its immediacy,
since it has not yet objectivized its Notion, transfigured that   5
which it has within itself in an immediate manner by
positing it, transformed its actuality so that it is adequate
to its Notion, is not spirit in its truth. The whole development   +
of spirit is nothing but its raising itself into its truth, and it is
only as the stages of this that the so-called psychic powers   10
have any significance. It is through this differentiation and
transfiguration of itself, and by the leading back of its
differences into its Notion, that spirit has truth, and is living,
organic, systematic. Similarly, it is only through the cognition of this, the nature of spirit, that the science of it is true,   15
living, organic, systematic. Neither rational nor empirical
psychology warrants these predicates however, for while the
former turns spirit into a dead essence, divorced from its   +
actualization, the latter destroys living spirit by tearing it
apart into a multiplicity of independent forces, underived   20
from and unrelated by the Notion.

As has already been noticed, animal magnetism has
contributed to the discrediting of the untrue and limited
conception of spirit postulated by the mere understanding.
It is particularly with regard to consideration of the natural   25
aspect of spirit that this extraordinary condition has had
effect. The understanding is at least capable of apprehending, in an external manner, the other conditions and natural
determinations, as well as the conscious activities of spirit.
It can also grasp what is called the natural course of things,   30
the external connection of cause and effect, by which, like
finite things, it is itself dominated. It is however evidently   +
incapable of ascribing even credibility to the phenomena of
animal magnetism, for in this instance it is no longer possible
for it to regard spirit as being completely fixed and bounded   35

bundenfein des Geiftes an Ort und Zeit, fo wie an den verftän-
bigen Zufammenhang von Urfache und Wirkung feinen Sinn verliert, und innerhalb des finnlichen Dafeyns felbft die dem Verftande
ein unglaubliches Wunder bleibende Erhabenheit des Geiftes über
das Außereinander und über deffen äußerliche Zufammenhänge zum
Vorfchein kommt. Obgleich es nun fehr thöricht wäre, in den
Erfcheinungen des thierifchen Magnetismus eine Erhebung des
Geiftes fogar über feine begreifende Vernunft zu fehen, und von
diefem Zuftande über das Ewige höhere Auffchlüffe als die von
der Philofophie ertheilten zu erwarten, — obgleich der magne-
tifche Zuftand vielmehr für eine Krankheit und für ein Herabfinken
des Geiftes felbft unter das gewöhnliche Bewußtfeyn infofern erklärt werden muß, als der Geift in jenem Zuftande fein in be-
ftimmten Unterfcheidungen fich bewegendes, der Natur fich gegen-
überftellendes Denken aufgiebt; — fo ift doch andererfeits das in
den Erfcheinungen jenes Magnetismus fichtbare Sichlosmachen
des Geiftes von den Schranken des Raums und der Zeit und
von allen endlichen Zufammenhängen Etwas, was mit der Philofophie eine Verwandtfchaft hat, und das, da es mit aller Bru-
talität einer ausgemachten Thatfache dem Skepticismus des Ver-
ftandes Trotz bietet, das Fortfchreiten von der gewöhnlichen Pfy-
chologie zum begreifenden Erkennen der fpeculativen Philofophie
nothwendig macht, für welche allein der thierifche Magnetismus
kein unbegreifliches Wunder ift.

§. 380.

Die concrete Natur des Geiftes bringt für die Betrachtung die eigenthümliche Schwierigkeit mit fich, daß
die befondern Stufen und Beftimmungen der Entwicklung
feines Begriffs nicht zugleich als befondere Exiftenzen zurück und feinen tiefern Geftaltungen gegenüber bleiben,
wie diß in der äußern Natur der Fall ift, wo die Materie
und Bewegung ihre freie Exiftenz als Sonnenfyftem hat,
die Beftimmungen der Sinne auch rückwärts als Eigen-
fchaften der Körper und noch freier als Elemente exifti-
ren u. f. f. Die Beftimmungen und Stufen des Geiftes
dagegen find wefentlich nur als Momente, Zuftände, Be-

13

in place and time as well as by the postulated connection of
cause with effect. It is therefore faced with what it cannot
regard as anything but an incredible miracle, the appearance
within sensuous existence of spirit's having raised itself above
extrinsicality and its external connections. Now it would be   5   +
quite absurd to regard the phenomena of animal magnetism
as a raising of spirit above its comprehending reason, and to
expect that this condition should yield revelations of the
eternal superior to those provided by philosophy. In so far
as spirit in the magnetic condition relinquishes thought   10
which, in positing itself in the face of nature, moves in
determinate differences, the condition is rather to be regarded as a distemper, in which spirit even sinks below
ordinary consciousness. In the phenomena of this magnetism   +
however, spirit visibly liberates itself from the limits of time   15
and space and from all finite connections, and the phenomena have, therefore, something of an affinity with philosophy. It is this which, by defying the scepticism of the
understanding with the blunt brutality of an undeniable
fact, necessitates the progression from ordinary psychology   20
to the comprehending cognition of speculative philosophy.
It is only for this cognition that animal magnetism is not an
incomprehensible miracle.

§ 380

**Observation of the concrete nature of spirit is
peculiarly difficult, in that the particular stages and   25
determinations in the development of its Notion do
not remain behind together as particular existences
confronting its profounder formations. In the case
of external nature they do however: matter and
motion have their free existence as the solar system,   30
the determinations of the senses also exist retro-
gressively as the properties of bodies, and even more
freely as elements etc. The determinations and stages
of spirit occur in the higher stages of its development essentially only as moments, conditions, de-   35**

ſtimmungen an den höhern Entwicklungsſtufen. Es ge-
ſchieht dadurch, daß an einer niedrigern, abſtractern Beſtimmung das Höhere ſich ſchon empiriſch vorhanden zeigt,
wie z. B. in der Empfindung alles höhere Geiſtige als Inhalt oder Beſtimmtheit. Oberflächlicherweiſe kann daher in
der Empfindung, welche nur eine abſtracte Form iſt, jener
Inhalt, das Religiöſe, Sittliche u. ſ. f., weſentlich ſeine
Stelle und ſogar Wurzel zu haben, und ſeine Beſtimmun-
gen als beſondere Arten der Empfindung zu betrachten
nothwendig ſcheinen. Aber zugleich wird es indem niedrigere Stufen betrachtet werden, nöhig, um ſie nach ihrer
empiriſchen Exiſtenz bemerklich zu machen, an höhere zu erinnern, an welchen ſie nur als Formen vorhanden ſind,
und auf dieſe Weiſe einen Inhalt zu anticipiren, der erſt
ſpäter in der Entwicklung ſich darbietet (z. B. beim natürlichen Erwachen das Bewußtſeyn, bei der Verrücktheit
den Verſtand u. ſ. f.).

Begriff des Geiſtes.

§. 381.

Der Geiſt hat für uns die Natur zu ſeiner Vor-
ausſetzung, deren Wahrheit, und damit deren abſo-
lut Erſtes er iſt. In dieſer Wahrheit iſt die Natur ver-
ſchwunden, und der Geiſt hat ſich als die zu ihrem Für-
ſichſeyn gelangte Idee ergeben, deren Object eben ſowohl
als das Subject der Begriff iſt. Dieſe Identität iſt
abſolute Negativität, weil in der Natur der Begriff
ſeine vollkommene äußerliche Objectivität hat, dieſe ſeine
Entäußerung aber aufgehoben, und er in dieſer ſich iden-
tiſch mit ſich geworden iſt. Er iſt dieſe Identität ſomit
zugleich nur, als Zurückkommen aus der Natur.

Zuſatz. Bereits im Zuſatz zu §. 379 iſt der Begriff des
Geiſtes dahin angegeben worden, daß dieſer die ſich ſelbſt wiſ-
ſende wirkliche Idee ſey. Dieſen Begriff hat die Philoſophie, wie
alle ihre ſonſtigen Begriffe, als nothwendig zu erweiſen, das
heißt, als Reſultat der Entwicklung des allgemeinen Begriffs oder

terminations, so that what is higher already shows itself to be empirically present in a lower and more abstract determination, all higher spirituality, for example, being already in evidence as content or determinateness within sensation. Superficially, it might therefore seem necessary to regard that which is religious, ethical, etc. as having its essential placing and even root as the content of the simply abstract form of sensation, and to regard the determinations of it as particular kinds of sensation. If lower stages are regarded with reference to their empirical existence however, higher stages will have to be simultaneously recollected. Since they are only present within these higher stages as forms, this procedure gives rise to the anticipation of a content which only presents itself later in the development. Consciousness is anticipated in natural awakening for example, the understanding in derangement etc.

NOTION OF SPIRIT

§ 381

For us, spirit has *nature* as its *presupposition*. It is the *truth* of nature, **and, therefore, its absolute prius.** Nature has vanished in this truth, and **spirit** has yielded itself as the Idea **which has attained to its being-for-self,** the *object* of which, to the same extent as its *subject*, is *the Notion*. This identity is *absolute negativity*, for the Notion has its complete external objectivity in nature, and has become identical with itself in that this its externalization has been sublated. **At the same time** therefore, it is only as this return out of nature that the Notion constitutes this identity.

Addition. The Notion of spirit as the Idea in its self-knowing actuality has already been enunciated in the Addition to § 379. Philosophy has to demonstrate the necessity of this Notion, as it does of all its other Notions, that is to say, it has to cognize it as a result of the development of the

ber logifchen Jbee zu erfennen. Dem Geifte geht aber in biefer
Entwicklung nicht nur bie logifche Jbee, fonbern auch bie äußere
Natur vorher. Denn bas fchon in ber einfachen logifchen Jbee
enthaltene Erkennen ift nur ber von uns gebachte Begriff bes
Erkennens, nicht bas für fich felbft vorhanbene Erkennen, nicht
ber wirkliche Geift, fonbern bloß beffen Möglichkeit. Der wirkliche Geift, welcher allein in ber Wiffenfchaft vom Geifte unfer
Gegenftanb ift, hat bie äußere Natur zu feiner nächften, wie bie
logifche Jbee zu feiner erften Vorausfetzung. Zu ihrem Enbre-
fultate muß baher bie Naturphilofophie — unb mittelbar bie Logik
— ben Beweis ber Rothwenbigkeit bes Begriffs bes Geiftes haben. Die Wiffenfchaft vom Geift ihrerfeits hat biefen Begriff
burch feine Entwicklung unb Verwirklichung zu bewähren. Was
wir baher hier zu Anfang unferer Betrachtung bes Geiftes ver-
ficherungsweife von bemfelben fagen, kann nur burch bie ganze
Philofophie wiffenfchaftlich bewiefen werben. Zunächft können wir
hier nichts Anberes thun, als ben Begriff bes Geiftes für bie
    * Vorftellung erläutern.

* *Kehler Ms.* SS. 10–11; vgl. *Griesheim Ms.* SS. 16–17: Vorläufig ist von
andern Weisen der Betrachtung des Geistes gesprochen; aber was wir zu
thun haben ist, den Geist nach (11) dem Begriff, an und für sich selbst zu
betrachten, die Philosophie des Geistes. Bei diesem Thema haben wir
anzufangen und das muß nothwendig der Begriff des Geistes sein. Daß wir
mit dem Geist anfangen, ist gebunden, daß die Philosophie des Geistes eine
einzelne philosophische Wissenschaft ist; die Natur der Philosophie ist aber,
daß sie nur ein Ganzes ist, ein Kreis, dessen Peripherie aus mehreren
Kreisen besteht; wir betrachten hier einen besonderen Kreis des Ganzen;
dieser hat aber ein Voraus, sein Anfang ist ein (*Griesheim*: das) Produkt eines
Vorhergehenden; für uns aber, die wir anfangen ist der Geist nicht ein
Produkt, sondern wir fangen unmittelbar an, und da fangen wir unphiloso-
phisch an mit etwas, das nicht bewiesen ist. Wie überhaupt in der Philosophie,
im Logischen ein wahrhafter Anfang gewonnen wird, gehört nicht hierher,
und wie nun auch der Anfang als Resultat erscheinen muß, ebenso wenig.
Wir fangen mit dem Geist an, sprechen unmittelbar davon, das ist voraussetzen,
versichern, was freilich oft als einzige Weise des Erkennens gilt. Wir fangen
an auf unberechtigste Weise, mit dem Begriff des Geistes. Wir können uns
zunächst Vorstellung von diesem Begriff machen, die sich hält, ihre Be-
währung dadurch nur finden kann, daß sich auf die gewöhnlichen Vorstellungsweisen, Weise des Bewußtseins, die wir sonst zugeben, berufen wird.
Damit wir wissen, auf welche Bestimmung es ankommt im bestimmten
Begriff des Geistes.

universal Notion, or of the logical Idea. In this development
however, spirit is preceded not only by the logical Idea, but
also by external nature. For the *cognition* already contained
in the simply *logical* Idea is not cognition present for itself,
but merely the Notion of cognition thought by us; it is not   5
actual spirit, but merely the possibility of it. In the science of   +
spirit we have only actual spirit as our general object, which
has nature as its proximate and the logical Idea as its
primary presupposition. The final result of the philosophy
of nature and mediatively of logic, is therefore the proof of   10
the necessity of the Notion of spirit. The science of spirit, for
its part, has to verify this Notion by means of its development and actualization. Consequently, what is stated
assertorically here at the beginning of our consideration of
spirit, can only be demonstrated scientifically by philosophy   15
in its entirety. All we can do here at the outset is to elucidate
the Notion of spirit for presentative thinking.*

---

* *Kehler* Ms. pp. 10–11; cf. *Griesheim* Ms. pp. 16–17: The provisional discussion has been concerned with other ways of considering spirit, but we
have to consider the philosophy of spirit, spirit in accordance with the (11)
Notion, as it is in and for itself. It is with this theme that we have to begin, and
it must of necessity be the Notion of spirit. Yet although our beginning with
spirit involves the philosophy of spirit as a single philosophic science, it is in
the nature of philosophy alone to be a whole, a cycle, the periphery of which
consists of various cycles. Here we are considering one particular cycle of
the whole, and it has a presupposition, its beginning being a (*Griesheim*: the)
product of what precedes it. For us as we begin however, spirit is not a
product, for we begin immediately and therefore unphilosophically with   +
something that is not proven. This is not the place for the way in which a
valid beginning is to be gained for philosophy in general within what is
logical, and the beginning's appearing as a result is equally out of place here.
We begin with spirit, speaking of it immediately and so presupposing,
giving assurance of, what one has to admit is often taken to be the only mode
of cognition. We begin in the most unjustified manner, with the Notion of
spirit. We can do so by making for ourselves a serviceable presentation of this
Notion, which can only find its confirmation through there being reference
to the ordinary modes of presentation and consciousness also admitted by
us. In this way we know the determination at issue in the determinate
Notion of spirit.

Um dieſen Begriff feſtzuſetzen, iſt nöthig, daß wir die Be=
ſtimmtheit angeben, durch welche die Idee als Geiſt iſt. Alle
Beſtimmtheit iſt aber Beſtimmtheit nur gegen eine andere Be=
ſtimmtheit; der des Geiſtes überhaupt, ſteht zunächſt die der Natur
gegenüber; jene iſt daher nur zugleich mit dieſer zu faſſen. Als
die unterſcheidende Beſtimmtheit des Begriffs des Geiſtes muß die
I d e a l i t ä t, das heißt, das Aufheben des Andersſeyns der Idee,
das aus ihrem Anderen in ſich Zurückkehren und Zurückgekehrtſeyn
derſelben bezeichnet werden, während dagegen für die logiſche Idee
das unmittelbare, e i n f a c h e I n ſ i c h ſ e y n, für die Natur aber
das A u ß e r ſ i c h ſ e y n der Idee das Unterſcheidende iſt. Eine aus=
führlichere Entwicklung des im Zuſatz zu §. 379 über die logiſche
Idee beiläufig Geſagten liegt uns hier zu fern; nothwendiger iſt
an dieſer Stelle eine Erläuterung Desjenigen, was als das Cha=
rakteriſtiſche der äußeren Natur angegeben worden iſt, da zu die=
ſer der Geiſt — wie ſchon bemerkt — ſeine nächſte Beziehung hat.

Auch die äußere Natur wie der Geiſt iſt vernünftig, göttlich,
eine Darſtellung der Idee. Aber in der Natur erſcheint die Idee
im Elemente des Außereinander, iſt nicht nur dem Geiſte äußerlich,
ſondern, — weil dieſem, weil der das Weſen des Geiſtes aus=
machenden an und für ſich ſeyenden Innerlichkeit, — eben deß=
halb auch ſich ſelber äußerlich. Dieſer ſchon von den Griechen
ausgeſprochene und ihnen ganz geläufige Begriff der Natur ſtimmt
vollkommen mit unſerer gewöhnlichen Vorſtellung von dieſer überein.
Wir wiſſen, daß das Natürliche räumlich und zeitlich iſt — daß
in der Natur Dieſes neben Dieſem beſteht, — Dieſes nach Die=
ſem folgt, — kurz, daß alles Natürliche in's Unendliche außer=
einander iſt; daß ferner die Materie, dieſe allgemeine Grundlage
aller daſeyenden Geſtaltungen der Natur, nicht bloß uns Wider=
ſtand leiſtet, außer unſerem Geiſte beſteht, ſondern gegen ſich
ſelber ſich auseinander hält, in concrete Punkte, in materielle
Atome ſich trennt, aus denen ſie zuſammengeſetzt iſt. Die Un=
terſchiede, zu welchen der Begriff der Natur ſich entfaltet, ſind
mehr oder weniger gegeneinander ſelbſtſtändige Exiſtenzen; durch
ihre urſprüngliche Einheit ſtehen ſie zwar mit einander in Be=
ziehung, ſo daß keine ohne die andre begriffen werden kann; aber

In order to explicate this Notion, it is necessary to specify the determinateness through which the Idea has being as spirit. All determinateness is however determinateness only with regard to another determinateness — that of spirit in general is in the first instance distinct from that of nature, and is therefore only to be grasped in conjunction with it. The distinctive determinateness of the Notion of spirit has to be characterized as *ideality*, that is to say, as the sublation of the otherness of the Idea, by which it returns and has returned into itself out of its other. The distinctive determinateness of the logical Idea is immediate and *simple being-in-self* however, while it is the *self-externality* of the Idea which characterizes nature. A more elaborate development of what was said in passing about the logical Idea in the Addition to § 379 would be out of place in the present context. At this juncture, an exposition of what has been presented as characteristic of external nature is more to the point, for as has already been noticed, it is to this that spirit is most closely related.

External nature resembles spirit in that it also is rational, divine, a representation of the Idea. In nature however, the Idea appears in the element of extrinsicality, being not only exterior to spirit, but also — and precisely because it is exterior to spirit, to the being-in-and-for-self of the internality constituting the essence of spirit — to itself. This Notion of nature was already enunciated by the Greeks, among whom it was widely recognized. It is moreover in complete agreement with our ordinary presentation of nature, for we know that what is natural is spatial and temporal, that in nature this subsists next to that, this follows that, in short, that the extrinsicality of everything natural is infinite. We also know that matter, the universal foundation of all the specific formations of nature, not only offers resistance to us in that it has subsistence external to our spirit, but also holds itself apart, in opposition to itself, dividing itself into the concrete points, the material atoms, of which it is composed. The differences into which the Notion of nature unfolds itself are more or less mutually independent existences with regard to one another. It is true that they stand in relation to one another on account of their original unity, so that one cannot be grasped without the other, but this relation is to a greater

diese Beziehung ist eine ihnen in höherem oder geringerem Grade äußerliche. Wir sagen daher mit Recht, daß in der Natur nicht die Freiheit, sondern die Nothwendigkeit herrsche; denn diese letztere ist eben, in ihrer eigentlichsten Bedeutung, die nur innerliche und deßhalb auch nur äußerliche Beziehung gegeneinander selbstständiger Existenzen. So erscheinen z. B. das Licht und die Elemente als gegeneinander selbstständig; so haben die Planeten, obgleich von der Sonne angezogen, trotz dieses Verhältnisses zu ihrem Centrum, den Schein der Selbstständigkeit gegen dasselbe und gegeneinander, welcher Widerspruch durch die Bewegung des Planeten um die Sonne dargestellt wird. — Im Lebendigen kommt allerdings eine höhere Nothwendigkeit zu Stande als die im Leblosen herrschende ist. Schon in der Pflanze zeigt sich ein in die Peripherie ergossenes Centrum, eine Concentration der Unterschiede, ein Sich=von=innen=Herausentwickeln, eine sich selbst unterscheidende und aus ihren Unterschieden in der Knospe sich selbst hervorbringende Einheit, somit Etwas, dem wir Trieb zuschreiben; aber diese Einheit bleibt eine unvollständige, weil der Gliederungsproceß der Pflanze ein Außersichkommen des vegetabilischen Subjects, jeder Theil die ganze Pflanze, eine Wiederholung derselben ist, die Glieder mithin nicht in vollkommener Unterwürfigkeit unter die Einheit des Subjects gehalten werden. — Eine noch vollständigere Ueberwindung der Aeußerlichkeit stellt der thierische Organismus dar; in diesem erzeugt nicht nur jedes Glied das andere, ist dessen Ursach und Wirkung, Mittel und Zweck, somit selbst zugleich sein Anderes, sondern das Ganze wird von seiner Einheit so durchdrungen, daß Nichts in ihm als selbstständig erscheint, jede Bestimmtheit zugleich eine ideelle ist, das Thier in jeder Bestimmtheit dasselbe Eine Allgemeine bleibt, daß somit am thierischen Körper das Außereinander sich in seiner ganzen Unwahrheit zeigt. Durch dieß Beisichseyn in der Bestimmtheit, durch dieß in= und aus seiner Aeußerlichkeit unmittelbar in sich Reflectirtseyn ist das Thier für sich seyende Subjectivität und hat es Empfindung; die Empfindung ist eben diese Allgegenwart der Einheit des Thieres in allen seinen Gliedern, die jeden Eindruck unmittelbar dem Einen Ganzen mittheilen, welches im Thiere

or lesser extent external to them. We are therefore justified
in saying that necessity and not freedom holds sway in
nature, for it is precisely the simply internal and consequently also the simply external relation of independent
existences to one another which constitutes the most intrinsic   5
significance of necessity. Light and the elements appear as
independent of one another for example. The planets are
drawn by the sun, but despite this relationship to their
centre, appear to be independent of it and of one another, a
contradiction which is exhibited in the planet's movement   10
about the sun. — Living being certainly exhibits a higher   +
necessity than that which dominates inanimate being. In the
plant there is already a display of a centre effused into the
periphery, a concentration of differences, a self-development
outwards from within, a unity which differentiates itself and   15
brings itself forth out of its differences into the bud, and
consequently into something to which we ascribe a drive.
This unity remains incomplete however, for in the plant's
process of formation, the vegetable subject comes forth from
itself, each part being the whole plant, or rather a repetition   20
of the same. The members are therefore not held in complete
subjection to the unity of the subject. — In the animal   +
organism externality is more completely overcome, for each
member engenders the other, being its cause and effect,
means and end. Each member is therefore simultaneously   25
its own other. What is more, the whole of the animal
organism is so pervaded by its unity, that nothing within it
appears as independent. Since each determinateness is at
the same time of an ideal nature, the animal remaining the
same single universal within each determinateness, it is in   30
the animal body that extrinsicality shows the full extent of
its lack of truth. It is on account of this being with itself in   +
determinateness, of this immediate return from its externality
into intro-reflectedness, that the animal is a subjectivity
which is for itself and has sensation. Sensation is precisely   35
this ubiquity of the unity of the animal in all its members,
which immediately communicate each impression to the
single whole. It is in the animal that the being-for-self of this

für sich zu werden beginnt. In dieser subjectiven Innerlichkeit liegt, daß das Thier durch sich selbst, von innen heraus, nicht bloß von außen bestimmt ist, das heißt, daß es Trieb und Instinkt hat. Die Subjectivität des Thieres enthält einen Widerspruch und den Trieb, durch Aufhebung dieses Widerspruchs sich selbst zu erhalten; welche Selbsterhaltung das Vorrecht des Lebendigen und in noch höherem Grade das des Geistes ist. Das Empfindende ist bestimmt, hat einen Inhalt und damit eine Unterscheidung in sich; dieser Unterschied ist zunächst noch ein ganz ideeller, einfacher, in der Einheit des Empfindens aufgehobener; der aufgehobene, in der Einheit bestehende Unterschied ist ein Widerspruch, der dadurch aufgehoben wird, daß der Unterschied sich als Unterschied setzt. Das Thier wird also aus seiner einfachen Beziehung auf sich in den Gegensatz gegen die äußerliche Natur hineingetrieben. Durch diesen Gegensatz verfällt das Thier in einen neuen Widerspruch; denn nun ist der Unterschied auf eine, der Einheit des Begriffs widersprechende Weise gesetzt; er muß daher ebenso aufgehoben werden, wie zuerst die ununterschiedene Einheit. Diese Aufhebung des Unterschieds geschieht dadurch, daß das Thier das in der äußerlichen Natur für dasselbe Bestimmte verzehrt und durch das Verzehrte sich erhält. So ist durch Vernichtung des dem Thiere gegenüberstehenden Anderen die ursprüngliche einfache Beziehung auf sich und der darin enthaltene Widerspruch von Neuem gesetzt. Zur wahrhaften Lösung dieses Widerspruchs ist nöthig, daß das Andere, zu welchem das Thier sich verhält, diesem gleich sey. Dies findet im Geschlechtsverhältniß statt; hier empfindet jedes der beiden Geschlechter im Anderen nicht eine fremde Aeußerlichkeit, sondern sich selbst oder die beiden gemeinsame Gattung. Das Geschlechtsverhältniß ist daher der höchste Punkt der lebenden Natur; auf dieser Stufe ist sie der äußeren Nothwendigkeit im vollsten Maaße entnommen; da die auf einander bezogenen unterschiedenen Existenzen nicht mehr einander äußerlich sind, sondern die Empfindung ihrer Einheit haben. Dennoch ist die thierische Seele noch nicht frei; denn sie erscheint immer als Eins mit der Bestimmtheit der Empfindung oder Erregung, als an Eine Bestimmtheit gebunden; nur in der Form

whole is initiated. This subjective inwardness involves the animal's being determined by means of itself, from within outwards, and not simply from without. It is the source, that is to say, of its drive and instinct. The subjectivity of the animal contains a contradiction, and the drive to preserve itself by sublating this contradiction. It is this self-preservation which constitutes the prerogative of living being, and in an even higher degree, of spirit. The sentient being is determinate, has a content, and therefore a difference within itself. In the first instance the difference here is still of a completely ideal nature, simple, and sublated within the unity of sentience. Sublated, subsisting within the unity, it is a contradiction, and this contradiction is sublated in that the difference posits itself as difference. The animal is therefore driven out of its simple self-relatedness into opposition to external nature. On account of this opposition the animal falls into a fresh contradiction, for the difference is not posited so as to contradict the unity of the Notion, and must therefore also be sublated, just as the initial undifferentiated unity was. This sublation of difference is brought about by the animal's consuming that within external nature which has determinate being for it, and preserving itself by means of what is consumed. The original simple self-relatedness, and the contradiction contained within it, are therefore posited afresh by the animal's annihilation of that with which it is confronted. In order that this contradiction should be truly resolved, it is necessary that the other to which the animal relates itself should be its equal. This is the case in the *sex-relationship*, in which each of the two sexes finds in the other not an alien externality, but itself, or rather the genus common to both of them. The sex-relationship is therefore the culminating point of animate nature. At this level nature is withdrawn from external necessity to the maximum extent, for the different interrelated existences have their unity in sensation, and are no longer external to one another. The animal soul is not yet free however, for it always appears as one with determinateness of sensation or stimulation, as bound to a single determinateness. It is only

der Einzelnheit ist die Gattung für das Thier; dieß empfindet nur die Gattung, weiß nicht von ihr; im Thiere ist noch nicht die Seele für die Seele, das Allgemeine als solches für das Allgemeine. Durch das im Gattungsprozeß stattfindende Aufheben der Besonderheit der Geschlechter kommt das Thier nicht zum Erzeugen der Gattung; das durch diesen Proceß Hervorgebrachte ist wieder nur ein Einzelnes. So fällt die Natur, selbst auf der höchsten Spitze ihrer Erhebung über die Endlichkeit, immer wieder in diese zurück und stellt auf diese Weise einen beständigen Kreislauf dar. Auch der durch den Widerspruch der Einzelnheit und der Gattung nothwendig herbeigeführte Tod bringt, — da er nur die leere, selbst in der Form der unmittelbaren Einzelnheit erscheinende, vernichtende Negation der Einzelnheit, nicht deren erhaltende Aufhebung ist, — gleichfalls nicht die an-und-für-sich-seyende Allgemeinheit oder die an - und - für - sich allgemeine Einzelnheit, die sich selbst zum Gegenstande habende Subjectivität hervor. Auch in der vollendetsten Gestalt also, zu welcher die Natur sich erhebt, — im thierischen Leben — gelangt der Begriff nicht zu einer seinem seelenhaften Wesen gleichen Wirklichkeit, zur völligen Ueberwindung der Aeußerlichkeit und Endlichkeit seines Daseyns. Dieß geschieht erst im Geiste, der eben durch diese in ihm zu Stande kommende Ueberwindung sich selber von der Natur unterscheidet, so daß diese Unterscheidung nicht bloß
* das Thun einer äußeren Reflexion über das Wesen des Geistes ist.

* *Kehler Ms.* SS. 12–14; vgl. *Griesheim Ms.* SS. 19–20:… der Geist unterscheidet sich von der Natur, wir vergleichen den Geist, wie er uns bestimmt ist, mit der Natur, wie wir es wissen. *Wir* thun dies, unterscheiden den Geist, aber es hat Beziehung auf das Gesagte, nicht nur wir sind es, die den Geist von der Natur unterscheiden, sondern der Geist ist wesentlich dies, sich selbst von der Natur zu unterscheiden, er ist dies, sich von der Natur zu scheiden, nicht nur, das ist seine Bestimmung, sondern das ist seine That, seine Substanz, nur dies ist er, und ist nur insofern, sich von der Natur zu scheiden, und für sich, bei sich selbst zu sein. Daher zwei Verhältnisse des Geistes zur Natur ausgedrückt: daß wir ihn mit der Natur vergleichen und sein sich scheiden von der Natur ist das Hervorgehen desselben aus dem Sinnlichen, das sich hervorbringen des Geistes, das ist dann das, was eben die Auflösung ist in dem Geist, dazu, Geist zu sein, zu seiner Wahrheit zu kommen, diese Wahrheit des Natürlichen ist der Geist…

(13) In der Natur hat der Begriff seine vollkommene äußerliche Objektivi-

in the form of individuality that the genus has being for the animal, which senses the genus without knowing it, and within which neither the soul nor the universal as such is for itself. The animal does not attain to the engendering of the genus through the sublation of the particularity of the   5
sexes which takes place in the generic process, since that which is brought forth by means of this process is once again merely an individual being. Even at the highest point of its elevation above finitude, nature is therefore perpetually falling back into it, and by doing so it therefore exhibits a   10
perpetual cycle. *Death* is necessarily brought about by the contradiction of individuality and the genus, but even death, since it itself appears in the form of immediate individuality, and since it is not the conserving sublation of individuality but simply the empty and annihilating negation of it, also   15
fails to bring forth the universality or the universal individuality which is in and for itself, the subjectivity which has itself as general object. Even in animal life therefore, which   +
is the most perfect shape to which nature raises itself, the Notion fails to attain to an actuality equal to the soul-like   20
nature of its essence, to completely overcome the externality and finitude of its determinate being. This is first accomplished in *spirit*, which distinguishes itself from nature precisely on account of the occurrence within it of this triumph over determinate being. This distinction is not, therefore,   25
simply the act of reflecting externally upon the essence of spirit.*   +

---

* *Kehler Ms.* pp. 12–14: cf. *Griesheim Ms.* pp. 19–20: Spirit distinguishes itself from nature, we compare spirit, as it is determined for us, with nature as we know it. *We* do this, we distinguish spirit, although it is related to nature. It is not only we who distinguish spirit from nature however, for it is of the essence of spirit to distinguish itself from it. Spirit is that which divides itself from nature, to do so being not only its determination but its act, its substance; it is only this, and has being only in so far as it does this and is for and with itself. Two relationships of spirit with nature are therefore expressed, our comparing it with nature, and its dividing itself from nature, its proceeding forth from what is sensuous. The precise nature of spirit's bringing itself forth is therefore that of dissolution within spirit in order to be spirit, in order to enter into truth. Spirit is this truth of what is natural...

(13) In nature, the Notion has its complete external objectivity, — the determination within which nature in general has being, being that of

Diese zum Begriff des Geistes gehörende Aufhebung der Aeußerlichkeit ist Das, was wir die Idealität desselben genannt haben. Alle Thätigkeiten des Geistes sind nichts als verschiedene Weisen der Zurückführung des Aeußerlichen zu der Innerlichkeit, welche der Geist selbst ist, und nur durch diese Zurückführung, durch diese Idealisirung oder Assimilation des Aeußerlichen wird und ist er Geist. — Betrachten wir den Geist etwas näher, so

tät; oder die Bestimmung, in der die Natur überhaupt ist, ist das Außereinander. Zunächst haben wir in unserer Vorstellung, daß der Geist die Natur äußerlich sich gegenüber hat. Die Natur ist dem Geist äußerlich, und der Geist kann nicht sein ohne Natur, der selbstbewußte Geist ist gebunden an eine äußerliche Natur; (*Griesheim*: Vom Geiste geben wir dieß leicht zu, aber wir meinen die Natur könne wohl sein wenn keine Menschen wären, Steine, Thiere, Pflanzen könnten ohne Menschen ihr Wesen fort treiben, ihre Existenz würde im Gegentheil ihnen nicht so mannigfach verkümmert werden.) Wir stellen uns vor die Natur auf die Weise daß sie unmittelbar ist; die Natur brauche den Geist nicht, aber der Geist die Natur, er habe nur Bewußtsein, indem ihm eine Natur gegenübersteht. Die Natur ist zunächst in der Form der Unmittelbarkeit bestimmt, und hat ihre Bedingung, die erscheint nicht am Geist. Wir sagen, die Natur ist dem Geist äußerlich, der Geist ist auch wieder äußerlich für die Natur; insofern sind beide correlativ, aber das zweite ist, was wir wissen müssen, daß der Geist höher ist als die Natur, und insofern ist nicht die Natur das ursprüngliche, wahrhafte, ist nicht nur dem Geist äußerlich, sondern das an sich äußerliche. Daß die Natur nicht unmittelbar ist, haben wir in der religiösen Vorstellung; die Natur ist geschaffen, ein gesetztes, dies vergessen wir aber, indem wir sagen, nachdem sie geschaffen ist, so ist sie affirmativ, sagen wir, sie ist geschaffen, so ist sie nur kraft eines anderen, ihr Bestehen ist nur ein negatives. Wenn wir dies voraussetzen, daß der Geist höher ist, als die Natur, so kommt ihr die Äußerlichkeit zu gegen den Geist, dem die Innerlichkeit gegen sie zukommt. Wenn wir dies Verhältnis nicht so nach unserem Bewußtsein bestimmen wollen, sondern wahrnehmen wollen, wie sich beides unterscheidet, so werden wir finden, daß die Natur ist in der Bestimmung der Äußerlichkeit überhaupt, der Geist ist nur als Begriff des Insichseins. Alles Natürliche ist räumlich, Zeit ist schon höher, da beginnt schon Innerlichkeit; die Räumlichkeit ist nichts, als etwas außereinander, im Raum hat alles Platz, da ist alles affirmativ, alles Bestimmtheit, und thut dem anderen keinen Eintrag, der (14) Raum ist das Bestehen aller Dinge, wo jedes gleichgültig gegen die andern ist. Das ist die abstrakte absolute Bestimmtheit der Natur, außereinander, und daß die Natur nicht nur relativ gegen den Geist als äußerlich bestimmt ist, sondern an ihm selbst sich äußerlich ist. Dem Geist dagegen kommt zu dieses Insichsein, die einfache Beziehung auf sich selbst, dieses Negiren der Äußerlichkeit; was man auch die Einfachheit genannt hat; einfach, nur Beziehung auf sich.

It is this sublation of externality which pertains to the Notion of spirit that we have called its *ideality*. All the activities of spirit are nothing but the various modes in which that which is external is led back into the internality, to what spirit is itself, and it is only by means of this leading 5 back, this idealizing or assimilation of that which is external, that spirit becomes and is spirit. — If we consider it some-

---

extrinsicality. Initially, we have a presentation of spirit's having nature external to and opposed to itself. Nature is external to spirit, and spirit cannot be without nature, self-conscious spirit being bound to an external nature. (*Griesheim*: We readily admit this in the case of spirit, but are of the opinion that nature could very well be without there being any people, that stones, animals, plants, could persist in their essence without people, and, indeed, that if there were none, their existence would not be encroached upon in so many ways.) We present nature to ourselves as being immediate, as if nature did not need spirit while spirit needs nature, as if spirit only had consciousness in that a nature stands over against it. Initially, nature is + determined in the form of immediacy, and has its condition, which does not appear in spirit. We say that nature is external to spirit, so that spirit also, for its part, is external to nature. To this extent both are correlative, but we also have to bear in mind that spirit is higher than nature, and that in so far as it is so, nature is not what is original, what is true, but what is external, not only to spirit but implicitly. We have the religious presentation of nature's + being not immediate but created, something posited, but we forget this in that we say that after it is created it is affirmative. If we say it is created, it is so only on account of another, its subsisting being only a negative being. + If we presuppose spirit's being higher than nature, what pertains to nature as opposed to spirit is externality, and what pertains to spirit as opposed to nature is inwardness. If we do not want to determine this relationship thus, in accordance with our consciousness, but to perceive how each distinguishes itself, we shall find that the being of nature is in the general determination of externality, and that spirit has being only as the Notion of being-in-self. All that is natural is spatial, time already being higher, already initiating inwardness; spatiality is nothing other than something extrinsic, everything + having place in space, where everything is affirmative, determinate, and does not interfere with anything else. (14) Space is the subsisting of all things, where each is indifferent to the others. This is the abstract absolute determinateness of nature, extrinsicality, nature being not only determined as external relative to spirit, but being in itself self-external. What pertains to spirit is however this being-for-self, simple self-relation, this negating of externality; what has also been called simplicity; what is simple, mere self-relation.

19

finden wir als die erste und einfachste Bestimmung desselben die,
daß er Ich ist. Ich ist ein vollkommen Einfaches, Allgemeines.
Wenn wir Ich sagen, meinen wir wohl ein Einzelnes; da aber
Jeder Ich ist, sagen wir damit nur etwas ganz Allgemeines.
Die Allgemeinheit des Ich macht, daß es von Allem, selbst von
seinem Leben abstrahiren kann. Der Geist ist aber nicht bloß
dieß dem Lichte gleiche abstract Einfache, als welches er betrachtet wurde, wenn von der Einfachheit der Seele im Gegensatze
gegen die Zusammengesetztheit des Körpers die Rede war; vielmehr ist der Geist ein trotz seiner Einfachheit in sich Unte.schie-
denes; denn Ich setzt sich selbst sich gegenüber, macht sich zu
seinem Gegenstande und kehrt aus diesem, allerdings erst abstracten, noch nicht concreten Unterschiede zur Einheit mit sich
zurück. Dieß Beisichselbstseyn des Ich in seiner Unterscheidung ist
die Unendlichkeit oder Idealität desselben. Diese Idealität bewährt
sich aber erst in der Beziehung des Ich auf den ihm gegenüberstehenden unendlich mannigfaltigen Stoff. Indem das Ich diesen
Stoff erfaßt, wird derselbe von der Allgemeinheit des Ich zugleich
vergiftet und verklärt, verliert sein vereinzeltes, selbstständiges Bestehen und erhält ein geistiges Dasein. Durch die unendliche
Mannigfaltigkeit seiner Vorstellungen wird der Geist daher aus
seiner Einfachheit, aus seinem Beisichseyn, so wenig in ein räum-
liches Außereinander hineingerissen, daß vielmehr sein einfaches
Selbst sich in ungetrübter Klarheit durch jene Mannigfaltigkeit
hindurchzieht und dieselbe zu keinem selbstständigen Best•hen kommen läßt.

Der Geist begnügt sich aber nicht damit, als endlicher Geist,
durch seine vorstellende Thätigkeit die Dinge in den Raum seiner
Innerlichkeit zu versetzen und ihnen somit auf eine selbst noch äußerliche Weise ihre Aeußerlichkeit zu nehmen; sondern, als reli-
giöses Bewußtseyn, bringt er durch die scheinbar absolute Selbstständigkeit der Dinge bis zu der in ihrem Inneren wirksamen, Alles
zusammenhaltenden, Einen, unendlichen Macht Gottes hindurch;
und vollendet, als philosophisches Denken, jene Idealisirung
der Dinge dadurch, daß er die bestimmte Weise erkennt, wie die
ihr gemeinsames Princip bildende ewige Idee sich in ihnen dar-

20

what more closely, we find that the ego is its primary and
simplest determination. Ego is a completely simple, universal
being. When we speak of it we are certainly referring to an
individual being, but since every individual being is ego, we
are merely referring to something extremely universal. It is   5
on account of its universality that the ego is able to abstract
from everything, even its life. Spirit is not merely this   +
abstract and simple being however, it does not resemble
light, as it was thought to when reference was made to the
simplicity of the soul as opposed to the compositeness of the   10
body. It is, rather, internally differentiated in spite of its   +
simplicity, for ego posits itself in opposition to itself, turns
itself into its general object, and out of this difference, which
in the first instance is certainly abstract and still lacking in
concreteness, returns into unity with itself. It is this, the ego's 15
being with itself in its differentiation, which constitutes its
infinity or ideality. However, this ideality first gives proof of
itself in the relation of ego to the infinite multiplicity of the
material confronting it. This material, in that it is grasped
by the ego, is at the same time infected and transfigured by   20
the ego's universality, so that it loses its individualized and
independent subsistence and assumes a spiritual existence.
Spirit is therefore far from being torn out of its simplicity and
being with itself, drawn into a spatial extrinsicality through
the infinite multiplicity of its presentations, for in the   25
simplicity and undimmed clarity of its self, it pervades this
multiplicity and denies it the attaining of any independent
subsistence.   +

Spirit is not content with its *finitude* however, with trans-
posing things in space into its inwardness through its   30
presentative activity, and so depriving them of their externality in a manner which is itself still external. As *religious*
consciousness, it penetrates from the apparently absolute
independence of things, to the singleness which is effective
within them and which maintains the coherence of every-   35
thing, to the infinite power of God. As *philosophic* thinking it   +
perfects this idealization of things by cognizing the precise
manner in which the eternal Idea, which constitutes their

stellt. Durch diese Erkenntniß kommt die schon im endlichen Geiste sich bethätigende idealistische Natur des Geistes zu ihrer vollendeten, concretesten Gestalt, macht sich der Geist zu der sich selbst vollkommen erfassenden wirklichen Idee und damit zum absoluten Geiste. Schon im endlichen Geiste hat die Idealität den Sinn einer in ihren Anfang zurückkehrenden Bewegung, durch welche der Geist aus seiner Ununterschiedenheit, — als der ersten Position — zu einem Anderen, — zur Negation jener Position — fortschreitend, und vermittelst der Negation jener Negation zu sich selber zurückkommend, sich als absolute Negativität, als die unendliche Affirmation seiner selbst erweist; und dieser seiner Natur gemäß, haben wir den endlichen Geist, erstens, in seiner unmittelbaren Einheit mit der Natur, dann, in seinem Gegensatze gegen dieselbe, und zuletzt, in seiner, jenen Gegensatz als einen aufgehobenen in sich enthaltenden, durch denselben vermittelten Einheit mit der Natur zu betrachten. So aufgefaßt wird der endliche Geist als Totalität, als Idee und zwar als die für sich seyende, aus jenem Gegensatze zu sich selbst zurückkehrende wirkliche Idee erkannt. Im endlichen Geiste aber hat diese Rückkehr nur ihren Beginn, erst im absoluten Geiste wird sie vollendet; denn erst in diesem erfaßt die Idee sich, weder nur in der einseitigen Form des Begriffs oder der Subjectivität, noch auch nur in der ebenso einseitigen Form der Objectivität oder der Wirklichkeit, sondern in der vollkommenen Einheit dieser ihrer unterschiedenen Momente, das heißt, in ihrer absoluten Wahrheit.

Was wir im Obigen über die Natur des Geistes gesagt haben, ist etwas allein durch die Philosophie zu Erweisendes und Erwiesenes, der Bestätigung durch unser gewöhnliches Bewußtseyn Nichtbedürftiges. Insofern aber unser nichtphilosophisches Denken seinerseits einer Vorstelligmachung des entwickelten Begriffs des Geistes bedarf, kann daran erinnert werden, daß auch die christliche Theologie Gott, d. h., die Wahrheit, als Geist auffaßt, und diesen nicht als ein Ruhendes, in leerer Einerleiheit Verbleibendes, sondern als ein Solches betrachtet, das nothwendig in den Proceß des Sich-von-sich-selbst-unterscheidens, des Setzens seines Anderen eingeht, und erst durch dieß Andere und durch die

21

common principle, exhibits itself within them. It is through this cognition that the idealistic nature of spirit, which is already operative in finite spirit, attains its most perfect and most concrete shape in that spirit makes itself absolute as the actual and perfectly self-comprehending Idea. Already in finite spirit, ideality signifies a movement returning into its beginning, through which spirit progresses out of the initial position of its lack of difference into the negation of this first position by another, and by means of the negation of this negation, returns to itself and shows itself to be absolute negativity, the infinite affirmation of itself. It is in conformity with this, the nature of spirit, that we have to consider finite spirit: firstly in its immediate unity with nature, then in its opposition to this, and finally in its unity with nature, which contains this opposition as a sublation, and which is mediated by it. Grasped in this manner, finite spirit is known as totality, Idea, and indeed as the actual Idea which, in its being-for-self, returns to itself out of this opposition. In finite spirit this return is however simply initiated. It is in absolute spirit that it is first perfected. In this, for the first time, the Idea apprehends itself neither in the simply one-sided form of the Notion or subjectivity, nor in the equally onesided form of objectivity or actuality, but in the perfect unity of these its different moments, that is to say, in its absolute truth.

What we have said above about the nature of spirit is to be demonstrated and has been demonstrated almost exclusively by means of philosophy, and has no need of confirmation by means of our ordinary consciousness. However, in so far as our non-philosophical thinking requires, for its part, that the developed Notion of spirit should be more presentative, it can be pointed out that Christian theology also conceives of God or truth as spirit, and that it regards it not as a quiescence, a persistence in empty uniformity, but as necessarily entering into the process of differentiating itself from itself, of positing its other, and as first coming to itself not

erhaltende Aufhebung — nicht durch Verlassung — desselben zu sich selber kommt. Die Theologie drückt, in der Weise der Vorstellung, diesen Proceß bekanntlich so aus, daß Gott der Vater (dieß einfach Allgemeine, Insichseyende), seine Einsamkeit aufgebend, die Natur (das Sichselbstäußerliche, Außersichseyende) erschafft, einen Sohn (sein anderes Ich) erzeugt, in diesem Anderen aber, kraft seiner unendlichen Liebe, sich selbst anschaut, darin sein Ebenbild erkennt, und in demselben zur Einheit mit sich zurückkehrt; welche nicht mehr abstracte, unmittelbare, sondern concrete, durch den Unterschied vermittelte Einheit der vom Vater und vom Sohne ausgehende, in der christlichen Gemeine zu seiner vollkommenen Wirklichkeit und Wahrheit gelangende heilige Geist ist, als welcher Gott erkannt werden muß, wenn er in seiner absoluten Wahrheit, — wenn er als an‑und‑für‑sich‑seyende wirkliche Idee, — und nicht — entweder nur in der Form des bloßen Begriffs, des abstracten Insichseyns — oder in der ebenso unwahren Form einer mit der Allgemeinheit seines Begriffs nicht übereinstimmenden einzelnen Wirklichkeit, — sondern in der vollen Uebereinstimmung seines Begriffs und seiner Wirklichkeit erfaßt werden soll.

Soviel über die unterscheidenden Bestimmtheiten der äußeren Natur und des Geistes überhaupt. Durch den entwickelten Unterschied ist zugleich die Beziehung angedeutet worden, in welcher die Natur und der Geist zu einander stehen. Da diese Beziehung häufig mißverstanden wird, ist eine Erläuterung derselben hier an ihrer Stelle. Wir haben gesagt, der Geist negire die Aeußerlichkeit der Natur, assimilire sich die Natur und idealisire sie dadurch. Diese Idealisirung hat im endlichen, die Natur außer sich setzenden Geiste eine einseitige Gestalt; hier steht der Thätigkeit unseres Willens wie unseres Denkens ein äußerlicher Stoff gegenüber, der gegen die Veränderung, welche wir mit ihm vornehmen, gleichgültig, die ihm dadurch zu Theil werdende Idealisirung durchaus leidend erfährt. — Bei dem die Weltgeschichte hervorbringenden Geiste aber findet ein anderes Verhältniß statt. Da steht nicht mehr auf der einen Seite eine dem Gegenstande äußerliche Thätigkeit, auf der anderen ein bloß leidender Gegenstand,

22

through the relinquishment but through the preserving
sublation of this other. As is well known, theology expresses
this process in the following presentative manner: God the
Father (this simple universal being-in-self), by putting aside
His solitariness, creates nature (that which is external to   5
itself, self-externality), begets a Son (His other ego), but in
the power of His infinite love, beholds Himself in this other,
recognizes His likeness therein, and returns in this to unity
with Himself. This unity is no longer abstract or immediate,
but concrete, being mediated through difference in that it   10
proceeds from the Father and the Son. Attaining its perfect
actuality and truth in the Christian community, it is the Holy
Spirit. If God is to be grasped in His absolute truth as the
actual being-in-and-for-self of the Idea, it is as such that he
has to be known, not simply in the form of the mere Notion,   15
of abstract being-in-self, or in the equally untrue form of an
individual actuality in non-agreement with the universality
of its Notion, but in the full agreement of His Notion with
His actuality.   +
So much for the differentiating determinatenesses of   20
external nature and of spirit in general. The relation in
which spirit and nature stand to one another has been
indicated through the development of their difference.
Since this relation is often misunderstood, this is the place for
an exposition of it. We have said that spirit negates the   25
externality of nature, assimilates nature, and so idealizes it.
This idealization has a onesided shape in finite spirit; by
which nature is posited as external. The activity of our will,
like that of our thought, is here confronted with an external
material, which, since it is indifferent to the alteration which   30
we undertake in respect of it, undergoes its resultant idealization with complete passivity. — In the spirit which brings
forth world history however, another relationship occurs,
for in this case an activity external to the general object no
longer stands on one side, with a simply passive general   35

sondern die geistige Thätigkeit richtet sich gegen einen in sich selber thätigen Gegenstand, — gegen einen solchen, der sich zu Dem, was durch jene Thätigkeit hervorgebracht werden soll, selbst heraufgearbeitet hat, so daß in der Thätigkeit und im Gegenstande ein und derselbe Inhalt vorhanden ist. So waren z. B. das Volk und die Zeit, auf welche die Thätigkeit Alexander's und Cäsar's als auf ihren Gegenstand wirkte, durch sich selber zu dem von jenen Individuen zu vollbringenden Werke fähig geworden; die Zeit schaffte sich ebenso sehr jene Männer, wie sie von ihnen geschaffen wurde; diese waren ebenso die Werkzeuge des Geistes ihrer Zeit und ihres Volkes, wie umgekehrt jenen Heroen ihr Volk zum Werkzeug für die Vollbringung ihrer Thaten diente. — Dem so eben geschilderten Verhältniß ähnlich ist die Weise, wie der philosophirende Geist sich zur äußeren Natur verhält. Das philosophische Denken erkennt nämlich, daß die Natur nicht bloß von uns idealisirt wird, — daß das Außereinander derselben nicht ein für sie selber, für ihren Begriff, durchaus Unüberwindliches ist, — sondern daß die der Natur inwohnende ewige Idee, oder — was das Nämliche ist — der in ihrem Inneren arbeitende an-sich-seyende Geist selber die Idealisirung, die Aufhebung des Außereinander bewirkt, weil diese Form seines Daseyns mit der Innerlichkeit seines Wesens in Widerspruch steht. Die Philosophie hat also gewissermaaßen nur zuzusehen, wie die Natur selber ihre Aeußerlichkeit aufhebt, das Sichselbstäußerliche in das Centrum der Idee zurücknimmt, oder dieß Centrum im Aeußerlichen hervortreten läßt, den in ihr verborgenen Begriff von der Decke der Aeußerlichkeit befreit und damit die äußerliche Nothwendigkeit überwindet. Dieser Uebergang von der Nothwendigkeit zur Freiheit ist nicht ein einfacher, sondern ein Stufengang von vielen Momenten, deren Darstellung die Naturphilosophie ausmacht. Auf der höchsten Stufe dieser Aufhebung des Außereinander, — in der Empfindung — kommt der in der Natur gefangen gehaltene an sich seyende Geist zum Beginn des Fürsichseyns und damit der Freiheit. Durch dieß selbst noch mit der Form der Einzelnheit und Aeußerlichkeit, folglich auch der Unfreiheit behaftete Fürsichseyn wird die Natur über sich hinaus zum Geiste als solchem,

object on the other. Spiritual activity directs itself toward a   +
general object which is in itself active, and which has
elaborated itself into that which is to be brought forth by
this activity. Consequently, one and the same content is
present in both the activity and the general object. For   5
example, the people and the time, the general object in
which the activity of Alexander or Caesar was effective, had
themselves become capable of the works accomplished by
these individuals. The time created these men to precisely
the same extent as it was created by them; these heroes   10
were as much the instruments of the spirit of their time and
people, as, conversely, their people served them as the
instrument for the accomplishment of their deeds. — The   +
manner in which spirit, in philosophizing, relates itself to
external nature, is similar to the relationship just described,   15
for philosophic thinking knows that nature is idealized not
simply by us, that nature or rather its Notion is not completely incapable of overcoming its extrinsicality, but that it
is the eternal Idea dwelling within nature, or rather the
implicit spirit working within it, which brings about the   20
idealizing or sublation of extrinsicality, and that it does so
because this form of spirit's determinate being stands in
contradiction to the inwardness of the essence of spirit.
Consequently, philosophy has as it were simply to observe
how nature itself sublates its externality, takes back that   25
which is self-external into the centrality of the Idea, or
allows this centre to come forth in that which is external, and
overcomes external necessity by freeing the Notion obscured
within it from the covering of externality. This is not a
simple transition from necessity to freedom however, but a   30
series of stages consisting of many moments, the exposition
of which constitutes the philosophy of nature. The highest
stage of this sublation of extrinsicality is sensation, within
which the spirit which is implicit and held captive in nature
reaches the initiation of being-for-self and so of freedom. It   35
is by means of this being-for-self, which is itself still burdened
with the lack of freedom involved in the form of individuality
and externality, that nature is driven forth beyond itself to
spirit as such, that is to say, to the being-for-self of spirit,

das heißt, zu dem, durch das Denken, in der Form der Allge=
meinheit für=sich=seyenden, wirklich freien Geiste fortgetrieben.

Aus unserer bisherigen Auseinandersetzung erhellt aber schon,
daß das Hervorgehen des Geistes aus der Natur nicht so gefaßt
werden darf, als ob die Natur das absolut Unmittelbare, Erste,
ursprünglich Setzende, — der Geist dagegen nur ein von ihr Ge=
setztes wäre; vielmehr ist die Natur vom Geiste gesetzt, — und
dieser das absolut Erste. Der an=und=für=sich=seyende Geist ist
nicht das bloße Resultat der Natur, sondern in Wahrheit sein
eigenes Resultat; er bringt sich selber aus den Voraussetzungen,
die er sich macht, — aus der logischen Idee und der äußeren
Natur hervor, und ist die Wahrheit sowohl jener als dieser, d. h.,
die wahre Gestalt des nur in sich und des nur außer sich seyenden
Geistes. Der Schein, als ob der Geist durch ein Anderes ver=
mittelt sey, wird vom Geiste selber aufgehoben, da dieser — so
zu sagen — die souveräne Undankbarkeit hat, Dasjenige, durch
welches er vermittelt scheint, aufzuheben, zu mediatisiren, zu einem
nur durch ihn Bestehenden herabzusetzen und sich auf diese Weise
vollkommen selbstständig zu machen. — In dem Gesagten liegt
schon, daß der Uebergang der Natur zum Geiste nicht ein Ueber=
gang zu etwas durchaus Anderem, sondern nur ein Zusichselber=
kommen des in der Natur außer sich seyenden Geistes ist. Ebenso
24   wenig wird aber durch diesen Uebergang der bestimmte Unterschied
der Natur und des Geistes aufgehoben; denn der Geist geht nicht
auf natürliche Weise aus der Natur hervor. Wenn §. 222 ge=
sagt wurde, der Tod der nur unmittelbaren einzelnen Lebendig=
keit sey das Hervorgehen des Geistes, so ist dieß Hervorgehen
nicht fleischlich, sondern geistig, — nicht als ein natürliches Her=
vorgehen, sondern als eine Entwicklung des Begriffs zu verste=
hen, welcher die Einseitigkeit der nicht zu adäquater Verwirklichung
kommenden, vielmehr im Tode sich als negative Macht gegen jene
Wirklichkeit erweisenden Gattung, und die jener gegenüberstehen=
den Einseitigkeit des an die Einzelheit gebundenen thierischen Da=
seyns, in der an=und=für=sich allgemeinen Einzelnheit, oder —
was Dasselbe — in dem auf allgemeine Weise für sich seyenden
Allgemeinen aufhebt, welches der Geist ist.

which in that it has the form of universality through thought, has actual freedom.

It will already have become apparent from the preceding exposition, that in a consideration of spirit's issuing forth from it, nature is not to be regarded as the absolute imme-   5
diacy, the primary and original positing factor, and spirit simply as that which is posited by it. It is, rather, nature which is posited by spirit, spirit being the absolute prius.   +
Spirit which is in and for itself is not the mere result of nature, but is in truth its own result. It brings itself forth   10
from the logical Idea and external nature, the presupposi-tions it constitutes for itself, and is as much the truth of one as it is of the other i.e. it is the true shape of spirit which is simply within itself, and of spirit which is simply self-external. The apparent mediation of spirit by an other is   15   +
sublated by spirit itself. Spirit displays, so to speak, the sovereign ingratitude of sublating, mediating, that by which it appears to be mediated, of degrading it to nothing but a dependent subsistence, and of thereby establishing itself as completely independent. — What has already been said   20
implies that the transition from nature to spirit is not a shift into something entirely distinct, but simply a return into self of the spirit which in nature is self-external. It should be added however, that the specific difference between nature and spirit is not sublated by this transition, for spirit does   25
not proceed forth from nature in a natural manner. When it was stated in § 222 that it is the death of simply immediate and singular animation which constitutes the proceeding forth of spirit, reference was made to a procedure not of the flesh but of the spirit. It is to be understood not as a natural   30
proceeding forth, but as a development of the Notion. The   +
Notion sublates both the onesidedness of the genus, which rather than reaching adequate actualization displays itself in death as a power negative to this actuality, and the oppo-site onesidedness of animal existence, which is bound to   35
singularity. It sublates them within spirit, which is the being-in-and-for-self of universal singularity, the universal in the universal mode of its being-for-self.

Die Natur als solche kommt in ihrer Selbstverinnerlichung nicht zu diesem Fürsichseyn, zum Bewußtseyn ihrer selbst; das Thier — die vollendetste Form dieser Verinnerlichung — stellt nur die geistlose Dialektik des Uebergehens von einer einzelnen, seine ganze Seele ausfüllenden Empfindung zu einer anderen, ebenso ausschließlich in ihm herrschenden einzelnen Empfindung dar; erst der Mensch erhebt sich über die Einzelnheit der Empfindung zur Allgemeinheit des Gedankens, zum Wissen von sich selbst, zum Erfassen seiner Subjectivität, seines Ichs, — mit Einem Worte — erst der Mensch ist der denkende Geist und dadurch und zwar allein dadurch wesentlich von der Natur unterschieden. Was der Natur als solcher angehört, liegt hinter dem Geiste; er hat zwar in sich selbst den ganzen Gehalt der Natur; aber die Naturbestimmungen sind am Geiste auf eine durchaus andere Weise als in der äußeren Natur.

§. 382.

Das Wesen des Geistes ist deswegen formell die **Freiheit**, die absolute Negativität des Begriffes als Identität mit sich. Nach dieser formellen Bestimmung kann er von allem Aeußerlichen und seiner eigenen Aeußerlichkeit, seinem Daseyn selbst abstrahiren; er kann die Negation seiner individuellen Unmittelbarkeit, den unendlichen Schmerz ertragen, d. i. in dieser Negativität affirmativ sich erhalten und identisch für sich seyn. Diese Möglichkeit ist seine abstracte für-sich-seyende Allgemeinheit in sich.

Zusatz. Die Substanz des Geistes ist die Freiheit, d. h. das Nichtabhängigseyn von einem Anderen, das Sichauffichselbstbeziehen. Der Geist ist der für-sich-seyende, sich selbst zum Gegenstand habende verwirklichte Begriff. In dieser in ih.. vorhandenen Einheit des Begriffs und der Objectivität besteht zugleich seine Wahrheit und seine Freiheit. Die Wahrheit macht den Geist, — wie schon Christus gesagt hat — frei; die Freiheit macht ihn wahr. Die Freiheit des Geistes ist aber nicht bloß eine außerhalb des Anderen, sondern eine im Anderen errungene

Nature as such, in interiorizing itself, does not attain to this being-for-self, to consciousness of itself. The animal, the most perfect form of this interioration, exhibits merely the spiritless dialectic of the transition from one sensation, which occupies the whole of its soul, to another, which dominates 5 it with equal exclusiveness. It is man who first raises himself above the singleness of sensation to the universality of thought, to self-knowledge, to the grasping of his subjectivity or ego. In other words, it is man who first constitutes thinking spirit, and it is by means of this, and indeed solely 10 by means of it, that he is essentially distinct from nature. What belongs to nature as such lies behind spirit. Spirit certainly has within itself the entire capacity of nature, but the manner in which determinations of nature occur in spirit is quite different from that in which they occur in 15 external nature.

§ 382

The **formal** *essence* of spirit is therefore *freedom*, the absolute negativity of the Notion as self-identity. **On account of this formal determination,** spirit *can* abstract from all that is external **and** 20 **even** from its own externality, its **determinate being. It can** bear the infinite *pain* of the negation of its individual immediacy i.e. **maintain itself affirmatively in this negativity and** have identity as a being-for-self. This possibility is the **abstract** 25 **being-for-self** of the universality **within it.**

Addition. The substance of spirit is freedom, that is to say lack of dependence upon an other, the relating of itself to itself. Spirit is the Notion actualized into being-for-self in that it has itself as its general object. This unity of the 30 Notion with objectivity, which is present within spirit, constitutes spirit's truth and freedom. Truth, as Christ has already said, makes spirit free; freedom makes it true. Spirit is not free however, simply in that it is independent of its other in being external to it, for it achieves this independence 35

Unabhängigkeit vom Anderen, — kommt nicht durch die Flucht vor dem Anderen, sondern durch dessen Ueberwindung zur Wirklichkeit. Der Geist kann aus seiner abstracten für-sich-seyenden Allgemeinheit, aus seiner einfachen Beziehung auf sich heraustreten, einen bestimmten, wirklichen Unterschied, ein Anderes, als das einfache Ich ist, somit ein Negatives in sich selbst setzen; und diese Beziehung auf das Andere ist dem Geiste nicht bloß möglich, sondern nothwendig, weil er durch das Andere und durch Aufhebung desselben dahin kommt, sich als Dasjenige zu bewähren und in der That Dasjenige zu seyn, was er seinem Begriffe nach seyn soll, nämlich die Idealität des Aeußerlichen, die aus ihrem Andersseyn in sich zurückkehrende Idee, oder — abstracter ausgedrückt — das sich selbst unterscheidende und in seinem Unterschiede bei-und-für-sich-seyende Allgemeine. Das Andere, das Negative, der Widerspruch, die Entzweiung gehört also zur Natur des Geistes. In dieser Entzweiung liegt die Möglichkeit des S ch m e r z e s. Der Schmerz ist daher nicht von außen an den Geist gekommen, wie man sich einbildete, wenn man die Frage aufwarf, auf welche Weise der Schmerz in die Welt gekommen sey. Ebenso wenig wie der Schmerz kommt das B ö s e — das Negative des an-und-für-sich-seyenden unendlichen Geistes — von außen an den Geist; es ist im Gegentheil nichts Anderes, als der sich auf die Spitze seiner Einzelnheit stellende Geist. Selbst in dieser seiner höchsten Entzweiung, in diesem Sichlosreißen von der Wurzel seiner an-sich-seyenden sittlichen Natur, in diesem vollsten Widerspruche mit sich selbst, bleibt daher der Geist doch mit sich identisch und daher frei. Das der äußeren Natur Angehörende geht durch den Widerspruch unter; würde z. B. in das Gold eine andere specifische Schwere gesetzt, als es hat, so müßte es als Gold untergehen. Der Geist aber hat die Kraft sich im Widerspruche, folglich im Schmerz — sowohl über das Böse wie über das Uebele — zu erhalten. Die gewöhnliche Logik irrt daher, indem sie meint, der Geist sey ein den Widerspruch gänzlich von sich Ausschließendes. Alles Bewußtseyn enthält vielmehr eine Einheit und eine Getrenntheit, somit einen Widerspruch; so ist z. B. die Vorstellung des Hauses ein meinem Ich völlig Widersprechendes und dennoch von

26

within this other. Its freedom is actualized not through the withdrawal from this other, but through the overcoming of it. Spirit can come forth from the abstract being-for-self of its universality, from its abstract self-relatedness, and by positing within itself a determinate and actual difference, something 5 other than the simple ego, posit a negative within itself. This relation with the other is, moreover, not simply possible for spirit but necessary to it, since it is through the other and the sublation of the other that it proves and constitutes itself as that which it should be in its accordance with its Notion. In 10 this respect it is the ideality of that which is external, the Idea returning into itself from its otherness. Expressed more abstractly, it is the self-differentiating universal which is with and for itself in its difference. The other, the negative, the contradiction, the disunion, is therefore inherent in the 15 nature of spirit. This disunion carries with it the possibility of *pain*. Pain therefore is not derived by spirit from without, as it was imagined to be when men enquired into the manner of its having come into the world. *Wickedness*, the negative   +
of the being-in-and-for-self of finite spirit, is no more derived 20 by spirit from without than is pain, for it is nothing other than spirit putting its singularity before all else. Conse- quently, even in this its supreme disunion, its divorcing itself from the root of its implicitly ethical nature, its most complete self-contradiction, spirit is still free in that it 25 retains its self-identity. Contradiction destroys whatever   + belongs to external nature; gold for example would perish if any other than its own specific gravity were posited within it. Spirit however is able to maintain itself in contradiction and consequently in pain, to survive wickedness as well as 30 evil. All consciousness contains a unity and a dividedness, and hence a contradiction, and ordinary logic is therefore in error in that it supposes spirit to exclude from itself all contradiction. Although, for example, the presentation of   + the house is in complete contradiction of my ego, the ego 35

diesem Ertragenes. Der Widerspruch wird aber vom Geiste ertragen, weil dieser keine Bestimmung in sich hat, die er nicht
als eine von ihm gesetzte und folglich als eine solche wüßte, die
er auch wieder aufheben kann. Diese Macht über allen in ihm
vorhandenen Inhalt bildet die Grundlage der Freiheit des Geistes.
In seiner Unmittelbarkeit ist der Geist aber nur an sich, dem Begriffe oder der Möglichkeit nach, noch nicht der Wirklichkeit nach
frei; die wirkliche Freiheit ist also nicht etwas unmittelbar im Geiste
Seyendes, sondern etwas durch seine Thätigkeit Hervorzubringen-
des. So als den Hervorbringer seiner Freiheit haben wir in
der Wissenschaft den Geist zu betrachten. Die ganze Entwicklung
des Begriffs des Geistes stellt nur das Sichfreimachen des Geistes von allen, seinem Begriffe nicht entsprechenden Formen seines
Daseyns dar; eine Befreiung, welche dadurch zu Stande kommt,
daß diese Formen zu einer dem Begriffe des Geistes vollkommen
angemessenen Wirklichkeit umgebildet werden.

27

§. 383.

Diese Allgemeinheit ist auch sein Daseyn. Als für
sich seyend ist das Allgemeine sich besondernd und hierin
Identität mit sich. Die Bestimmtheit des Geistes ist daher
die Manifestation. Er ist nicht irgend eine Bestimmtheit oder Inhalt, dessen Aeußerung und Aeußerlichkeit nur
davon unterschiedene Form wäre; so daß er nicht Etwas
offenbart, sondern seine Bestimmtheit und Inhalt ist dieses
Offenbaren selbst. Seine Möglichkeit ist daher unmittelbar unendliche, absolute Wirklichkeit.

Zusatz. Wir haben früher die unterscheidende Bestimmtheit des Geistes in die Idealität, in das Aufheben des Andersseyns der Idee gesetzt. Wenn nun in dem obenstehenden §. 383
„die Manifestation" als die Bestimmtheit des Geistes angegeben
wird, so ist dieß keine neue, keine zweite Bestimmung desselben,
sondern nur eine Entwicklung der früher besprochenen. Denn
durch Aufhebung ihres Andersseyns wird die logische Idee, oder
der an-sich-seyende Geist, für sich, das heißt, sich offenbar. Der
für-sich-seyende Geist, oder der Geist als solcher, ist also, — im

endures it. It is however on account of spirit's containing no determination which it might not know to be posited by itself, and therefore also potentially susceptible of being spiritually sublated, that it endures the contradiction. It is this power over the whole content present within it which 5 constitutes the basis of the freedom of spirit. In its immediacy, spirit is however only free implicitly, in accordance not with actuality but with the Notion or possibility. Consequently, actual freedom is not something which occurs within spirit as an immediacy, but is to be brought forth through the 10 activity of spirit. In the science of spirit we have therefore to regard spirit as that which brings forth its freedom. The whole development of the Notion of spirit simply exhibits the manner in which spirit frees itself from all forms of its determinate being which do not correspond to its Notion. 15 + The liberation is accomplished in that these forms are transformed into an actuality entirely adequate to the Notion of spirit.

§ 383

This universality is also its *determinate being*. + The universal *particularizes* itself **in that it has** 20 **being-for-self** and **so** constitutes self-identity. The determinateness **of spirit is therefore** *manifestation*. Spirit is not a **certain** determinateness **or content, the expression or** exteriority **of which is merely a distinct form of it. Rather than reveal-** 25 **ing something therefore,** its determinateness and content is itself this revelation. Its possibility is therefore the immediacy of infinite, absolute *actuality*.

Addition. We have already taken the distinguishing 30 determinateness of spirit to be *ideality*, the sublating of the otherness of the Idea. In the Paragraph above (§ 383), 'manifestation' is said to constitute the determinateness of spirit. This is not however a new, a second determination, but simply a development of what has already been said, for the 35 logical Idea or the implicit being of spirit has being-for-self, is revealed to itself, through the sublation of its otherness. As distinct from the implicit spirit which is unknown to itself, revealed only to us, effused into the extrinsicality of nature, spirit which is for itself, or spirit as such, is therefore 40

Unterschiede von dem sich selber unbekannten, nur uns offenbaren, in das Außereinander der Natur ergossenen, an-sich-seyenden Geiste, — das nicht bloß einem Anderen, sondern sich selber Sichoffenbarende, oder — was auf Dasselbe hinauskommt — das in seinem eigenen Elemente, nicht in einem fremden Stoffe seine Offenbarung Vollbringende. Diese Bestimmung kommt dem Geiste als solchem zu; sie gilt daher von demselben, nicht nur, insofern er sich einfach auf sich bezieht, sich selbst zum Gegenstande habendes Ich ist, sondern auch insofern er aus seiner abstracten für-sich-seyenden Allgemeinheit heraustritt, eine bestimmte Unterscheidung, ein Anderes als er ist, in sich selbst setzt; denn der Geist verliert sich nicht in diesem Anderen, erhält und verwirklicht sich vielmehr darin, prägt darin sein Inneres aus, macht das Andere zu einem ihm entsprechenden Daseyn, kommt also durch diese Aufhebung des Anderen, des bestimmten wirklichen Unterschiedes, zum concreten Fürsichseyn, zum bestimmten Sichoffenbarwerden. Der Geist offenbart daher im Anderen nur sich selber, seine eigene Natur; diese besteht aber in der Selbstoffenbarung; das Sichselbstoffenbaren ist daher selbst der Inhalt des Geistes und nicht etwa nur eine äußerlich zum Inhalt desselben hinzukommende Form; durch seine Offenbarung offenbart folglich der Geist nicht einen von seiner Form verschiedenen Inhalt, sondern seine, den ganzen Inhalt des Geistes ausdrückende Form, nämlich seine Selbstoffenbarung. Form und Inhalt sind also im Geiste mit einander identisch. Gewöhnlich stellt man sich allerdings das Offenbaren als eine leere Form vor, zu welcher noch ein Inhalt von außen hinzukommen müsse; und versteht dabei unter Inhalt ein Insichseyendes, ein Sich = in = sich = haltendes, unter der Form dagegen die äußerliche Weise der Beziehung des Inhalts auf Anderes. In der speculativen Logik wird aber bewiesen, daß in Wahrheit der Inhalt nicht bloß ein Insichseyendes, sondern ein durch sich selbst mit Anderem in Beziehung Tretendes ist, wie umgekehrt in Wahrheit die Form nicht bloß als ein Unselbstständiges, dem Inhalte Aeußerliches, sondern vielmehr als Dasjenige gefaßt werden muß, was den Inhalt zum Inhalt, zu einem Insichseyenden, zu einem von Anderem Unterschiedenen macht. Der wahrhafte Inhalt enthält also in sich selbst

that which reveals itself to itself and not simply to another.
It is, one might say, that which consummates its revelation
in its own element, not in an alien material. This determination belongs to spirit as such and therefore holds good of it
not only in so far as it is simply self-relating as the ego which   5
has itself as its general object, but also in so far as it comes
forth from the abstract being-for-self of its universality by
positing within itself a determinate distinction, something
other than itself. For spirit, rather than losing itself in this
other, maintains and actualizes itself there, marking forth its   10
internality by turning the other into a determinate being
commensurate with it, and by thus sublating the other, the
determinate and actual difference, reaching concrete
being-for-self, determinate self-revelation. Consequently,
spirit reveals in the other only itself, only its own nature. Its   15
own nature consists however of self-revelation. The self-
revelation is therefore itself the content of spirit, and not in
some way merely a form in which spirit derives its content
from without. Consequently, spirit reveals in its revelation
not a content differing from its form, but the form in which   20
its entire content is expressed i.e. its self-revelation. In
spirit, form and content are therefore identical with one
another. It is of course usual to conceive of that which is
revealed as an empty form which still has to derive its
content from without, and so to regard content as a being-   25
in-self, something self-contained, and form on the other
hand as the external manner in which content relates to its
other. Speculative logic shows it to be demonstrably true
however, that content is not simply a being-in-self, but that
it enters into relation with another by means of itself;   30
conversely, that form is not simply a dependency external to
content, but has rather to be grasped as that which makes
the content what it is, i.e. a being-in-self distinct from the
other. The true content contains the form within itself   +

die Form, und die wahrhafte Form ist ihr eigener Inhalt. Den Geist aber haben wir als diesen wahrhaften Inhalt und als diese wahrhafte Form zu erkennen. — Um diese im Geiste vorhandene Einheit der Form und des Inhalts — der Offenbarung und des Geoffenbarten — für die Vorstellung zu erläutern, kann an die Lehre der christlichen Religion erinnert werden. Das Christenthum sagt: Gott habe sich durch Christus, seinen eingebornen Sohn, geoffenbart. Diesen Satz faßt die Vorstellung zunächst so auf, als ob Christus nur das Organ dieser Offenbarung, — als ob das auf diese Weise Geoffenbarte ein Anderes als das Offenbarende sei. Jener Satz hat aber in Wahrheit vielmehr d e n Sinn, Gott habe geoffenbart, daß seine Natur darin besteht, einen Sohn zu haben, d. h. sich zu unterscheiden, zu verendlichen, in seinem Unterschiede aber bei sich selbst zu bleiben, im Sohne sich selber anzuschauen und zu offenbaren, und durch diese Einheit mit dem Sohne, durch dieß Fürsichseyn im Anderen, absoluter Geist zu seyn; so daß der Sohn nicht das bloße Organ der Offenbarung, sondern selbst der Inhalt der Offenbarung ist.

Ebenso wie der Geist die Einheit der Form und des Inhalts darstellt, ist er auch die Einheit der Möglichkeit und Wirklichkeit. Wir verstehen unter dem Möglichen überhaupt das noch Innerliche, noch nicht zur Aeußerung, zur Offenbarung Gekommene. Nun haben wir aber gesehen, daß der Geist als solcher nur ist, insofern er sich selber sich offenbart. Die Wirklichkeit, die eben in seiner Offenbarung besteht, gehört daher zu seinem Begriffe. Im endlichen Geiste kommt allerdings der Begriff des Geistes noch nicht zu seiner absoluten Verwirklichung; der absolute Geist aber ist die absolute Einheit der Wirklichkeit und des Begriffs oder der Möglichkeit des Geistes.

§. 384.

Das Offenbaren, welches als die abstracte Idee unmittelbarer Uebergang, Werden der Natur ist, ist als Offenbaren des Geistes, der frei ist, Setzen der Natur als seiner Welt; ein Setzen, das als Reflexion zugleich Voraussetzen der Welt als selbstständiger Natur ist

therefore, and the true form is its own content. We have
therefore to know spirit as this true content and form. — In
order to help presentative thinking to understand this unity
of form and content, of revelation and revealed, present in
spirit, attention can be drawn to the doctrine of the Christian 5
religion. According to Christianity, God has revealed Himself through Christ, His only begotten Son. In the first
instance, presentative thinking takes this proposition to mean
that Christ is merely the organ of this revelation, as if that
which is revealed in this manner were something other than 10
that which reveals it. However, the true meaning of the +
proposition is rather that God has revealed, that it is in
His nature to have a Son i.e. to differentiate, to limit Himself,
yet to remain with Himself in His difference; to contemplate
and reveal Himself in the Son, and through this unity with 15
the Son, through this being-for-self in the other, to be absolute spirit. Consequently, the Son is not the mere organ, but
the very content of the revelation.

Just as spirit exhibits the unity of form and content, so too
is it the unity of possibility and actuality. That which is 20
possible we generally understand to be that which is still
internal, which has not yet attained to expression, to revelation. We have now seen, however, that spirit as such has
being only in so far as it reveals itself to itself. Actuality, the
precise constitution of which is the revelation of spirit, is 25
therefore involved in the Notion of spirit. In finite spirit of
course, the Notion of spirit does not yet reach its absolute
actualization; absolute spirit is however the absolute unity
of actuality and of the Notion or possibility of spirit.

§ 384

Revelation, which in that the *abstract* Idea is 30 +
revealed is the immediate transition of nature's
coming into being, is, in that the freedom of spirit
is revealed, the *positing* of nature as the world
of spirit; as reflection, this positing is at the
same time the *presupposing* of the world as inde- 35

Das Offenbaren im Begriffe ift Erſchaffen derſelben als ſeines Seyns, in welchem er die Affirmation und Wahrheit ſeiner Freiheit ſich gibt.

Das Abſolute iſt der Geiſt; diß iſt die höchſte Definition des Abſoluten. — Dieſe Definition zu finden und ihren Sinn und Inhalt zu begreifen, diß kann man ſagen, war die abſolute Tendenz aller Bildung und Phi= loſophie, auf dieſen Punkt hat ſich alle Religion und Wiſſenſchaft gedrängt; aus dieſem Drang allein iſt die Weltgeſchichte zu begreifen. — Das Wort und die Vor= ſtellung des Geiſtes iſt früh gefunden, und der In= halt der chriſtlichen Religion iſt, Gott als Geiſt zu er= kennen zu geben. Diß was hier der Vorſtellung gege= ben, und was an ſich das Weſen iſt, in ſeinem eige= nen Elemente, dem Begriffe, zu faſſen, iſt die Aufgabe der Philoſophie, welche ſo lange nicht wahrhaft und immanent gelöſt iſt, als der Begriff und die Freiheit nicht ihr Gegenſtand und ihre Seele iſt.

Zuſatz. Das Sichoffenbaren iſt eine dem Geiſte überhaupt zukommende Beſtimmung; daſſelbe hat aber drei unterſchiedene Formen. Die erſte Weiſe, wie der an=ſich=ſeyende Geiſt oder die logiſche Idee ſich offenbart, beſteht in dem Umſchlagen der Idee in die Unmittelbarkeit äußerlichen und vereinzelten Daſeyns. Dieß Umſchlagen iſt das Werden der Natur. Auch die Natur iſt ein Geſetztes; aber ihr Geſetztſeyn hat die Form der Unmittelbarkeit, des Seyns außer der Idee. Dieſe Form widerſpricht der Inner= lichkeit der ſich ſelbſt ſetzenden, aus ihren Vorausſetzungen ſich ſelber hervorbringenden Idee. Die Idee, oder der in der Natur ſchlafende anſichſeyende Geiſt hebt deßhalb die Aeußerlichkeit, Ver= einzelung und Unmittelbarkeit der Natur auf, ſchafft ſich ein ſei= ner Innerlichkeit und Allgemeinheit gemäßes Daſeyn, und wird dadurch der in ſich reflectirte, für=ſich=ſeyende, ſelbſtbewußte, er= wachte Geiſt, oder der Geiſt als ſolcher. — Hiermit iſt die zweite Form der Offenbarung des Geiſtes gegeben. Auf dieſer Stufe ſtellt der nicht mehr in das Außereinander der Natur ergoſſene Geiſt ſich als das Fürſichſeyende, Sichoffenbare, der bewußtloſen,

pendent nature. Within the Notion, revelation is
the generation of nature as the being of the
Notion, in which the latter **yields to itself** the
affirmation and *truth* of its freedom.

The absolute is spirit; this is the supreme   5
definition of the absolute. It may be said that
the discovery of this definition and the grasping of its meaning and content was the ultimate purpose of all education and philosophy.
All religion and science has driven toward   10
this point, and world history is to be grasped
solely **from this drive.** — The word spirit, the
presentation of it, was an early discovery, and it **is**
the purport of the Christian religion to **make**
God **known** as spirit. That which is *rendered* here   15
presentatively, the *implicit* essence, is to be
grasped in its own element, the Notion. This
is the task of philosophy, and so long as the
Notion and freedom do not constitute the
general object and soul of philosophy, it is a   20
task which is not truly and immanently accomplished.

Addition. The self-revelation, although it is a determination
pertaining to spirit in general, has *three* distinct forms. The
first of these, the way in which implicit spirit or the logical   25
Idea reveals itself, consists of the Idea's switching into the
immediacy of external and singularized determinate being.   +
It is in this switch that nature comes into being. Nature is
also a posited being, although its positedness has the form of
immediacy, of being external to the Idea. In that the Idea   30
posits itself, bringing itself forth from its presuppositions, its
internality is contradicted by this form. The externality,   +
individuation and immediacy of nature is therefore sublated
by the Idea, by the spirit implicit and dormant in nature,
which creates for itself a determinate being corresponding   35
to its internality and universality, and so becomes the
intro-reflected being-for-self of self-consciousness and
awakened spirit i.e. spirit as such. — It is this that yields the   +
second form of the revelation of spirit. At this level, the spirit
which is no longer effused into the extrinsicality of nature   40
ranges itself as self-revealing being-for-self in the face of

ihn ebensosehr verhüllenden wie offenbarenden Natur gegenüber, macht dieselbe sich zum Gegenstande, reflectirt über sie, nimmt die Aeußerlichkeit der Natur in seine Innerlichkeit zurück, idealisirt die Natur, und wird so in seinem Gegenstande für sich. Aber dieß erste Fürsichseyn des Geistes ist selbst noch ein unmittelbares, abstractes, nicht absolutes; durch dasselbe wird das Außersichselbstseyn des Geistes nicht absolut aufgehoben. Der erwachende Geist erkennt hier noch nicht seine Einheit mit dem in der Natur verborgenen an-sich-seyenden Geiste, steht daher zur Natur in äußerlicher Beziehung, erscheint nicht als Alles in Allem, sondern nur als die Eine Seite des Verhältnisses, ist zwar in seinem Verhältniß zu Anderem auch in sich reflectirt und somit Selbstbewußtseyn, läßt aber diese Einheit des Bewußtseyns und des Selbstbewußtseyns noch als eine so äußerliche, leere, oberflächliche bestehen, daß das Selbstbewußtseyn und das Bewußtseyn zugleich auch noch außereinanderfallen, und daß der Geist, trotz seines Beisichselbstseyns, zugleich nicht bei sich selber, sondern bei einem Anderen ist, und seine Einheit mit dem im Anderen wirksamen ansichseyenden Geiste noch nicht für ihn wird. Der Geist setzt hier die Natur als ein Insichreflectirtes, als seine Welt, nimmt der Natur die Form eines gegen ihn Anderen, macht das ihm gegenüberstehende Andere zu einem von ihm Gesetzten; zugleich aber bleibt dies Andere noch ein von ihm Unabhängiges, ein unmittelbar Vorhandenes, vom Geiste nicht Gesetztes, sondern nur Vorausgesetztes, also ein Solches, dessen Gesetztwerden dem reflectirenden Denken vorhergeht. Das Gesetztseyn der Natur durch den Geist ist auf diesem Standpunkte somit noch kein absolutes, sondern nur ein im reflectirenden Bewußtseyn zu Stande kommendes; die Natur wird daher noch nicht als nur durch den unendlichen Geist bestehend, als seine Schöpfung begriffen. Der Geist hat folglich hier noch eine Schranke an der Natur, und ist eben durch diese Schranke endlicher Geist. — Diese Schranke wird nun vom absoluten Wissen aufgehoben, welches die dritte und höchste Form der Offenbarung des Geistes ist. Auf dieser Stufe verschwindet der Dualismus einer selbstständigen Natur oder des in das Außereinander ergossenen Geistes einerseits, und des erst für sich zu

31

unconscious nature, which conceals as much as it reveals of
it. It makes this a general object, reflects upon it, takes back
into its internality the externality of nature, idealizes nature,
and so gains being-for-self in its general object. This primary
being-for-self of spirit is however not absolute, but is itself   5
still an abstract immediacy, for the self-externality of spirit
is not absolutely sublated by means of it. At this juncture the
awakening spirit does not yet know its unity with the implicit
spirit concealed within nature, so that it stands in external
relation to nature, appearing not as all in all, but merely as   10
one side of the relationship. In its relationship with the other,
it is certainly also intro-reflected and therefore self-conscious,
but it still allows this unity of consciousness and self-consciousness to subsist in such an external, empty, superficial
manner, that at the same time there is also a falling apart   15
of the two terms. Spirit, despite its being with itself, is
therefore at the same time not with itself but with another,
since its unity with the spirit implicitly active within the
other has not yet become its being-for-self. At this juncture
spirit posits nature as an intro-reflectedness, as its world, and   20
by denying it the form of an other distinct from itself, turns
that by which it is opposed into that which it posits. At the
same time however, this other remains independent of it as
an immediate presence, not posited but merely presupposed
by it. As such it is therefore a becoming posited which pre-   25
cedes reflecting thought. From this standpoint, nature's
being posited by spirit is therefore still not absolute, since
it merely takes place in reflecting consciousness. Consequently, nature is not yet grasped as subsisting only
by means of the infinite spirit which creates it. At this   30
juncture spirit is therefore still limited by nature, and it is
precisely on account of this limit that it is finite. — The   +
limit is now sublated by absolute knowledge, the *third* and
highest form of the revelation of spirit. The dualism, one
side of which consists of an independent nature or spirit   35
effused into extrinsicality, and the other of spirit merely

werden beginnenden, aber seine Einheit mit jenem noch nicht begreifenden Geistes andererseits. Der absolute Geist erfaßt sich
als selber das Seyn setzend, als selber sein Anderes, die Natur
und den endlichen Geist hervorbringend, so daß dieß Andere jeden
Schein der Selbstständigkeit gegen ihn verliert, vollkommen auf-
hört, eine Schranke für ihn zu seyn, und nur als das Mittel
erscheint, durch welches der Geist zum absoluten Fürsichseyn, zur
absoluten Einheit seines Ansichseyns und seines Fürsichseyns, seines Begriffs und seiner Wirklichkeit gelangt.

Die höchste Definition des Absoluten ist die, daß dasselbe
nicht bloß überhaupt der Geist, sondern daß es der sich absolut
offenbare, der selbstbewußte, unendlich schöpferische Geist ist, welchen wir so eben als die dritte Form des Offenbarens bezeichnet
haben. Wie wir in der Wissenschaft von den geschilderten unvoll-
kommnen Formen der Offenbarung des Geistes zur höchsten Form
derselben fortschreiten, so stellt auch die Weltgeschichte eine Reihe
von Auffassungen des Ewigen dar, an deren Schluß erst der Begriff des absoluten Geistes hervortritt. Die orientalischen Re-
ligionen — auch die jüdische — bleiben noch beim abstracten
Begriff Gottes und des Geistes stehen; was sogar die nur von
Gott dem Vater wissen wollende Aufklärung thut; denn Gott der
Vater für sich ist das in sich Verschloßne, Abstracte, also noch
nicht der geistige, noch nicht der wahrhaftige Gott. In der griechischen Religion hat Gott allerdings angefangen, auf bestimmte
Weise offenbar zu werden. Die Darstellung der Griechischen Götter
hatte zum Gesetz die Schönheit, die zum Geistigen gesteigerte Natur.
Das Schöne bleibt nicht ein abstract Ideelles, sondern in seiner
Idealität ist es zugleich vollkommen bestimmt, individualisirt. Die
griechischen Götter sind jedoch zunächst nur für die sinnliche Anschauung oder auch für die Vorstellung dargestellt, noch nicht im
Gedanken aufgefaßt. Das sinnliche Element aber kann die Totalität des Geistes nur als ein Außereinander, als einen Kreis
individueller geistiger Gestalten darstellen; die alle diese Gestalten
zusammenfassende Einheit bleibt daher eine den Göttern gegenüberstehende, ganz unbestimmte fremde Macht. Erst durch die
christliche Religion ist die in sich selber unterschiedene Eine Natur

initiating its being-for-self but not yet grasping its unity with its counterpart, disappears at this level. Absolute spirit apprehends itself as itself positing being, as itself its other, as bringing forth nature and finite spirit so that this other loses every appearance of being independent of it, ceases com- 5 pletely to limit it, and appears merely as the means by which spirit achieves absolute being-for-self, the absolute unity of its implicitness and being-for-self, of its Notion and its actuality. +

The supreme definition of the absolute is not that it is 10 spirit in general, but that it is the absolutely self-revealing, self-conscious, infinitely creative spirit we have just indicated as constituting the third form of revelation. In science we progress from the imperfect forms of the revelation of spirit described above to the highest form of it, just as world history 15 presents a series of conceptions of the eternal, the Notion of absolute spirit coming forth only at its conclusion. The + oriental religions, like the Judaic do not progress beyond the abstract Notion of God and spirit. The enlightenment, which + wants to know only of God the Father, resembles them, for 20 + God the Father is for Himself, confined to Himself, abstract, and consequently not yet the God of spirit and of truth. In the Greek religion there is, it is true, the initiation of revealing God in a determinate manner. The representation of the Greek gods had beauty, nature heightened to spirituality, as 25 its law. The beautiful does not remain of an abstractly ideal nature, but in its ideality is at the same time completely determined, individualized. Nevertheless, the Greek gods are at first only represented for sensuous intuition or for this and presentation, they are not yet conceived through thought. 30 The sensuous element can, moreover, only represent the totality of spirit as an extrinsicality, as a sphere of individual spiritual shapes, so that the unity embracing all these shapes remains a wholly indeterminate and alien power, dwelling apart from the gods. It is in the Christian religion that the 35 + single internally self-differentiated nature of God, the

Gottes, die Totalität des göttlichen Geistes in der Form der Einheit geoffenbart worden. Diesen in der Weise der Vorstellung

33   gegebenen Inhalt hat die Philosophie in die Form des Begriffs oder des absoluten Wissens zu erheben, welches, wie gesagt, die höchste Offenbarung jenes Inhalts ist.

Eintheilung.

§. 385.

Die Entwicklung des Geistes ist:

daß er.

   I. in der Form der Beziehung auf sich selbst ist, innerhalb seiner ihm die ideelle Totalität der Idee, d. i. daß das, was sein Begriff ist, für ihn wird, und ihm sein Seyn diß ist, bei sich, d. i. frey zu seyn, — subjectiver Geist;

   II. in der Form der Realität als einer von ihm hervorzubringenden und hervorgebrachten Welt, in welcher die Freiheit als vorhandene Nothwendigkeit ist, — objectiver Geist;

   III. in an und für sich seyender und ewig sich hervorbringender Einheit der Objectivität des Geistes und seiner Idealität oder seines Begriffs, der Geist in seiner absoluten Wahrheit, — der absolute Geist.

*

---

* *Diktiert, Sommer 1818* ('Hegel-Studien' Bd. 5, S. 29, 1969) : Die Idee des Geistes, welche als Begriff *an sich* oder *für uns* ist, muß für ihn selbst werden, oder vielmehr er muß sie für sich hervorbringen. Alle Thätigkeit des Geistes und alle Veranstaltung seiner Welt und Geschichte sind Momente dieser Arbeit sich selbst zu erfassen. Die erste Stufe, der subjective Geist, ist der *theoretische Prozeß* desselben, der Prozeß innerhalb seiner selbst, daß er die Gewißheit seiner selbst, als subjectives, für sich werde, welches absolute Objectivität sey, daß er sich als *freyes* erfasse, d.i. zunächst als abstrakt freyes. Die 2te Stufe[,] der objective Geist geht von dieser selbstbewußten Freyheit aus, welche sich zu einer wirklichen Welt als einer vorhandenen Natur macht, die rechtliche und sittliche Welt. Die 3te ist das Erfassen dieses objektiven

+ Geistes durch sich selbst in seiner Idealität, Allgemeinheit und Wahrheit, das Wesen des absoluten Geistes.

totality of the Divine spirit in the form of unity, has had its
initial revelation. Philosophy has to raise the content   +
yielded here in the mode of presentation into the form of the
Notion or of absolute knowledge. As has been noted, this
is the supreme revelation of this content.   5   +

Division

§ 385

In its development, spirit occurs:

**I. in the form of its being related to itself, the
ideal nature of the totality of the Idea being within
it and for it, i.e. that which constitutes its Notion is
for it, and it has being in that it is with itself i.e. free,   10
— subjective spirit;**

**II. in the form of reality, as a world it is to bring
and has brought forth, freedom being present within
this world as necessity, — objective spirit;**

**III. in its absolute truth, the unity of its objectivity   15
and its ideality or Notion, a unity which is in and
for itself and is eternally bringing itself forth, —
absolute spirit.***

* *Dictated, Summer 1818* ('Hegel-Studien' vol. 5, p. 29, 1969): The Idea of   +
spirit, which as Notion is *implicit* or *for us*, must come to be for itself, or rather
spirit must bring forth the Idea for itself. All spirit's activity, and all the
bringing about of its world and history, are moments of this work of apprehending itself. The first stage, that of subjective spirit, is the *theoretical process*
of this, the process within itself by which spirit becomes certain of itself as
what is subjective, becomes for itself. This is absolute objectivity, its apprehending itself as *free* i.e. initially as abstractly free. The second stage is
objective spirit's proceeding forth from this self-conscious freedom, which
makes itself into an actual world as a nature which is present, — the world
of right and ethics. The third is the apprehending of this objective spirit, by
means of itself, in its ideality, universality and truth, — the essence of
absolute spirit.

Zuſatz. Der Geiſt iſt immer Idee; zunächſt aber iſt er nur der **Begriff** der Idee, oder die Idee in ihrer Unbeſtimmt= heit, in der abſtracteſten Weiſe der Realität, das heißt, in der Weiſe des Seyns. Im Anfange haben wir nur die ganz allge= meine, unentwickelte Beſtimmung des Geiſtes, nicht ſchon das Beſondre deſſelben; dieß bekommen wir erſt, wenn wir von Einem zu Anderem übergehen; — denn das Beſondere enthält Eines und ein Anderes; — dieſen Uebergang haben wir aber eben zu Anfang noch nicht gemacht. Die Realität des Geiſtes iſt alſo zunächſt noch eine ganz allgemeine, nicht beſonderte: die Ent= wicklung dieſer Realität wird erſt durch die ganze Philoſophie des Geiſtes vollendet. Die noch ganz abſtracte, unmittelbare Realität iſt aber die Natürlichkeit, die Ungeiſtigkeit. Aus dieſem Grunde

34

iſt das Kind noch in der Natürlichkeit befangen, hat nur natür= liche Triebe, iſt noch nicht der Wirklichkeit, ſondern nur der Mög= lichkeit oder dem Begriffe nach, geiſtiger Menſch. Die erſte Rea= lität des Begriffs des Geiſtes muß demnach, eben weil ſie noch eine abſtracte, unmittelbare, der Natürlichkeit angehörende iſt, als die dem Geiſte unangemeſſenſte bezeichnet, die wahrhafte Realität aber als die Totalität der entwickelten Momente des Begriffs be= ſtimmt werden, welcher die Seele, die Einheit dieſer Momente bleibt. Zu dieſer Entwicklung ſeiner Realität geht der Begriff des Geiſtes nothwendig fort; denn die Form der Unmittelbarkeit, der Unbeſtimmtheit, welche ſeine Realität zunächſt hat, iſt eine ihm widerſprechende; das unmittelbar im Geiſt vorhanden zu ſeyn Scheinende iſt nicht ein wahrhaft Unmittelbares, ſondern an ſich ein Geſetztes, Vermitteltes. Durch dieſen Widerſpruch wird der Geiſt getrieben, das Unmittelbare, das Andere, als welches er ſich ſelber vorausſetzt, aufzuheben. Durch dieſe Aufhebung kommt er erſt zu ſich ſelbſt, tritt er als Geiſt hervor. Man kann da= her nicht mit dem Geiſte als ſolchem, — ſondern muß von ſei= ner unangemeſſenſten Realität anfangen. Der Geiſt iſt zwar ſchon im Anfange der Geiſt, aber er weiß noch nicht, daß er dieß iſt. Nicht er ſelber hat zu Anfange ſchon ſeinen Begriff erfaßt, ſon= dern nur wir, die wir ihn betrachten, ſind es, die ſeinen Begriff erkennen. Daß der Geiſt dazu kommt, zu wiſſen, was er iſt,

Addition. Spirit is always Idea. Initially however it is merely the *Notion* of the Idea, or the Idea in its indeterminateness, in the mode of being, the most abstract mode of reality. In the first instance, we do not have particularity of spirit, only the entirely universal, undeveloped determination 5 of it. We first have particularity when we pass from one to an other, for the particular contains both these moments. At the outset we have not yet made this transition however, so that the reality of spirit is still entirely universal, unparticularized. The development of this reality is only completed 10 through the whole philosophy of spirit. This still quite abstract, immediate reality, is moreover naturality, non-spirituality. This is why the child is still immersed in naturality, having only natural impulses, being a spiritual human being only according to possibility or the Notion, not in 15 accordance with actuality. Consequently, it is precisely on account of its abstraction, its immediacy, its affinity with naturality, that the initial reality of the Notion of spirit has to be regarded as being the most inadequate to it. Its true reality has to be determined as the totality of the developed 20 moments of the Notion, and it is the soul which remains the unity of these moments. The Notion of spirit necessarily progresses into this development of its reality, since it is contradicted by the immediacy and indeterminateness of the initial form of its reality. That which appears as being 25 immediately present in spirit is implicitly mediated, a positedness, not a true immediacy. This contradiction drives spirit to sublate the immediacy, the otherness it attributes to that which it presupposes. It is through this sublation that it first comes to itself, emerging as spirit. Consequently, 30 a beginning has to be made not with spirit as such, but with that of its reality which is the least adequate to it. Although spirit is certainly already spirit at the outset, it does not yet know this. In the first instance, only we who are considering spirit know its Notion, it has not apprehended it itself. The 35 realization of spirit consists of its coming to know what it is.

Dieß macht seine Realisation aus. Der Geist ist wesentlich nur Das, was er von sich selber weiß. Zunächst ist er nur an sich Geist; sein Fürsichwerden bildet seine Verwirklichung. Für sich aber wird er nur dadurch, daß er sich besondert, sich bestimmt, oder sich zu seiner Voraussetzung, zu dem Anderen seiner selber macht, sich zunächst auf dieß Andere als auf seine Unmittelbarkeit bezieht, dasselbe aber als Anderes aufhebt. So lange der Geist in der Beziehung auf sich als auf ein Anderes steht, ist er nur der subjective, der von der Natur herkommende Geist und zunächst selbst Naturgeist. Die ganze Thätigkeit des subjectiven Geistes geht aber darauf aus, sich als sich selbst zu erfassen, sich als Idealität seiner unmittelbaren Realität zu erweisen. Hat er sich zum Fürsichseyn gebracht, so ist er nicht mehr bloß subjectiver, sondern objectiver Geist. Während der subjective Geist wegen seiner Beziehung auf ein Anderes noch unfrei, oder — was dasselbe — nur an sich frei ist, kommt im objectiven Geiste die Freiheit, das Wissen des Geistes von sich als freiem zum Daseyn. Der objective Geist ist Person, und hat, als solche, im Eigenthum eine Realität seiner Freiheit; denn im Eigenthum wird die Sache als Das, was sie ist, nämlich als ein Unselbstständiges, und als ein Solches gesetzt, das wesentlich nur die Bedeutung hat, die Realität des freien Willens einer Person und darum für jede andere Person ein Unantastbares zu seyn. Hier sehen wir ein Subjectives, das sich frei weiß, und zugleich eine äußerliche Realität dieser Freiheit; der Geist kommt daher hier zum Fürsichsein, die Objectivität des Geistes zu ihrem Rechte. So ist der Geist aus der Form der bloßen Subjectivität herausgetreten. Die volle Verwirklichung jener im Eigenthum noch unvollkommenen, noch formellen Freiheit, die Vollendung der Realisation des Begriffs des objectiven Geistes wird aber erst im Staate erreicht, in welchem der Geist seine Freiheit zu einer von ihm gesetzten Welt, zur sittlichen Welt entwickelt. Doch auch diese Stufe muß der Geist überschreiten. Der Mangel dieser Objectivität des Geistes besteht darin, daß sie nur eine gesetzte ist. Die Welt muß vom Geiste wieder frei entlassen, das vom Geist Gesetzte zugleich als ein unmittelbar Seyendes gefaßt werden. Dieß

Essentially, spirit is only what it knows of itself. Initially, it + is only implicitly spirit, its actualization being formed by the becoming of its being-for-self. However, it only has being-for-self through its particularizing, determining itself, through its making itself the presupposition, the other of 5 itself, and while initially relating itself to this other as to its immediacy, sublating its otherness. Spirit is merely *subjective*, merely the spirit deriving from nature, initially it is itself merely the nature-spirit, so long as it stands in relation to itself as to an other. The whole activity of subjective spirit is 10 + directed toward apprehending itself, toward proving itself as the ideality of its immediate reality. If it has achieved being-for-self, it is no longer merely subjective, but *objective* spirit. On account of its relation to an other, subjective spirit is still not free, or — and this amounts to the same 15 thing — is only *implicitly* so. In objective spirit however, freedom attains determinate being through spirit's know-ledge of itself as being free. Objective spirit is a person, and it is as such that its freedom has a reality in property, for in property the matter is posited as what it is, i.e. as something 20 which lacks independence, and which, since it has essential significance only as the reality of a person's freewill, is not to be touched by anyone else. We see here a subjective being which knows itself to be free, and at the same time an external reality of this freedom. It is here therefore that spirit attains 25 to being-for-self, its objectivity coming into its own, and it is thus that it comes forth from the form of mere subjectivity. + However, the full actualization of this still incomplete and formal freedom in property, the perfection of the realization of the Notion of objective spirit, is first achieved in the state, 30 within which spirit develops its freedom into a world which it posits, the world of ethics. Spirit must, nevertheless, also + advance beyond this stage. The deficiency in this objectivity of spirit lies in its being merely posited. Spirit must let the world free once again, that which is posited by it must at 35 the same time be grasped as an immediate being. This occurs

geschieht auf der dritten Stufe des Geistes, auf dem Standpunkt des absoluten Geistes, d. h. der Kunst, der Religion und
der Philosophie.

§. 386.

Die zwei ersten Theile der Geisteslehre befassen
den endlichen Geist. Der Geist ist die unendliche Idee,
und die Endlichkeit hat hier ihre Bedeutung der Unangemessenheit des Begriffs und der Realität mit der Bestimmung, daß sie das Scheinen innerhalb seiner ist, — ein
Schein, den an sich der Geist sich als eine Schranke setzt,
um durch Aufheben derselben für sich die Freiheit als
sein Wesen zu haben und zu wissen, d. i. schlechthin ma-
nifestirt zu seyn. Die verschiedenen Stufen dieser Thätigkeit auf welchen als dem Scheine zu verweilen und welche
zu durchlaufen die Bestimmung des endlichen Geistes ist,
sind Stufen seiner Befreiung, in deren absoluten Wahrheit das Vorfinden einer Welt als einer vorausgesetzten, das Erzeugen derselben als eines von ihm gesetzten, und die Befreiung von ihr und in ihr eins und
dasselbe sind, — einer Wahrheit, zu deren unendlichen Form der Schein als zum Wissen derselben sich
reinigt.

Die Bestimmung der Endlichkeit wird vornehmlich vom Verstande in der Beziehung auf den Geist
und die Vernunft fixirt; es gilt dabei nicht nur für
eine Sache des Verstandes, sondern auch für eine mora-
lische und religiöse Angelegenheit, den Standpunkt
der Endlichkeit als einen letzten festzuhalten, so wie
dagegen für eine Vermessenheit des Denkens, ja für eine
Verrücktheit desselben über ihn hinausgehen zu wollen.
— Es ist aber wohl vielmehr die schlechteste der Tugen-
den, eine solche Bescheidenheit des Denkens, welche
das Endliche zu einem schlechthin Festen, einem Absoluten macht, und die ungründlichste der Erkenntnisse,
in dem, was seinen Grund nicht in sich selbst hat, stehen
zu bleiben. Die Bestimmung der Endlichkeit ist längst

at the third stage of spirit, the standpoint of *absolute* spirit,
i.e. of art, religion and philosophy.   +

§ 386

The first two parts of the *doctrine of spirit* deal   +
with *finite* spirit. Spirit is the infinite Idea, and
the disproportion between the Notion and real-   5
ity, the meaning of finitude, has here the added
determination of its constituting the appearance
within spirit. Spirit posits this appearance as an
implicit limit to itself, in order that it may sublate it, and as *being-for-self* possess and know   10
freedom as *its* essence, i.e. **simply be manifest.
The determination of finite spirit consists of its
treating** the various stages **of this** activity **as appearance, by dwelling upon and traversing them.** They
are stages in its liberation, in the absolute truth   15
of which, the *ascertaining* of a presupposed world,
the *generation* of the same as being posited by
spirit and the liberation from **and within this world,**
are one and the same. **Appearance purifies itself into
the infinite freedom of this truth in order to attain   20
knowledge of it.**

It is mainly with regard to *spirit* and *reason*
that the *understanding* fixes upon the determination of *finitude*, it being regarded not only as a
matter of understanding but also of moral and   25
religious concern that the *supremacy* of **the
standpoint of finitude** should be firmly maintained. It is, moreover, regarded as presumpt-
uous and even deranged for thought to want
to pass beyond this standpoint. — Such *modesty*   30
of thought is itself the worst of virtues how-
ever, since it treats the *finite* **as something
simply fixed, as an** *absolute*, and to remain involved
in what does not contain its own ground is to have
the worst basis for knowledge. The deter-   35
mination of *finitude* has been elucidated and

an seinem Orte, in der Logik, beleuchtet und erörtert
worden; diese ist dann ferner für die weiter bestimmten
aber noch immer noch einfachen Gedankenformen der
Endlichkeit, wie die übrige Philosophie für die concreten
Formen derselben nur diß Aufzeigen, das Endliche nicht
ist, d. i. nicht das Wahre, sondern schlechthin nur ein
Uebergehen und Ueber sich hinausgehen ist. —
Dieses Endliche der bisherigen Sphären ist die Dialek-
tik, sein Vergehen durch ein Anderes und in einem
Andern zu haben; der Geist aber, der Begriff und das
an sich Ewige, ist es selbst, dieses Vernichtigen des
Nichtigen, das Vereiteln des Eiteln in sich selbst zu
vollbringen. — Die erwähnte Bescheidenheit ist das
Festhalten dieses Eiteln, des Endlichen, gegen das
Wahre, und darum selbst das Eitle. Diese Ekelkeit
wird sich in der Entwicklung des Geistes selbst als seine
höchste Vertiefung in seine Subjectivität und innerster
Widerspruch und damit Wendepunkt, als das Böse,
ergeben.

Zusatz. Der subjective und der objective Geist sind noch
endlich. Es ist aber nothwendig, zu wissen, welchen Sinn die
Endlichkeit des Geistes hat. Gewöhnlich stellt man sich dieselbe
als eine absolute Schranke vor, — als eine feste Qualität, nach
deren Wegnahme der Geist aufhörte, Geist zu seyn; wie das Wesen der natürlichen Dinge an eine bestimmte Qualität gebunden
ist, — wie, zum Beispiel, das Gold nicht von seiner specifischen
Schwere getrennt werden, dieß und jenes Thier nicht ohne Klauen,
ohne Schneidezähne u. s. w. seyn kann. In Wahrheit aber darf
die Endlichkeit des Geistes nicht als eine feste Bestimmung betrachtet, sondern muß als ein bloßes Moment erkannt werden;
denn der Geist ist, wie schon früher gesagt, wesentlich die Idee
in der Form der Idealität, das heißt, des Regirtseyns des Endlichen. Das Endliche hat also im Geiste nur die Bedeutung eines
Aufgehobenen, nicht die eines Seyenden. Die eigentliche Qualität des Geistes ist daher vielmehr die wahrhafte Unendlichkeit,
das heißt, diejenige Unendlichkeit, welche dem Endlichen nicht ein-
seitig gegenübersteht, sondern in sich selber das Endliche als ein

37

examined **long since at its place in** the logic. **The** +
further import of this exposition for the **more fully**
determined but always still simple thoughtforms
of finitude, like that of the rest of philosophy
for the concrete forms of the same, is merely 5
that the finite is not, i.e. that it is **not that**
which is true, but simply a mere *passing over* **which**
passes beyond itself. — **That which is finite in**
the preceding spheres constitutes the dialectic, in
that it passes away by means of an other and into 10
an other. It is however precisely the bringing +
about **within itself** of this nullification of nullity,
the making vain of that which is vain, that
constitutes spirit, the Notion and the *implicitly*
eternal. — The modesty mentioned, the holding 15
fast **to this vainness, to that which is finite** as
opposed to that which is true, **is therefore itself**
vain. In the development of spirit, this vanity
will yield itself as *evil,* spirit's supreme immersion **in its subjectivity,** its innermost **contra-** 20
diction, and therefore its turning point. +

Addition. Subjective and objective spirit are still finite. It
is necessary to know in what sense spirit is finite however. Its
finitude is usually regarded as an absolute limit, a fixed
quality, the removal of which would give rise to its ceasing 25
to be spirit. The essence of natural things is certainly bound
to a definite quality — gold, for example, may not be
separated from its specific gravity, this or that animal cannot
be what it is without claws, incisors etc. The finitude of
spirit cannot in fact be regarded as a fixed determination 30
however. It has to be cognized as a mere moment, for as has
already been pointed out, spirit is essentially the Idea in the
form of ideality, that is to say, the negatedness of that which
is finite. In spirit therefore, that which is finite is of significance only as a sublatedness, not as a being. Consequently, 35
it is, rather, true infinity that constitutes the specific
quality of spirit, that is to say, not the infinity which stands
onesidedly opposed to that which is finite, but that which

Moment enthält. Es ist deßhalb ein leerer Ausdruck, wenn man sagt: Es gibt endliche Geister. Der Geist als Geist ist nicht endlich, er hat die Endlichkeit in sich, aber nur als eine aufzuhebende und aufgehobene. Die hier nicht genauer zu erörternde echte Bestimmung der Endlichkeit muß dahin angegeben werden, daß das Endliche eine ihrem Begriffe nicht gemäße Realität ist. So ist die Sonne ein Endliches, da sie nicht ohne Anderes gedacht werden kann, — da zur Realität ihres Begriffs nicht bloß sie selbst, sondern das ganze Sonnensystem gehört. Ja, das ganze Sonnensystem ist ein Endliches, weil jeder Himmelskörper in ihm gegen den anderen den Schein der Selbstständigkeit hat, folglich diese gesammte Realität ihrem Begriff noch nicht entspricht, noch nicht dieselbe Idealität darstellt, welche das Wesen des Begriffes ist. Erst die Realität des Geistes ist selber Idealität, erst im Geiste findet also absolute Einheit des Begriffs und der Realität, somit die wahrhafte Unendlichkeit statt. Schon, daß wir von einer Schranke wissen, ist Beweis unseres Hinausseyns über dieselbe, unserer Unbeschränktheit. Die natürlichen Dinge sind eben darum endlich, weil ihre Schranke nicht für sie selber, sondern nur für uns vorhanden ist, die wir dieselben mit einander vergleichen. Zu einem Endlichen machen wir uns dadurch, daß wir ein Anderes in unser Bewußtseyn aufnehmen. Aber eben, indem wir von diesem Anderen wissen, sind wir über diese Schranke hinaus. Nur der Unwissende ist beschränkt; denn er weiß nicht von seiner Schranke; wer dagegen von der Schranke weiß, Der weiß von ihr nicht als von einer Schranke seines Wissens, sondern als von einem Gewußten, als von einem zu seinem Wissen Gehörenden; nur das Ungewußte wäre eine Schranke des Wissens; die gewußte Schranke dagegen ist keine Schranke desselben; von seiner Schranke wissen, heißt daher, von seiner Unbeschränktheit wissen. Wenn aber der Geist für unbeschränkt, für wahrhaft unendlich erklärt wird, so soll damit nicht gesagt seyn, daß die Schranke ganz und gar nicht im Geiste sey, vielmehr haben wir zu erkennen, daß der Geist sich bestimmen, somit verendlichen, beschränken muß. Aber der Verstand hat Unrecht, diese Endlichkeit als eine starre, — den

contains finitude as a moment. The statement that there are
finite spirits is therefore meaningless. Spirit as spirit *is* not
finite, but *has* finitude within itself, although only as that
which is to be and has been sublated. Consequently, the
proper determination of finitude, which is not to be explained   5
more precisely here, is that that which is finite is a reality
which is inadequate to its Notion. The sun for example is
finite, since it cannot be conceived of without reference to
an other, the reality of its Notion involving not only itself
but the entire solar system. What is more, the entire solar   10
system is finite, since each celestial body within it appears to
be independent of the others. This collective reality does not
yet correspond to its Notion therefore, does not yet exhibit
the same ideality as that which constitutes the essence of the
Notion. It is only in *spirit* that reality is itself ideality, and   15   +
only in spirit therefore that true infinity occurs as an absolute
unity of Notion and reality. Our knowing of a limit is
already evidence of our being beyond it, of our unlimitedness.
Natural things are finite precisely because their limit is
present not for them themselves, but only for us, in that we   20
compare them with one another. We turn ourselves into a
finitude by taking up an other into our consciousness. It is
however precisely through our knowing of this other that we
overcome this limit. Only he who does not know is limited,
for he does not know his limit. Whoever knows of the limit   25
however, knows of it not as a limit to his knowledge, but as
known, as belonging to his knowledge. Only the unknown
could constitute a limit to knowledge, the known limit does
not limit it. To know one's limit is therefore to know of
one's unlimitedness. When spirit is expounded as being   30   +
unlimited however, as truly infinite, this does not imply that
limit is entirely absent from it. We have rather to recognize,
that spirit has to determine and therefore introduce finitude,
limit itself. However, the understanding is not justified in
regarding this finitude as inflexible — the difference between   35

39

Unterschied der Schranke und der Unendlichkeit als einen absolut festen zu betrachten, und demgemäß zu behaupten, der Geist sey entweder beschränkt oder unbeschränkt. Die Endlichkeit, wahrhaft aufgefaßt, ist, wie gesagt, in der Unendlichkeit, die Schranke im Unbeschränkten enthalten. Der Geist ist daher sowohl unendlich als endlich, und weder nur das Eine noch nur das Andere; er bleibt in seiner Verendlichung unendlich; denn er hebt die Endlichkeit in sich auf; Nichts ist in ihm ein Festes, ein Seyendes, Alles vielmehr nur ein Ideelles, ein nur Erscheinendes. So muß Gott, weil er Geist ist, sich bestimmen, Endlichkeit in sich setzen, (sonst wäre er nur eine todte, leere Abstraction); da aber die Realität, die er sich durch sein Sichbestimmen giebt, eine ihm vollkommen gemäße ist, wird Gott durch dieselbe nicht zu einem Endlichen. Die Schranke ist also nicht in Gott und im Geiste, sondern sie wird vom Geiste nur gesetzt, um aufgehoben zu werden. Nur momentan kann der Geist in einer Endlichkeit zu bleiben scheinen; durch seine Idealität ist er über dieselbe erhaben, weiß er von der Schranke, daß sie keine feste Schranke ist. Daher geht er über dieselbe hinaus, befreit sich von ihr; und diese Befreiung ist nicht — wie der Verstand meint — eine niemals vollendete, eine in's Unendliche immer nur erstrebte, sondern der Geist entreißt sich diesem Progreß in's Unendliche, befreit sich absolut von der Schranke, von seinem Anderen, und kommt somit zum absoluten Fürsichseyn, macht sich wahrhaft unendlich.

the limit and infinitude as being absolutely fixed, and on the basis of this to assert that spirit is *either* limited *or* unlimited. As has been pointed out, the true conception of finitide is that it is contained in infinitude, and the true conception of limit that it is contained in the unlimited. Spirit is finite *as*   5 *well as* infinite, and *not* simply one *or* the other. It remains infinite in its being made finite, for it sublates the finitude within itself. Nothing within it is a fixed being, all is simply of an ideal nature, simply what is appearing. Since God is spirit, He must therefore determine Himself, posit finitude   10 within Himself. If He did not He would be simply a dead, empty abstraction. The reality He gives Himself through His determining Himself is completely adequate to Him however, so that it does not endue Him with finitude. God and spirit are therefore devoid of limit, which is only posited by   15 spirit in order that it may be sublated. Spirit can only appear to remain in a finitude momentarily; it is raised above it through its ideality, through its knowing the limit to be no limit. It passes beyond finitude therefore, freeing itself from it, and this liberation is not, as the understanding supposes   20 it to be, never accomplished, perpetually merely striven for in that which is infinite. Spirit wrests itself from this progress into infinity, frees itself absolutely from limit, from its other, and so attains to absolute being-for-self, makes itself truly infinite.   25   +

40

Erste Abtheilung der Philosophie des Geistes.

Der subjective Geist.

---

§. 387.

Der Geist in seiner Idealität sich entwickelnd ist der Geist als erkennend. Aber das Erkennen wird hier nicht blos aufgefaßt, wie es die Bestimmtheit der Idee als logischer ist (§. 223.), sondern wie der concrete Geist sich zu demselben bestimmt.

Der subjective Geist ist:

A. An sich oder unmittelbar; so ist er Seele oder Naturgeist; — Gegenstand der

Anthropologie.

B. Für sich oder vermittelt, noch als identische Reflexion in sich und in Anderes; der Geist im Verhältniß oder Besonderung; Bewußtseyn, — der Gegenstand der

Phänomenologie des Geistes.

C. Der sich in sich bestimmende Geist, als Subject für sich, der Gegenstand der

Psychologie.

In der Seele erwacht das Bewußtseyn; das Bewußtseyn setzt sich als Vernunft, die unmittelbar zur sich wissenden Vernunft erwacht ist, welche sich durch ihre Thätigkeit zur Objectivität, zum Bewußtseyn ihres Begriffs befreit.

Wie im Begriffe überhaupt die Bestimmtheit, die an ihm vorkommt, Fortgang der Entwicklung ist, so ist auch an dem Geiste jede Bestimmtheit, in der er sich zeigt, Moment der Entwicklung und in der Fortbestimmung, Vorwärtsgehen seinem Ziele zu, sich zu

Section One of the Philosophy of Spirit

Subjective Spirit

---

§ 387

In developing itself in its ideality, spirit has being as   +
being **cognitive**. At this juncture, however, cognition
is taken up not simply as a determinateness of the
logical aspect of the Idea (§ 223), but as that into
which **concrete** spirit determines itself.   5
    Subjective spirit is:
A. **Implicit or immediate, as which it is soul
    or natural spirit, the** general object of
                    anthropology.
B. **For itself or mediated, still** as identical   10
    reflection into self and into an other; **spirit** in
    relationship or particularization, — *conscious-
    ness*, the general object of the
                    phenomenology of spirit.
C. **Internally self-determining spirit,** as   15
    subject or **being-for-self,** — the general object
    of
                    psychology.
Consciousness awakens in the *soul.* It *posits itself as
reason*, **which is immediately aroused into self-**   20
knowing reason, and which, by means of its activity,
liberates itself into objectivity, **into consciousness
of its Notion.**

    **In the Notion in general, the determinateness
which occurs is a progressive development.**   25
Similarly in spirit, each determinateness in which
it exhibits itself is a moment of development,
of the further determination by which it moves
forward to its goal of making itself into that which

41

dem zu machen und für sich zu werden das was er
an sich ist. Jede Stuffe ist innerhalb ihrer dieser Proceß, und das Product derselben, daß für den Geist
(d. i. die Form desselben, die er in ihr hat) das ist, was
er im Beginn derselben an sich oder damit nur für
* uns war. — Die psychologische sonst gewöhnliche Betrachtungsweise gibt an, erzählungsweise, was der Geist
oder die Seele ist, was ihr geschieht, was sie thut;
so daß die Seele als fertiges Subject vorausgesetzt ist,
an dem dergleichen Bestimmungen nur als Aeußerungen zum Vorschein kommen, aus denen soll erkannt
werden, was sie ist, — in sich für Vermögen und Kräfte
besitzt; ohne Bewußtseyn darüber, daß die Aeußerung
dessen, was sie ist, im Begriffe dasselbe für sie setzt,
wodurch sie eine höhere Bestimmung gewonnen hat. —
Von dem hier betrachtenden Fortschreiten ist dasjenige
zu unterscheiden und davon ausgeschlossen, welches Bildung und Erziehung ist. Dieser Kreis bezieht sich nur
auf die einzelnen Subjecte als solche, daß der allgemeine Geist in ihnen zur Existenz gebracht werde. In
der philosophischen Ansicht des Geistes als solchen wird
er selbst als in seinem Begriffe sich bildend und er-
ziehend betrachtet, und seine Aeußerungen als die Momente seines Sich-zu-sich-selbst-Hervorbringens, seines
Zusammenschließens mit sich, wodurch er erst wirklicher
Geist ist.

Zusatz. In §. 385 ist der Geist in seine drei Hauptformen,
den subjectiven, den objectiven und den absoluten Geist unterschieden, und zugleich die Nothwendigkeit des Fortgangs von dem
ersten zu dem zweiten und von diesem zum dritten angedeutet
worden. Wir haben diejenige Form des Geistes, welche wir zuerst betrachten müssen, den subjectiven Geist genannt, weil der
Geist hier noch in seinem unentwickelten Begriffe ist, sich seinen
Begriff noch nicht gegenständlich gemacht hat. In dieser seiner
Subjectivität ist der Geist aber zugleich objectiv, hat eine unmittelbare Realität, durch deren Aufhebung er erst für sich wird, zu
42   sich selbst, zum Erfassen seines Begriffs, seiner Subjectivität ge-

* Dieser Satz erstmals 1830.

it is implicitly, and becoming the being-for-self of it. Each stage constitutes this process within itself, and the product is that what spirit was implicitly at the initiation of the stage and so only for us, has being for it, or rather for the form 5
spirit possesses in the stage.* — The ordinary psychological approach makes statements as to what spirit or the soul is, what happens to it, what it does, presupposing it to be a ready-made subject within which such determinations appear 10
only as expressions. These expressions are supposed to make known what it is, i.e. what inner faculties and powers it has, it not being realized that in the Notion, in that it posits for itself the expression of what it is, the soul has gained a 15
higher determination. — That which constitutes instruction and education is to be distinguished and excluded from the progression to be considered here, for this is a sphere relating only to the bringing of universal spirit into existence within 20
individual subjects as such. In the philosophic
treatment of spirit as such, spirit is considered as
instructing and educating itself within its Notion,
and its expressions as the moments of its bringing
itself forth to itself, of the self-coincidence whereby 25
it initiates its actuality.

Addition. In § 385 the three principal forms of spirit, the
subjective, the objective and the absolute, have been
distinguished, and at the same time, the necessity of the
progression from the first to the second and from this to the 30
third has been indicated. We have called the first form of
spirit we have to consider *subjective*, since spirit here has not
yet objectivized its Notion, is still Notionally undeveloped.
In this its subjectivity, spirit is at the same time objective
however, for it has an immediate reality, in the sublation of 35
which it initiates its being-for-self by coming to itself,
apprehending its Notion, achieving its subjectivity. It is

---

* Sentence first published 1830.

langt. Man könnte daher ebensowohl sagen, der Geist sey zu=
nächst objectiv und solle subjectiv werden, wie umgekehrt, er sey
erst subjectiv und habe sich objectiv zu machen. Der Unterschied
des subjectiven und des objectiven Geistes ist folglich nicht als
ein starrer anzusehen. Schon im Anfange haben wir den Geist
nicht als bloßen Begriff, als ein bloß Subjectives, sondern als
Idee, als eine Einheit des Subjectiven und Objectiven zu fassen,
und jeder Fortgang von diesem Anfange ist ein Hinausgehen über
die erste einfache Subjectivität des Geistes, ein Fortschritt in der
Entwicklung der Realität oder Objectivität desselben. Diese Ent=
wicklung bringt eine Reihe von Gestaltungen hervor, die zwar
von der Empirie angegeben werden müssen, in der philosophi=
schen Betrachtung aber nicht äußerlich neben einander gestellt blei=
ben dürfen, sondern als der entsprechende Ausdruck einer noth=
wendigen Reihe bestimmter Begriffe zu erkennen sind, und für
das philosophische Denken nur insofern Interesse haben, als sie
eine solche Reihe von Begriffen ausdrücken. — Zunächst können
wir nun aber die unterschiedenen Gestaltungen des subjectiven
Geistes nur versicherungsweise angeben; erst durch die bestimmte
Entwicklung desselben wird deren Nothwendigkeit hervortreten.

Die drei Hauptformen des subjectiven Geistes sind 1) die
Seele, 2) das Bewußtseyn und 3) der Geist als sol=
cher. Als Seele hat der Geist die Form der abstracten Allge=
meinheit, als Bewußtseyn die der Besonderung, als für
sich seyender Geist die der Einzelnheit. So stellt sich in sei=
ner Entwicklung die Entwicklung des Begriffes dar. Warum die
jenen drei Formen des subjectiven Geistes entsprechenden Theile
der Wissenschaft in dem obenstehenden Paragraphen den Namen
Anthropologie, Phänomenologie und Psychologie er=
halten haben, wird aus einer näheren vorläufigen Angabe des
Inhalts der Wissenschaft vom subjectiven Geiste erhellen.

Den Anfang unserer Betrachtung muß der unmittelbare Geist
bilden; dieß aber ist der Naturgeist, die Seele. Wenn ge=
meint würde, es sey mit dem bloßen Begriff des Geistes zu be=
ginnen, so ist dieß ein Irrthum; denn, wie bereits gesagt, ist
der Geist immer Idee, also verwirklichter Begriff. Zu Anfang

therefore as true to say that spirit is initially objective and has to become subjective as it is to say the opposite i.e. that it is at first subjective and has to objectivize itself. Consequently, the difference between subjective and objective spirit is not to be regarded as a rigid one. We have to grasp spirit at the outset not as mere Notion, as mere subjectivity, but as Idea, as a unity of subjective and objective, and each progression from this goes beyond the initial simple subjectivity of spirit, since it is an advance in the development of its reality or objectivity. This development brings forth a series of figurations, which certainly have to be specified empirically. In philosophic consideration they may not remain externally juxtaposed however, for they are to be known as the corresponding expression of a necessary series of specific Notions, and it is only in so far as they express such a series that they are of interest to philosophic thinking. Here at the outset we can only specify these various figurations of subjective spirit assertorically; their necessity will only emerge from the determinate development of spirit.

The three principle forms of subjective spirit are 1) the *soul*, 2) *consciousness* and 3) *spirit as such*. As soul, spirit has the form of abstract *universality*, as consciousness that of *particularization*, as spiritual being-for-self that of *singularity*. It is thus that its development exhibits the development of the Notion. The reason why the parts of the science corresponding to these three forms of subjective spirit have been designated as *anthropology*, *phenomenology* and *psychology* in the Paragraph above, will become apparent through a fuller introductory account of the content of the science of subjective spirit.

Our consideration must begin with spirit in its immediacy; this however is *natural spirit*, the *soul*. It is an error to think that a beginning is made with the mere Notion of spirit, for as has already been pointed out, spirit, in that it is always Idea, is actualized Notion. At the outset however, the Notion

aber kann der Begriff des Geistes noch nicht die vermittelte Rea
lität haben, welche er im abstracten Denken erhält; seine Realität
muß zu Anfang zwar auch schon eine abstracte seyn — nur da
durch entspricht sie der Idealität des Geistes — sie ist aber noth
wendig eine noch unvermittelte, noch nicht gesetzte, folglich eine
seyende, ihm äußerliche, eine durch die Natur gegebene. Wir
müssen also von dem noch in der Natur befangenen, auf seine
Leiblichkeit bezogenen, noch nicht bei sich selbst seyenden, noch nicht
freien Geiste anfangen. Diese — wenn wir so sagen dürfen —
Grundlage des Menschen ist der Gegenstand der Anthropologie.
In diesem Theile der Wissenschaft vom subjectiven Geiste ist der
gedachte Begriff des Geistes nur in uns, den Betrachtenden, noch
nicht im Gegenstande selber; den Gegenstand unserer Betrachtung
bildet hier der erst bloß seyende Begriff des Geistes, der seinen
Begriff noch nicht erfaßt habende, noch außer = sich = seyende Geist.

Das Erste in der Anthropologie ist die qualitativ be
stimmte, an ihre Naturbestimmungen gebundene Seele (hierher ge
hören z. B. die Racenunterschiede). Aus diesem unmittelbaren
Einsseyns mit ihrer Natürlichkeit tritt die Seele in den Gegen
satz und Kampf mit derselben (dahin gehören die Zustände der
Verrücktheit und des Somnambulismus). Diesem Kampfe folgt
der Sieg der Seele über ihre Leiblichkeit, die Herabsetzung und
das Herabgesetztseyn dieser Leiblichkeit zu einem Zeichen, zur Dar
stellung der Seele. So tritt die Idealität der Seele in ihrer
Leiblichkeit hervor, wird diese Realität des Geistes auf eine, selbst
aber noch leibliche Weise ideell gesetzt.

In der Phänomenologie erhebt sich nun die Seele durch
die Negation ihrer Leiblichkeit zur reinen ideellen Identität mit
sich, wird Bewußtseyn, wird Ich, ist ihrem Anderen gegenüber
für sich. Aber dieß erste Fürsichseyn des Geistes ist noch be
dingt durch das Andere, von welchem der Geist herkommt. Das
Ich ist noch vollkommen leer, eine ganz abstracte Subjectivität,
setzt allen Inhalt des unmittelbaren Geistes außer sich, und be
zieht sich auf denselben als auf eine vorgefundene Welt. So
wird Dasjenige, was zunächst nur unser Gegenstand war, zwar
dem Geiste selber zum Gegenstande, das Ich weiß aber noch nicht,

44

of spirit cannot yet exhibit the mediated reality it assumes in
abstract thinking. It is true that even at the outset its reality
must also be abstract, for it is only as such that it corresponds
to the ideality of spirit, but it is of necessity a still unmediated,
unposited reality, and is therefore a being external to spirit, a   5
being rendered by nature. We have therefore to begin with
spirit which is as yet not with itself, not free but still involved
in nature, related to its corporeity. This foundation of
humanity — if we may so call it — is the general object of
anthropology. In this part of the science of subjective spirit   10
the contemplated Notion of spirit is merely in us the con-
templators, and not yet in the general object itself. At this
juncture the general object we are considering forms the
first simple being of the Notion of spirit, in which spirit in
its self-externality has not yet grasped its Notion.   15

It is the qualitatively determined soul, bound to its
determinations in nature, which constitutes the initiation of
anthropology. Racial differences, for example, belong here.
Out of this immediate union with its naturality, the soul
enters into opposition and conflict with it. It is here that the   20
conditions of derangement and somnambulism belong. This
conflict follows from the triumph of the soul over its corporeity i.e. from the reducing and reduction of this corporeity
to a sign for the representation of soul. It is thus that the
ideality of the soul comes forth in its corporeity, but although   25
the ideal nature of this reality of spirit is posited, this still
occurs in a corporeal mode.

In *phenomenology*, the soul now raises itself by means of the
negation of its corporeity into the purely ideal nature of self-
identity. It becomes consciousness, ego, has being-for-self   30
in the face of its other. This primary being-for-self of spirit is
however still conditioned by the other from which spirit
emerges. The ego is still completely empty, an entirely
abstract subjectivity, which posits the entire content of
immediate spirit as external to itself and relates itself to it as   35
to a world which it finds before it. Thus, although that
which in the first instance was merely our general object
certainly becomes the general object of spirit itself, the ego

daß das ihm Gegenüberstehende der natürliche Geist selber ist. Das Ich ist daher trotz seines Fürsichseyns doch zugleich nicht für sich, da es nur in Beziehung auf Anderes, auf ein Gegebenes ist. Die Freiheit des Ich ist folglich nur eine abstracte, bedingte, relative. Der Geist ist hier zwar nicht mehr in die Natur versenkt, sondern in sich reflectirt und auf dieselbe bezogen, erscheint aber nur, steht nur in Beziehung zur Wirklichkeit, ist noch nicht wirklicher Geist. Daher nennen wir den Theil der Wissenschaft, in welchem diese Form des Geistes betrachtet wird, die Phänomenologie. Indem nun aber das Ich sich aus seiner Beziehung auf Anderes in sich reflectirt, wird es Selbstbewußtseyn. In dieser Form weiß das Ich sich zunächst nur als das unerfüllte Ich, und allen concreten Inhalt als ein Anderes. Die Thätigkeit des Ich besteht hier darin, die Leere seiner abstracten Subjectivität zu erfüllen, das Objective in sich hinein zu bilden, das Subjective dagegen objectiv zu machen. Dadurch hebt das Selbstbewußtseyn die Einseitigkeit seiner Subjectivität auf, kommt es aus seiner Besonderheit, aus seinem Gegensatze gegen das Objective zu der beide Seiten umfassenden Allgemeinheit, und stellt in sich die Einheit seiner selbst mit dem Bewußtseyn dar; denn der Inhalt des Geistes wird hier ein objectiver, wie im Bewußtseyn, und zugleich, wie im Selbstbewußtseyn, ein subjectiver. Dieß allgemeine Selbstbewußtseyn ist an sich oder für uns Vernunft; aber erst im dritten Theil der Wissenschaft vom subjectiven Geiste wird die Vernunft sich selber gegenständlich.

45

Dieser dritte Theil, die Psychologie, betrachtet den Geist als solchen, den Geist, wie er im Gegenstande sich nur auf sich selber bezieht, darin nur mit seinen eigenen Bestimmungen zu thun hat, seinen eigenen Begriff erfaßt. So kommt der Geist zur Wahrheit; denn nun ist die in der bloßen Seele noch unmittelbare, noch abstracte Einheit des Subjectiven und Objectiven durch Aufhebung des im Bewußtseyn entstehenden Gegensatzes dieser Bestimmungen als eine vermittelte wieder hergestellt, die Idee des Geistes also aus der ihr widersprechenden Form des einfachen Begriffs und aus der ihr ebenso sehr widersprechenden Trennung ihrer Momente zur vermittelten Einheit und somit zur wahren

does not yet know that that which confronts it is itself
natural spirit. Consequently, despite its being-for-self, the
ego is at the same time not for itself, since it is merely related
to an other which is given to it. The freedom of the ego is
therefore simply abstract, conditioned, relative. Spirit at this   5
juncture is certainly no longer immersed within nature. In
that it is intro-reflected and related to nature however, it
simply *appears*, for it merely stands in relation to actuality
and is not yet *actual* spirit. This is why we call that part of
science in which this form of spirit is considered *phenomen-*   10
ology. In that the ego is now intro-reflected through its   +
relation to the other however, it becomes self-consciousness.
Initially the ego in this form knows itself only as lacking in
content, and all concrete content only as an other. Its
activity at this juncture consists of its filling the void of its   15
abstract subjectivity, and, while forming that which is
objective within itself, objectivizing that which is subjective.
It is through this that self-consciousness sublates the onesidedness of its subjectivity, and breaks out of its particularity, its opposition to that which is objective, into the   20
universality which embraces both sides and exhibits within
itself the unity of its self with consciousness. It does so here
because the content of spirit becomes objective, as in
consciousness, and at the same time subjective, as in self-
consciousness. Implicitly, or for us, this universal self-   25
consciousness is reason; it is however only in the third part
of the science of subjective spirit that reason becomes objective to itself.

   This *third* part, *psychology*, treats of spirit as such, of spirit
which in the general object merely relates itself to itself,   30
dealing there only with its own determinations, grasping
there only its own Notion. It is thus that spirit enters upon
truth, for the unity of the subjective and objective, which in
the mere soul is still immediate and abstract, is now, through
the sublation of the opposition between these determinations   35
which occurs in consciousness, re-established as mediated.
By thus superseding the form of the mere Notion and the
disunion of its moments, both of which contradict it to the

Wirklichkeit gelangt. In dieser Gestalt ist der Geist die für sich selbst seyende Vernunft. Geist und Vernunft stehen zu einander in solchem Verhältniß, wie Körper und Schwere, wie Wille und Freiheit; die Vernunft bildet die substantielle Natur des Geistes; sie ist nur ein anderer Ausdruck für die Wahrheit oder die Idee, welche das Wesen des Geistes ausmacht; aber erst der Geist als solcher weiß, daß seine Natur die Vernunft und die Wahrheit ist. Der beide Seiten, die Subjectivität und die Objectivität, befassende Geist setzt sich nun erstens in der Form der Subjectivität, — so ist er Intelligenz, — zweitens in der Form der Objectivität, — so ist er Wille. Die zunächst auch selbst noch unerfüllte Intelligenz hebt ihre dem Begriff des Geistes unangemessene Form der Subjectivität dadurch auf, daß sie den ihr gegenüberstehenden, noch mit der Form des Gegebenseyns und der Einzelnheit behafteten objectiven Inhalt nach dem absoluten Maaßstabe der Vernunft mißt, diesem Inhalt die Vernünftigkeit anthut, die Idee in ihn einbildet, ihn zu einem concret Allgemei=nen verwandelt und so in sich aufnimmt. Dadurch kommt die Intelligenz dahin, daß das, was sie weiß, nicht eine Abstraction, sondern der objective Begriff ist, und daß andererseits der Ge=genstand die Form eines Gegebenen verliert und die Gestalt eines dem Geiste selber angehörenden Inhalts bekommt. Indem die Intelligenz aber zu dem Bewußtseyn gelangt, daß sie den Inhalt aus sich selbst nimmt, wird sie zu dem nur sich selber zum Zwecke setzenden praktischen Geiste, dem Willen, der nicht, wie die In=telligenz, mit einem von außenher gegebenen Einzelnen, sondern mit einem solchen Einzelnen anfängt, das er als das Seinige weiß, — dann aus diesem Inhalt, den Trieben, Neigungen, sich in sich reflectirend, denselben auf ein Allgemeines bezieht, — und endlich zum Wollen des an und für sich Allgemeinen, der Frei=heit, seines Begriffes sich erhebt. Zu diesem Ziele gelangt, ist der Geist ebenso sehr zu seinem Anfange, zur Einheit mit sich zurückgekehrt, wie zur absoluten, zur wahrhaft in sich bestimmten Einheit mit sich fortgeschritten, einer Einheit, in welcher die Be=stimmungen nicht Naturbestimmungen, sondern Begriffsbestimmun=gen sind.

46

same extent, the Idea of spirit attains to mediated unity and
so to true actuality. Spirit in this shape is *reason which is for*   +
itself. The relationship in which spirit stands to reason is as
the relationship between body and gravity, will and freedom.   +
Reason constitutes the substantial nature of spirit and is   5
merely another expression for the truth or the Idea which
constitutes the essence of it. It is however only spirit as such
which knows its nature to be reason and truth. The spirit
involved in both the subjective and objective aspects now
posits itself, initially in the form of subjectivity as *intelligence,*   10
and then in the form of objectivity as *will.* In the first   +
instance intelligence itself is as yet also without content, but
it sublates its form of subjectivity, inadequate as this is to the
Notion of spirit, by dispensing, in accordance with the
absolute standard of reason, with the objective content with   15
which it is confronted, a content which is still affected with
the form of its being given, with singularity. It does this by
assimilating it, that is, by infusing it with rationality,
introducing into it the Idea, transforming it into a concrete
universal. It is thus that what is known to intelligence is not   20
an abstraction but the objective Notion, while on the other
hand the general object loses the form of its being given and
assumes the shape of a content belonging to spirit itself. In
that intelligence becomes conscious of its deriving the content
from itself however, it becomes the will, the practical spirit   25
which posits only itself as its purpose. The will is unlike the
intelligence in that it begins not with a singularity rendered
to it from without, but with one which it knows to be its
own. It then reflects into itself from this content of desires,
inclinations, and so relates the content to a universal.   30
Finally, it raises itself to volition of that which is universal in
and for itself, that is, of its Notion, of freedom. In attaining   +
this end spirit has returned to its beginning, its unity with
itself. It has however also progressed to absolute self-unity,
to a unity which is truly determined within itself, the   35
interior determinations of it being those of the Notion, not
nature.

ANHANG

Ein Fragment zur Philosophie des Geistes
(*1822/5*)

§

Die *Philosophie des Geistes* hat den Geist als *unser* inneres Selbst zum Gegenstande, — weder das uns und sich selbst Äusserliche — noch das sich selbst schlechthin Innerliche; — unsern Geist, der zwischen der natürlichen Welt und der ewigen Welt steht, und beyde als Extreme bezieht und zusammenknüpft.

§

Der Mensch wendet sein Bewußtseyn früher nach diesen beyden Seiten; er lebt, empfindet, schaut an, stellt vor, denkt, will und vollbringt, und hat in allem diesem, äussere Dinge oder seine Zwecke, andere und zwar *beschränktere* Gegenstände, als seine Thätigkeit in Allem diesem selbst, vor sich. Ebenso geht er zugleich über diesen seinen endlichen Boden hinaus zum Unendlichen, als einem ihm Fernern oder Nähern, aber einem solchen Andern, in welchem er verschwebt.

§

Sich selbst zu erkennen, diese Richtung auf das, was unmittelbar gegenwärtig ist, wie die endliche Gegenstände, und als ein Inneres, wie der unendliche Gegenstand, ist später.

Erkenne dich selbst, ist das bekannte Gebot des delphischen Apollo, und bezeichnet den eigenthümlichen Standpunkt der griechischen Bildung als der sich selbst individuellen Geistigkeit. Es macht dem griechischen Sinne Ehre, durch die Innschrifft, Γνωθι σεαυτον auf dem Tempel des höchsten Wissens diß wahrhafte Selbstbewußtseyn über die Eigenthümlichkeit des griechischen Geistes bewiesen zu haben. Die Auslegung jenes Gebotes im Verstande einer Selbstkenntniß, die nur auf die particulären Zufälligkeiten, Neigungen, Fehler, Schwächen u. s. f. des Individuums ginge, wäre, könnte man sagen, des delphischen Apollo, des *Wissenden*, unwürdig, weil solche

APPENDIX

A Fragment on the Philosophy of Spirit
(*1822/5*)

§

The *Philosophy of Spirit* has as its general object neither that which is   +
external to us and to itself, nor that which is simply within itself, but
spirit as *our* inner self; — our spirit, which stands between the natural
and the everlasting worlds, relating and linking them together as
extremes.   5

§

Man applies his consciousness to both these aspects prior to philosophizing; he lives, senses, looks about, presents, thinks, wills and
accomplishes, and in doing so he constantly has external things, his
purposes, before himself, — general objects other than his actual
activity within all this, and certainly *more limited* than this activity.   10
Consequently, as soon as he passes beyond his finite foundation to
what is infinite, he flounders about in a more or less alien otherness.

§

The *knowing of oneself*, the being orientated as an inwardness, as the   +
infinite general object, toward what is immediately present, as are
finite general objects, occurs later.   15
Know thyself is the celebrated injunction of the Delphic Apollo, and
characterizes the specific standpoint of Greek culture as that of a
spirituality which sees itself as being individualized. It says much
for the genius of Greece that through the inscription Γνωθι σεαυτον
on the temple of supreme knowledge, it should have proclaimed this   20
genuine self-consciousness over and above the peculiarity of the
Greek spirit. It might be said to be unworthy of the Delphic Apollo,
of *he who knows*, to interpret this injunction as a matter of self-
knowledge involving nothing more than the individual's particular
contingencies, inclinations, faults, weaknesses etc., for such subjective   25

subjective Menschenkennerey, dem griechischen Geiste noch fremde und ein späteres, modernes Erzeugniß ist.*

* *Parallelfassung zum Anfang des ersten Bruchstücks.* Den *Geist* zum Gegenstande der Betrachtung zu machen, setzt — können wir sagen — ein Bedürfniß voraus, denselben kennen zu lernen. Was wir so ein Bedürfniß nennen mögen, darüber können wir uns erinnern, daß es den Griechen als ein *Gebot* des delphischen *Apollos* ausgesprochen worden ist. *Erkenne dich selbst*, war die berühmte Aufschrift an dem Tempel des *wissenden* Gottes. Um so viel höher der Himmel über der Erde, um so viel höher, ja unendlich hoch ist der Geist über der Natur, und die Erkenntniß desselben ist schon durch ihren Gegenstand die würdigste.

Dem Griechen war das *Menschliche* zu seinem Antheil gegeben, das ist, der *freye* Geist, der aber seine *Unendlichkeit* noch nicht erfaßt hat. — Es ist nicht der absolute, der *heilige* Geist, der über die griechische Welt ausgegossen wäre und zu dessen Erkenntniß er kommen könnte. Es ist der Mensch, als *frey innerhalb* der Natur, so daß er an ihr das Organ seines Bewußtseyns behält, in ihr befangen bleibt, und indem er zwar in der Philosophie und nur in ihr, nicht in der Religion zum reinen Gedanken fortgeht, dieser selbst sich von der Abstraction — dem der Unmittelbarkeit im Gedanken entsprechenden Befangenseyn — nicht losmachen kann, nicht zum *Begriffe* des Geistes selbst kommt.

Die Aufgabe, den Geist zu erkennen, ist auf diese Weise, an und für sich beschränkt. Auf dieselbe Stuffe begränzt sich auch die Erkenntniß, welche das Ziel dieser Wissenschaft ist. Aber zugleich bestimmt sich uns die Aufgabe auf vielfache Weise anders, ebendadurch daß unser allgemeiner Standpunkt durch die Erhebung unseres Bewußtseyns zum Bewußtseyn des unendlichen Geistes, — eine Erhebung, die in der Religion begonnen hat, höher gestellt ist. Durch diesen Standpunkt ist dem Geist, welcher zunächst unter dem Menschlichen verstanden zu werden pflegt, nunmehr der absolute Geist gegenüber getreten, und jener wird durch diese Vergleichung zu einem *Endlichen*, d. i. in der Natur beschränkten einerseits herabgedrükt. Andererseits aber hat der Mensch durch die Beziehung selbst, welche mit jener Vergleichung zugleich zu Stande kommt, in sich einen ganz freyen Boden gewonnen, und sich ein anderes Verhältniß gegen die Natur, das Verhältniß der Unabhängigkeit von ihr, gegeben.

So ist uns der Geist, den wir hier betrachten, sogleich als eine Mitte zwischen zwey Extreme, die *Natur* und *Gott*, gestellt, — zwischen einen Ausgangspunkt und zwischen einen Endzweck und Ziel. Die Frage, was der Geist *ist*, schließt damit sogleich die zwey Fragen in sich, *wo* der Geist *her*kommt, und *wo* der Geist *hingeht!* Und wenn diß zunächst zwey *weitere* Betrachtungen zu seyn scheinen über die, was er *ist*, so wird sich bald zeigen, daß sie es allein wahrhaftig sind, durch welche erkannt wird, was er *ist*.

Wo er herkommt, — es ist von der Natur; wo er hingeht, — es ist zu seiner Freyheit. Was er *ist*, ist eben diese Bewegung selbst von der Natur sich zu befreyen. Diß ist sosehr seine Substanz selbst, daß man von ihm nicht als einem so feststehenden Subjecte sprechen darf, welches diß oder jenes thue und wirke, als ob solche Thätigkeit eine Zufälligkeit, eine Art von Zustand wäre, ausser welchem es bestehe, sondern seine Thätigkeit ist seine Substantialität, die Actuosität ist sein Seyn.

information about man is a later, a modern production, and is still alien to the spirit of Greece.*

* *Parallel version of these opening Paragraphs.* Being acquainted with *spirit* may be said to be the prerequisite of making it the general object of consideration. And it can be borne in mind, that what we may therefore refer to as a requirement, was declared to the Greeks as a *commandment* of the Delphic *Apollo*. The celebrated inscription on the temple of the god of *knowledge* was *Know thyself*. As the heavens are higher than the earth, so is spirit higher, and indeed infinitely higher, than nature. It is already on account of its general object therefore, that knowledge of spirit is knowledge of the worthiest kind.

Although the Greek has as his portion that which is *human*, that is to say, *free* spirit, it was spirit which had not yet grasped its *infinity*. — What would be poured forth over the Greek world, what the Greek might gain knowledge of, is not absolute spirit, spirit in its *holiness*, but man as *free within nature*, as retaining in nature the organ of his consciousness; as remaining confined within nature however, and while undoubtedly progressing into pure thought in philosophy, although only in philosophy, not in religion, being himself unable to rid even this thought of abstraction, of being bogged down in what corresponds in thought to immediacy,— to attain to the *Notion* of spirit itself.

It is on account of this that the task of cognizing spirit is limited in and for itself, and the cognition constituting the goal of this science also confines itself to this level. At the same time, however, the task presents itself to us in a manner which is in many respects different, for it is precisely through the raising of our consciousness to consciousness of infinite spirit, — a raising which began in religion, that our general standpoint is a higher one. It is on account of this standpoint that absolute spirit has confronted the spirit which used at first to be understood as subsumed under that which is human, and which in one respect has been reduced through this comparison to a *finitude*, i.e. to limited, natural spirit. In another respect, however, it is through the very relation which comes into being together with this comparison, that man has won a wholly free foundation within himself, and established for himself another relationship with nature, that of being independent of it.

As we are considering it here, therefore, the precise placing of spirit is that of a middle between the two extremes of *nature* and *God*, — between a point of departure and a final purpose and goal. To ask what spirit *is*, is therefore to pose the component questions of *whence* it comes and *whither* it tends! Although these two questions appear at first to be *supplementary* to the consideration of what spirit *is*, it will soon become apparent that it is only through them that there can be any true cognition of *this*.

Where does it come from, — nature; whither does it tend, — to its freedom. Its *being* is this motion of freeing itself from nature. This is its very substance to such an extent, that it is not permissible to speak of it as a fixed subject, which does and brings about this or that, as if such activity were a matter of chance, a kind of condition, by means of which the subject is constituted. The activity of spirit is its substantiality, its being is actuosity.

§

Der Geist als in der § 1 angegebnen Stellung ein unterschiedenes Besonderes gegen die natürliche und gegen die ewige Welt, ist *endlicher* Geist. Indem aber die Philosophie einen Gegenstand in seiner Wahrheit betrachtet, hat sie den Geist in seiner von der Schranke unabhängigen Unendlichkeit zu betrachten. Weil der Geist sich auf die Natur und auf die göttliche Idee zugleich *bezieht,* somit beydes zugleich in seiner Bestimmung liegen muß, so liegt hierin schon daß die Endlichkeit nicht seine allgemeine Bestimmung ist.

§

Es können hier zunächst die endlichen Betrachtungsweisen des Geistes erwähnt werden, welche sonst die Philosophie des Geistes ausmachten und mit ihr verwechselt werden können.

§

a) Die *Menschenkenntniß* und *Selbsterkenntniß* bezieht sich auf das Zufällige und Besondere der Charaktere, ihre Neigungen, Leidenschaften, Gewohnheiten, Ansichten, Vorurtheile, Launen, Schwächen, Fehler u. s. f. — eine Kenntniß *der Menschen,* die oft mit der Kenntniß *des Menschen,* und deren Interesse und Wichtigkeit eben so häuffig mit dem Interesse und der Gewalt der Sache verwechselt wird. Die *Selbsterkenntniß* hat ihr Interesse für den moralischen Zweck in Rüksicht auf das particuläre Individuum, und führt, wenn sie nicht das Substantielle und Gründliche der Moralität und Religiosität mehr vor Augen hat, als die subjectiven Particularitäten, leicht zu einer grüblerischen Ängstlichkeit, vornemlich aber zu einer einbilderischen Selbstsucht. — Die sogenannte *Menschenkenntniß,* für welche man vorzüglich auch auf Romane, Schauspiele, ferner gemeine Gesellschaft, u. s. f. angewiesen hat, fällt nach der Seite der *Klugheit* im Leben vornehmlich hin, und erlangt um so mehr Wichtigkeit in denjenigen, die desto weniger eigenen Gehalt des Charakters besitzen und sich auf Zwecke richten, die sie nicht durch die Sache selbst, sondern durch die Zufälligkeiten und Particularitäten Anderer zu erreichen hoffen, oder deren Geschäffte mit Andern es mehr mit deren Zufälligkeiten zu thun haben (wie z. B. die Kammerdiener). — Die *Zufälligkeiten, Particularitäten,* und noch mehr die blossen Leidenschaften der Menschen können leicht mit dem verwechselt werden und das übersehen machen, *was ihr substantieller Charakter und Wille ist.* So geschieht es in einer psychologisch-

§

The context presented in § 1 is that of a distinct particular in respect of the natural and the everlasting world, and within it spirit is *finite*. Yet in that philosophy considers a general object in its truth, it has to consider spirit in the infinity of its being independent of limit. Finitude's not being the universal determination of spirit is already involved in 5 spirit's so *relating* itself to nature and the divine Idea, that both must at the same time lie within its determination.

§

A beginning can be made here by mentioning the finite ways of considering spirit, which have usually constituted the Philosophy of Spirit, and which can be mistaken for it. 10

§

a) *Knowledge of human nature* and of ourselves is concerned with the contingency and particularity of characters, — their inclinations, passions, habits, opinions, prejudices, moods, weaknesses, faults etc. — This knowledge *of men* is frequently mistaken for knowledge *of man*, just as its interest and importance are often mistaken for the authority of 15 the matter itself.
Self-knowledge is of interest in respect of the particular individual's moral purpose, but if it is not concerned primarily with what is substantial and basic, with morality and religiousness rather than subjective particularities, it can easily give rise to a brooding anxiety, 20 and even more readily to conceited self-centredness. — What is called *knowledge of human nature*, which is also said to be acquired from novels and plays, as well as society in general etc., is an aspect of *prudence*, and therefore largely a matter of how one lives. Its importance is enhanced to the extent that people are lacking in character 25 of their own, and either pursue purposes which they hope to achieve not by means of the matter itself, but through the chance peculiarities of others, or have predominantly fortuitous dealings with others, as does a valet for example. — The *fortuitous particularities* of people, and to an even greater extent simply their passions, can easily be mistaken 30 for *that which constitutes their substantial character and will*, and lead to its being overlooked. This happens when history is treated from the

pragmatischen Geschichtsansicht, daß die grossen Begebenheiten nur als Producte kleiner oder mächtigerer Leidenschaften, und die Individuen in ihren Handlungen nur als von subjectiven Interessen regiert betrachtet werden, so daß die Geschichte auf diese Weise zu einem Spiele gehaltloser Thätigkeit und zufälligen Ereignisses herabsinkt.

§

b) Die *Psychologie* ist ihrem Fundamente nach gleichfalls empirisch, bringt aber die Erscheinungen in allgemeine Classen, und beschreibt dieselben unter dem Nahmen von *Seelenkräften, Vermögen* u. s. f. und betrachtet den Geist nach den *Besonderheiten,* in die er auf diese Weise zerlegt ist, so daß er als eine *Sammlung (ein Aggregat)* solcher Vermögen und Kräffte vorgestellt wird, deren jede für sich nach ihrer Beschränktheit wirkt, und mit den andern nur in Wechselwirkung und somit äusserliche Beziehung tritt.

Alle Erkenntniß fängt subjectiv von Wahrnehmungen und Beobachtungen an, und die Kenntniß *der Erscheinungen* ist von höchster Wichtigkeit, ja eine durchaus unentbehrliche Kenntniß. Aber sowohl für die Wissenschaft als unmittelbar auch für einen solchen Gegenstand, wie der Geist ist, wird etwas ganz anderes erfordert, als die Hererzählung von einer Reihe von Vermögen, und die Darstellung derselben als einer unorganischen Menge. Die Foderung des *harmonischen Zusammenhangs,* — (was ein Schlagwort in dieser Materie und ein so unbestimmtes ist, als sonst die *Vollkommenheit* war) in welchen jene Vermögen und deren Ausbildung gebracht werden *soll,* zeigt wohl die Erinnerung an eine wesentliche Einheit an, aber nur als eine seyn sollende, nicht als die ursprüngliche Einheit des Begriffs, die doch jeder Mensch vor sich hat, wenn er den Geist sich vorstellt — nemlich als ein wesentlich an sich Eines, als eine Monade; diese Harmonie bleibt dann darum auch eine leere, und sich nur in leeren Redensarten etwa .amplificirende, Vorstellung, weil der Begriff, die ursprüngliche Einheit, nicht als das Princip, vielmehr das Gegentheil: die unorganische Vielheit und Besonderheit der Geisteskräfte vorausgesetzt ist.

§

c) Die *rationelle Psychologie, Pneumatologie* betrachtet den Geist in ganz abstracter Allgemeinheit, und ist die alte Metaphysik über den Geist, welche denselben oder die Seele als *Ding* und nach abstracten *Verstandesbestimmungen,* wie *einfach* oder zusammengesetzt, nach der

psychologico-pragmatic point of view, — great events are regarded as only the products of various degrees of passion, and the actions of individuals as dominated solely by subjective interests, so that history is degraded to an interplay of futile activity and adventitious occurrence.

§

b) Although *psychology* is also fundamentally empirical, it sorts out the appearances into general classes, describes them under the headings of *psychic powers, faculties* etc., and considers spirit in accordance with the *particularities* into which this procedure dissects it. Spirit is therefore presented as being a *collection* or *aggregate* of such faculties and powers, each of which is effective in its own limited manner, and only enters into reciprocal action, and so into external relation, with the others.

All cognition derives subjectively from perceptions and observations, and the cognition of the *appearances* is not only of the utmost importance, but completely indispensable. Science, however, as well as the immediate comprehension of such a general object as spirit, requires something wholly distinct from the enumeration of a series of faculties and the representation of it as an inorganic mixture. To require that these faculties and their cultivation *should* be *harmoniously combined*, — (a catch-phrase in this context, and as vague as *perfection* used to be), certainly indicates some awareness of an essential unity. It is only a unity which ought to be however, it is not the original unity of the Notion, although it is this that everyone has in mind when picturing spirit to himself as essentially and implicitly one, as a monad. This harmony also remains empty therefore, a presentation which amplifies itself in nothing but empty turns of phrase, for it is not the Notion, the original unity which is presupposed as the principle, but rather the opposite, — the inorganic plurality and particularity of spiritual powers.

§

c) *Rational psychology, pneumatology,* considers spirit in a wholly abstract and general way, and is the old metaphysics of spirit, which took spirit or the soul to be a *thing,* and in accordance with the abstract *determinations of the understanding,* to be *simple* or composite, related to the

Beziehung auf den Körper, als auf ein schlechthin Selbstständiges, u. s. f. faßte. In solcher Betrachtungsweise tritt das, wodurch der Geist Geist ist, nicht ein.

§

Es sind vornemlich *zwey Umstände*, wodurch diese Betrachtungsweisen verdrängt worden sind: der *eine* ist die völlige Veränderung des *Begriffs* der Philosophie, welcher für die Wissenschaft weder *empirische Erkenntnisse* und Erscheinungen oder sogenannte Thatsachen des Bewußtseyns, noch deren Erhebung zu *Gattungen* und *Classification*, noch abstracte Verstandesbestimmungen, überhaupt nicht die *endliche* Betrachtungsweise unseres gewöhnlichen Bewußtseyns und reflectirenden Denkens für hinreichend und adäquat hält, sondern zum Gegenstand der Wissenschaft vom Geiste *nur den lebendigen Geist*, und zur *Form* des Erkennens nur dessen eigenen *Begriff* und nach der Nothwendigkeit seiner immanenten Entwicklung, haben kann.

§

Der *andere* Umstand kommt von der empirischen Seite selbst, und ist der *animalische Magnetismus*, welcher in der Welt des Geistes ein Gebiet von *Wundern* entdekt, und uns damit bekannt gemacht hat. Für die Auffassung der verschiedenen Zustände und sonstiger natürlicher Bestimmungen des Geistes, welche den Zusammenhang der Natur und des Geistes enthalten, wie für die Auffassung seines Bewußtseyns und seiner geistigen Thätigkeit, reicht, wenn man bey den Erscheinungen stehen bleibt, nothdürftig die *gewöhnliche endliche Betrachtungsweise* hin, und der *verständige* Zusammenhang von *Ursachen* und *Wirkung*, den man den *natürlichen Gang* der *Dinge* nennt, findet in diesem äusserlichen Gebiete sein Auskommen. Aber in den Erfahrungen des thierischen Magnetismus ist es die *Region* der *äusserlichen Erscheinungen* selbst, in welcher der verständige Zusammenhang von Ursachen und Wirkungen, mit seinen Bedingungen von den räumlichen und zeitlichen Bestimmungen seinen Sinn verliert, und innerhalb des sinnlichen Daseyns selbst und seiner Bedingtheit die höhere Natur des Geistes sich geltend macht und zum Vorschein kommt. Es wird sich späterhin zeigen, daß die Erscheinungen des animalischen Magnetismus nicht aus dem Begriffe des Geistes, namentlich nicht über sein Denken und seine Vernunft, hinausgehen, daß sie im Gegentheil nur einem Zustande und einer Stuffe angehören, in der er krank und in ein niedrigeres Daseyn unter die Kraft seiner wahrhaften Würde herabgesunken ist. So thörigt und eine so falsche Hoffnung es daher ist, in den Erscheinungen dieses Magnetismus eine Erhöhung des Geistes und eine Eröffnung von

body as to a simply independent entity etc. That whereby spirit is
spirit eluded such a manner of consideration.   +

§

Two principal *factors* have contributed to the discrediting of these
ways of considering spirit. The first is the complete change in the
Notion of philosophy. The Notion rejects as being insufficient for and   5
inadequate to science the whole *finite* manner of consideration of our
ordinary consciousness and reflecting thought, — the *empirical cognitions* and appearances or the so-called facts of consciousness, the raising
of these to the status of *genera* and *classification*, the abstract determinations of the understanding. As the general object of the science of spirit   10
it can have *only living spirit*, and as the form of cognition only cognition's
own *Notion* in the necessity of its immanent development.

§

The *other* factor is *animal magnetism*, which derives from the empirical   +
aspect itself, opens up a field of *wonders* in the world of spirit, and has
made us familiar with it. Even if one gets no further than appearances,   15
nothing more than the *ordinary finite manner of consideration* is enough for
the conception of the various conditions and other natural determinations containing the connection between nature and spirit, as it is for
the conception of its consciousness and spiritual activity. In this external
field, the *understanding's* connection between *cause* and *effect*, which is   20
called the *natural course of things*, enjoys a certain validity. In the experiences of animal magnetism however, it is within this very *region* of
external appearances that the understanding's connection between causes
and effects, with its conditions of spatial and temporal determinations,
loses its validity, and in sensuous determinate being itself, together   25
with its conditionality, that the higher nature of spirit makes itself
effective and becomes apparent. It will subsequently become apparent
that the appearances of animal magnetism do not transcend the thought
and reason of the Notion of spirit, and that rather than breaking loose
from it, they simply belong to a state and a stage of spiritual disease,   30
of spirit's having sunk beneath the power of its true dignity to a lower
determinate being. Consequently, although it is certainly ridiculous   +
and pointless to attempt to see in such magnetic appearances an
elevation of spirit, an opening up of depths profounder than the thinking

Tieffen, die weiter gingen als sein denkender Begriff, sehen zu wollen, so sind es dagegen diese Erscheinungen, welche *im Felde des Erscheinens* selbst nöthigen, den *Begriff des Geistes* herbeyzuruffen, und nicht gestatten, bey dem begrifflosen Auffassen des Geistes, nach der *gewöhnlichen Psychologie* und nach *dem sogenannten natürlichen Gange der Dinge*, mehr stehen zu bleiben. Die an diesen Erscheinungen sich beweisende *Idealität* der sinnlichen und verständigen, überhaupt der endlichen Bestimmungen, ist es, wodurch dieses Gebiet für sich eine Verwandschaft zur Philosophie hat, so wie es auch für die Geschichte, in welcher so vieles unter dem Nahmen des Wunderbaren, von dem Verstand, der den *Zusammenhang äusserlicher Ursachen und Wirkungen und die Bedingtheiten* des *sinnlichen Daseyns*, zum *Maßstabe* der *Wahrheit* nimmt, so Vieles, Ereignisse und Individuen, mishandelt und verworfen worden ist, eine *versöhnende* Wichtigkeit hat.

Von schriftstellerischen Werken über die Natur des Geistes, welche von einem höhern Standpunkte der Philosophie ausgehen, als aus welchem die § ff. genannten Ansichten und Wissenschaften entsprangen, sind zwey zu nennen:

Eschenmayers Psychologie in drey Theilen, als empirische, reine und angewandte. Stuttg. u. Tüb. 1817. Der zweyte Theil enthält eine Logik, Ästhetik und Ethik, der dritte eine Kosmologie oder Physik; diese beyden Theile gehören also nicht hieher. Der erste, die Psychologie macht sogleich als empirische, für sich keinen Anspruch auf Wissenschaftlichkeit; der zweyte Theil, die reine Psychologie, soll die Bestimmung haben, die Principien jenes empirischen Materials auf zu stellen, und von dem dabey nur vorausgesetzten Schema die Construction gefunden und seine Abkunft aufgezeigt zu haben. *Eschenmayer* setzt aber sogleich (§ 289) die speculative Erkenntniß, die hier eintreten soll, bloß 1) in Reflexionen durch Begriffe, Urtheile und Schlüsse und 2) in ideale Anschauung. So findet sich in diesem zweyten Theil die gewöhnliche Methode, eine Voraussetzung zu analysiren, darüber zu reflectiren und das hiebey unentbehrliche, in der That ganz empirische und willkührliche Verfahren, den Inhalt ganz beliebig herzuerzählen und zu bestimmen, — der gebrauchte Nahme: ideale Anschauung thut nichts zur Sache; so spricht jeder, der seine Kenntniße und Vorstellungen in einer beliebigen Ordnung abhandelt, aus idealer Anschauung. Es ist in dieser Darstellung daher gerade die speculative Erkenntnißweise, welche man gänzlich vermißt; und an deren Stelle dagegen die bekannte Manier, ein Schema vorauszusetzen, und die vorhandenen Materialien unter dasselbe zu rubriciren, in Verbindung mit einem Herrn *Eschenmayern* eigenthümlichen Formalismus, mathematische Terminologie an die Stelle von Gedanken zu setzen, herrschend.

Notion of it, it is these appearances which, *in the field of appearance* itself, make it necessary to invoke *this Notion*, and no longer permissible to fall in with ordinary *psychology* and *the so-called natural course of things*, in failing to advance beyond the Notionless conception of spirit. Since it is in these appearances that the *ideality* of what pertains to the under-   5
standing, of sensuous and generally finite determinations gives proof of itself, this is a field which has its own affinity with philosophy. It is relevant to history in a *reconciliatory* capacity, — so much that has been regarded as miraculous in respect of events and individuals having been mishandled and rejected by the understanding, which takes the   10
connection between external causes and effects and the conditionalities of *sensuous determinate being* to be the *yardstick* of *truth*.   +

With regard to literary works on the nature of spirit founded on a higher philosophical standpoint than that which gave rise to the views and sciences mentioned in §   ff., mention might be made of   15
two: *Eschenmayer's* 'Psychology in three parts, *empirical, pure and applied*' (Stuttg. and Tüb., 1817). The second part contains a Logic, Aesthetics and Ethics, the third a Cosmology or Physics, so that both these parts do not belong here. The first part, the Psychology, in that it is empirical, makes no intrinsic pretension to being scientific.   20
The second part, the Pure Psychology, ought to have been devoted to enunciating the principles of this empirical material, and to searching out the construction and indicating the derivation of the model which it simply presupposes. But *Eschenmayer* (§ 289) merely sets straight about positing the speculative cognition which ought to   25
emerge here, 1) in reflections by means of notions, judgements and syllogisms, and 2) in ideal intuition. This second part therefore exhibits the ordinary method of analyzing a presupposition and reflecting upon it, together with what is indispensable to this — the in fact wholly empirical and arbitrary procedure of discussing and   30
determining the content just as one pleases. There is no point in the invoking of ideal intuition, for the phrase is used by anyone discussing his presentative cognitions in a sequence which pleases him. It is, therefore, precisely the speculative manner of cognition which is entirely absent from this exposition, and it is the well-known pro-   35
cedure which predominates in place of it, — a model is presupposed, and the materials available are rubricated beneath it, — together with a formalism peculiar to Mr. *Eschenmayer*, — that of replacing thoughts with mathematical terminology.   +

Steffens Anthropologie in 2 Bänden. Breslau 1822. verflicht Geologie sosehr mit Anthropologie, daß auf die letztere etwa der 10te oder 12te Theil des Ganzen kommt. Da das Ganze aus empirischem Stoffe, aus Abstractionen und aus Combinationen der Phantasie erzeugt, dagegen das, wodurch Wissenschaft constituirt wird, Gedanke, Begriff und Methode verbannt ist, so hat solches Werk wenigstens für die Philosophie kein Interesse.

Die speculative Betrachtung und Erkenntniß der Natur und Thätigkeit des Geistes ist in neuern Zeiten bis auf die Ahndung davon so sehr untergegangen, daß noch immer die Schrifften des *Aristoteles* über diesen Theil der Philosophie, beynahe, oder da die tieffen Ansichten des *Spinoza* doch nur ein Anfang sind, und weil sie, wie seine ganze Philosophie, nur Anfang sind, auf einer nur einseitigen Metaphysik beruhen, *Leibnitzens* Betrachtungen aber einerseits gleichfalls nur metaphysisch, andererseits nur empirisch sind, — so bleiben also durchaus die *Aristotelischen* Schriften die einzigen, welche wahrhaft speculative Entwicklungen über das Seyn und die Thätigkeit des Geistes enthalten, obgleich nichts so sehr misverstanden worden ist, als die Aristotelische Ansicht von der allgemeinen Natur des Erkennens, daß man sogar den *Aristoteles* an die Spitze der Empiriker gesetzt hat, und diese Ansicht seiner Lehre in allen Geschichten der Philosophie als ein festes Vorurtheil, zu finden ist; die Aristotelischen Speculationen aber über die Empfindung und überhaupt über die besondern Wirksamkeiten des Geistes, sind für die Psychologie ganz unbeachtet geblieben.

§

Die Philosophie des Geistes kann weder empirisch noch metaphysisch seyn, sondern hat den *Begriff* des Geistes in seiner immanenten, nothwendigen Entwicklung aus sich selbst zu einem Systeme seiner Thätigkeit zu betrachten.

Die empirische Betrachtungsweise des Geistes bleibt bey der Kenntniß der Erscheinung des Geistes stehen, ohne den Begriff desselben; die metaphysische Betrachtungsweise will es nur mit dem Begriffe zu thun haben, ohne seine Erscheinung; der Begriff wird so nur ein Abstractum, und die Bestimmungen desselben ein todter Begriff. Der Geist ist diß wesentlich, thätig zu seyn, das heißt, sich und zwar nur seinen Begriff zur Erscheinung zu bringen, ihn zu offenbaren.

In jeder besondern philosophischen Wissenschaft ist das Logische, als die reine allgemeine Wissenschaft, hiemit als das Wissenschaftliche in aller Wissenschaft vorausgesetzt.

Steffens' 'Anthropology' (2 vols. Breslau, 1822), mixes Geology with Anthropology to such an extent, that only about a tenth or twelfth of the whole is concerned with the latter. Since the whole of it is rustled up out of empirical material, abstractions and fantastic combinations, no attention being paid to the thought, Notion and   5 method of science, such a work is of no interest, at least to philosophy.   +

Even the remote awareness of the speculative consideration and cognition of the nature and activity of spirit has become so dim in recent times, that by and large it is only what *Aristotle* has written on this part of philosophy which contains a truly speculative develop-   10 ment of the being and activity of spirit. *Spinoza's* views are profound, but they are only a beginning, like the whole of his philosophy, based as it is upon a simply one-sided metaphysics. While in one respect   + *Leibnitz's* observations too are merely metaphysical, their other aspect is merely empirical. — Nothing has been so misunderstood   15   + as the *Aristotelian* view of the nature of cognition. *Aristotle* has even been regarded as the epitome of empiricism, and while this prejudiced view of his doctrine is to be found firmly embedded in all histories of philosophy, no attention at all has been paid to the importance to Psychology of the Aristotelian speculations concerning sensation   20 and the other particular operations of spirit.   +

§

The Philosophy of Spirit can be neither empirical nor metaphysical,   + but has to consider the *Notion* of Spirit in its immanent, necessary development from out of itself into a system of its activity.   25

The empirical manner of considering spirit does not progress beyond the cognition of its appearance to its Notion. Since the metaphysical manner of consideration is concerned only with the Notion, to the exclusion of its appearance, the Notion becomes simply an abstraction, and its determinations a Notion without life. Spirit is essentially   30 active, that is to say that it brings itself forth into appearance, revealing only its Notion.

In each particular philosophical science, what is logical is presupposed as the purely universal science, and so as the scientific factor in all science.   35   +

Begriff des Geistes
und
Eintheilung der Wissenschaft

§

Den Begriff des Geistes festzusetzen, dazu, ist nöthig, die *Bestimmtheit* anzugeben, wodurch er die Idee als Geist ist. Alle Bestimmtheit ist aber Bestimmtheit nur gegen eine andere Bestimmtheit; der des Geistes überhaupt steht zunächst die der Natur gegenüber, und jene ist daher nur zugleich mit dieser zu fassen. Indem dieser Unterschied des Geistes und der Natur zunächst *für uns*, für die subjective Reflexion ist, so wird sich dann an ihm selbst zeigen, daß und wie Natur und Geist sich durch sich selbst aufeinander beziehen.

§

Die Bestimmtheit, in welcher die Idee als *Natur* ist, ist, daß sie die Idee als *unmittelbar* ist; die sich entwickelnden Bestimmungen aber, oder was dasselbe ist, der Inhalt der Idee, in der Form der Unmittelbarkeit sind für sichseyende *Vereinzelungen*, die *sind*, d. i. als *gleichgültig* gegen einander bestehend erscheinen. Das *Aussereinander* macht daher die allgemeine, abstracte Bestimmtheit der Idee als Natur aus. Der Natur wird darum die *Realität* zugeschrieben.

§

Die *Unmittelbarkeit* und damit das, was die Realität der Natur heißt, ist zugleich nur eine Form, vielmehr ein *Vermitteltes*, und diese wesentliche Seite, nemlich die Beziehung der Natur auf den Geist, stellt dieselbe Bestimmtheit der Natur von ihrer andern Seite dar. Sie ist nemlich das dem Geiste *Andre* oder *Äusserliche*; aber was sie gegen den Geist ist, diß ist ihre wahrhafte Bestimmung an ihr selbst, weil der Geist ihre Wahrheit ist. Sie ist deßwegen das *sich selbst Andre*, das ihr selbst Äusserliche, und ihre Realität begründet sich somit wesentlich auf das Verhältniß zum Geiste. — Diß Aussereinander macht, in seiner ganz unmittelbaren, abstracten Form genommen, und zwar selbst sogleich in zwey Bestimmungen den *Raum* und die *Zeit* aus. Alles Natürliche ist räumlich und zeitlich. Aber die *Unterschiedenheit* oder die Grenze in Raum und Zeit ist nicht nur das abstracte Eins, Raum- und Zeit*punkt*, sondern das concretere Eins, das *Atom* als *materielles*, wornach das

The Notion of Spirit
and
Division of the Science

§

In order to establish the Notion of spirit, it is necessary to specify the   +
determinateness whereby it constitutes the Idea as spirit. All determinateness, however, is only determinateness in respect of another determinateness; that of spirit in general is initially in opposition to that of
nature, so that the former is only to be grasped together with the latter.   5
Since at first it is *for us*, for subjective reflection, that this difference
between spirit and nature has being, the difference itself will subsequently make apparent that nature and spirit are themselves inter-
relative, and how they interrelate.

§

Although the determinateness in which the Idea has being as *nature*   10   +
is that of nature's being the immediacy of the Idea, the self-developing
determinations which are the content of the Idea in the form of immediacy are *singularizations* which are for themselves. Since these determinations *are* i.e. appear as subsisting in a state of mutual *indifference*,
the abstract determinateness of the Idea as nature is *extrinsicality*. This   15
is why *reality* is ascribed to nature.

§

The *immediacy*, together with what is called the reality of nature, is at   +
the same time only a form, or rather a *mediatedness*, and this essential
aspect, that is to say nature's relation to spirit, represents the other
aspect of the same determinateness of nature. Nature is indeed that   20
which is *other than* or *external* to spirit; since spirit is its truth however,
what it is in respect of spirit is its true determination in itself. Nature is
therefore *that which* is *other than* itself, external to itself, so that its reality
bases itself essentially on the relationship to spirit. — This extrinsicality,
taken in its wholly immediate, abstract form, and indeed directly as it   25
is in itself, constitutes the two determinations of *space* and *time*. Although
all that is natural is spatial and temporal, the *state of difference*, or the
limitation in space and time, is not only the abstract unit, the spatial
and temporal *point*, but the more concrete unit or *material being* of the
atom, whereby the extrinsicality of nature and with it the general basis   30   +

Aussereinander der Natur, und damit die allgemeine Grundlage aller ihrer daseyenden Gestaltungen, sich zur Materie bestimmt, welche, weil jenes Atom, als nur Eins für sich, selbst nur ein abstractes Moment ist, wesentlich nicht als solches Eins, sondern nur als ein Aussereinander derselben, als *zusammengesetzt* existirt.

§

Diese erste Bestimmtheit, das *Aussereinander*, führt die andere Bestimmtheit der natürlichen Dinge mit sich. Die Materie ist ausserdem, daß sie ein sich äusserliches überhaupt ist, vielfach bestimmt und beschaffen, und die Beziehung...

§

Die *Endlichkeit* des Geistes ist eine für sich, aber auch darum vornemlich wichtige Bestimmung, weil von ihrem wahrhaften Verhältniß nur eine speculative Erkenntniß möglich ist, diese aber, weil die Endlichkeit für eine bekannte für sich sich verstehende, und schlechthin feste Bestimmung genommen wird, sosehr den Misverständnißen ausgesetzt ist. Obgleich die Unwahrheit solcher Bestimmung, wie die Endlichkeit überhaupt, aus der Logik vorauszusetzen ist, so ist sie in der concreten Bedeutung als Endlichkeit des Geistes, und um des besondern Interesses, das sie insofern hat, hier näher zu erörtern.

§

Die *Endlichkeit* ist zunächst die qualitative überhaupt, so daß die Qualität als Bestimmtheit mit dem Seyn, der Gattung des Gegenstandes identisch, von ihr untrennbar, und daß sie an einer andern von diesem Subjecte ausgeschlossenen Qualität ihre Bestimmtheit und Schranke hat. Diese Endlichkeit ist die der natürlichen Dinge, wie die specifische Schwere des Goldes vom Seyn des Goldes untrennbar und an einer andern ausser dem Golde ihren Unterschied und Bestimmtheit hat, so diese Form der Zähne, der Klauen, u. s. f. eines Thieres u. s. w. Die Endlichkeit in ihrem Begriffe aber ist die Unangemessenheit des Begriffes und seiner Realität, so daß diese seine Realität an dem Begriffe ihre Bestimmtheit oder Schranke hat, und für den Begriff eines endlichen Gegenstandes bedarf es um dieser Unangemessenheit willen, weil der Begriff ganz und ungetrennt ist, noch anderer Gegenstände — wie für den Begriff der Sonne nicht bloß der Sonne, sondern auch der Planeten und so ferner.

of the determinate being of all its formations, determines itself as matter. However, since this atom itself only has being as a unit which is for itself, matter itself is simply an abstract moment, and exists essentially not as such a unit, but only as an extrinsicality of it, a *composite* being.

§

This initial determinateness of *extrinsicality* carries the other deter-   5   +
minateness of natural things with it. Matter is , moreover, a general self-externality, variously determined and constituted, and the relation…

§

Although the *finitude* of spirit is a determination which is important   +
for itself, it is also of special importance in that the only possible cognition of its true relationship is a speculative one. The cognition of   10
finitude is open to misunderstandings however, in so far as finitude is taken to be a simply fixed and evidently self-explanatory determination. Although the untruth of such a determination, like finitude in general, is to be presupposed from the Logic, it is to be examined more closely here in its concrete significance as the finitude of spirit, and on account   15
of the particular interest which this entails.

§

Initially, *finitude* is what in general is qualitative, the determinateness   +
of the quality being identical with or inseparable from the being, the genus of the general object, and having its determinateness and limit in another quality, which is excluded from this subject. This is the finitude   20
of natural things, — as in the case of the specific gravity of gold, which is inseparable from the being of gold, and different and determinate on account of something other than and external to gold, or the form of an animal's teeth, claws etc., and so on. In its Notion however, finitude is the lack of conformity between the Notion and its reality, so that this   25
reality of the Notion has its determinateness or limit in the Notion, and on account of this lack of conformity, since the Notion is whole and undivided, the Notion of a finite general object requires still further general objects, — the Notion of the Sun not only that of the Sun, but also that of the planets and so on.   30   +

§

Die Idealität, welche die Qualität des Geistes ausmacht, ist ein solches, worin alle Qualität als solche sich aufhebt, das Qualitätslose — und die Endlichkeit des Geistes ist daher so zu fassen, daß, indem er in der Idealität aller Schranken der zur Existenz gekommene freye unendliche Begriff ist, seine Endlichkeit nur in die ihm unangemessene Weise der Realität fällt.

§

Weil der Geist die zur Existenz gekommene Freyheit des Begriffes ist, so ist jene ihm unangemessene Realität, die *Schranke, für ihn.* Eben darin, daß sie *für ihn* ist, steht er über derselben, und die Beschränktheit des Geistes hat damit eben diesen ganz andern Sinn, als die der natürlichen Dinge; daß er sich als beschränkt weiß, ist der Beweis seiner Unbeschränktheit.

Die Schranken der Vernunft, die *Beschränktheit des Geistes* sind Vorstellungen, welche ebenso für ein Letztes, ein für sich gewisses Factum als für etwas Bekanntes und für sich Verständliches gelten. Sie sind aber so wenig ein für sich Verständliches, daß die Natur des Endlichen und Unbeschränkten, und ebendamit sein Verhältniß zum Unendlichen den schwersten Punkt, man könnte sagen, den einzigen Gegenstand der Philosophie ausmachen; ebenso ist die Schranke nicht das Letzte, sondern vielmehr indem und weil der bewußte Mensch von der Schranke weiß und spricht, ist sie Gegenstand für ihn und er hinaus über sie. Diese einfache Reflexion liegt ganz nahe, und sie ist es, die nicht gemacht wird, indem von den Schranken der Vernunft und des Geistes gesprochen wird. — Der Misverstand beruht auf der Verwechslung der qualitativen Schranken der natürlichen Dinge, und der nur im Geiste nur ideellen, wesentlich zum Scheine herabgesetzten Schranke. Die natürlichen Dinge sind eben insofern natürliche Dinge, als ihre Schranke *nicht für sie selbst* ist; sie ist es nur für den Geist. Die natürlichen Dinge *sind* beschränkt, und sie sind es *für uns*, in *Vergleichung* mit *andern* Dingen und ohnehin mit dem Geiste. — Diese Vergleichung aber machen die natürlichen Dinge nicht, nur *wir* machen sie, machen die Vergleichung des Geistes in sich mit demselben, wie er als fühlender, verständiger, wollender u. s. f. beschränkt ist, aber eben dieses Vergleichen, diß Aussprechen seines Beschränktseyns ist selbst die Erhebung über sein Beschränktseyn. — Zur Schranke gehören *zwey*; die Schranke ist eine Negation überhaupt; daß *Etwas* beschränkt sey, dazu gehört das *Andere* desselben; jedes der beyden ist beschränkt, und die Schranke

§

The ideality which constitutes the quality of spirit is one in which all + quality as such sublates itself, being that which is without quality. Since spirit is in the ideality of all the limits of the free and infinite Notion which has come into existence, its finitude is to be grasped as falling only within the mode of reality which does not conform to it.   5

§

Since spirit is the freedom of the Notion which has come into exist- + ence, this reality which does not conform to it is the *limit* which has being *for it*. It is precisely on account of this limit's being *for it* that spirit stands above it, and it is precisely on account of this that spirit's limitation is so entirely different from that of natural things; the proof   10 of its limitlessness is that it knows itself to be limited.

The limits of reason, the *limitedness of spirit*, are presentations which are not only taken to be an irreducibility, an intrinsically settled fact, but also something which is known and implicitly understandable. They are so far from being implicitly understandable however, that   15 the nature of what is finite and unlimited, together with its relationship to what is infinite, might be said to constitute the most difficult point in philosophy, the sole general object of it. What is more, the limit is not an irreducibility, for since the conscious person knows of and speaks about it, it is a general object for him, and he is therefore   20 beyond it. It is this simple reflection, which arises of its own accord, which is overlooked when there is talk of the limits of reason and of spirit. — The misunderstanding derives from confusing the qualitative limits of natural things with the limit which is solely of an ideal nature only in spirit, and which in essence is reduced to an appear-   25 ance. Natural things are natural things precisely in so far as their limit is *not for themselves*; it is this only for spirit. Natural things *are* limited, and they are *for us*, in *comparison* with *other* things as well as with spirit. — Natural things do not make this comparison however, it is only *we* who do so, comparing spirit in itself with spirit, investigat-   30 ing the way in which it is limited as feeling, understanding, willing spirit etc. It is, however, precisely this comparing, this expressing of the limitedness of spirit, which itself constitutes the surpassing of the limitation. — Limit involves *two*, being in general a negation; the limitation of *something* involves its *other*; each of them is limited,   35

ist wenn man will, das Gemeinschaftliche beyder, oder vielmehr das Allgemeine derselben. Indem aber der Geist von der Schranke weiß indem sie *für ihn* ist, darin schon ist sie selbst als Gegenstand, als das *Andere* gesetzt; diß Andere der Schranke aber zunächst ist das Unbeschränkte, das Andere des Endlichen ist das Unendliche. — So hat dann das Endliche seine Beschränktheit an dem Unendlichen, das Beschränkte hat das Unbeschränkte zu seiner Schranke oder Grenze. Allein diesen beyden ist so die Schranke das Gemeinschaftliche, und in der That ist das Unbeschränkte, die Unendlichkeit, welchen das Beschränkte, die Endlichkeit das gegenüberstehende Andere ist, selbst nur eine endliche. An den Misverstand über die Natur der Schranken des Geistes, und die Endlichkeit überhaupt, schließt der Verstand sogleich diesen andern Misverstand an, eine solche Unendlichkeit, welcher das Endliche gegenüber stehen bleibt, für etwas mehr als ein bloßes Abstractum des Verstandes, für etwas Wahrhaftes zu halten. — Hier wo es sich von einer concreten Idee, dem Geiste handelt, müssen alle diese Gewohnheiten des abstracten Verstandes, längst aufgegeben seyn.

§

Der Begriff oder die wahrhafte Unendlichkeit überhaupt, und damit die des Geistes ist daß die Schranke *als* Schranke *für ihn* sey, daß er sich in seiner Allgemeinheit *bestimme*, d. i. sich eine Schranke setze, aber daß sie als ein *Schein* sey; er *ist* diß, ewig sich diesen Schein zu setzen, die Endlichkeit nur als ein Scheinen in ihm zu haben, d. i. sich Begriff so zu seyn, wie der Begriff in der Philosophie ist. Das *Seyn* des Geistes ist nicht Seyn, insofern es von der Thätigkeit unterschieden wird, sondern sein Seyn ist eben diese Bewegung, sich als *Anderes* seiner selbst zu setzen, und diß Andre seiner aufzuheben, zum Scheine herabzusetzen, und so in sich zurükzukehren; diese sich hervorbringende Idealität...

Racenverschiedenheit

§

Das allgemeine Naturleben der *Bewegung*, der freye Mechanismus des Sonnensystems und darin der individuellere der Erde ist im anthropologischen Leben jener ganz untergeordnete Unbestimmtheit, dieser theils für sich noch nicht weiter concrete Veränderung, theils ganz unbestimmte und untergeordnete Stimmung.

Erst die Erde kann als physicalische Individualität eine Besonderung

and the limit may be said to be what is common to them both, or rather their universal. Yet in that spirit knows of the limit, in that the limit has being *for it*, it is a limit which is itself already posited as general object, as the *other*. This other of the limit is what is primarily unlimited however, the other of what is finite being what is infinite. — 5
It is thus, therefore, that what is finite has its limitation in what is infinite, what is limited having what is unlimited as its limit or boundary. It is therefore the limit alone which is common to both, and the unlimited, the infinitude, which is the other opposed to the limited or finitude, is in fact itself only a finitude. To the misunderstanding of 10
the nature of the limits of spirit and of finitude in general, the understanding immediately adds this further misunderstanding of regarding an infinitude such as this, which remains opposed to what is finite, as something more than a mere abstraction of the understanding, as something which is true. — Since what is being dealt 15
with here is a concrete Idea, spirit, all these habits of the abstract understanding have to have been long since abandoned.

§

The Notion, which since it is true infinitude in general is also the +
infinitude of spirit, has being in that limit is *for it as* limit, in that it *determines* itself in its universality, i.e. in that the limit it posits itself 20
has being as an *apparency*. The Notion *is* the eternal positing of this apparency to itself, the internal possession of finitude as nothing but an apparency; i.e. being its own Notion, as it is in philosophy. The *being* of spirit is not being in so far as it differs from activity, but precisely this motion of returning into itself by positing itself as the *other* of itself, 25
and sublating this its other, reducing it to apparency. This self-producing ideality...

Racial variety

§

In anthropological life, the universal natural life of *motion* is a wholly +
subordinate indeterminateness, while that of the free mechanism of the solar system, and within this of the more individual mechanism of the 30
Earth, is partly for itself in not yet involving further concrete change, and partly a wholly indeterminate and subordinate mood.
The physical individuality of the Earth is the first capable of sustain-

zu qualitativ unterschiedenen Massen an ihr haben, und die physicalische Unterscheidung dieses Bodens des Menschen als anthropologische Besonderung an dessen allgemeiner Natur zur Existenz kommend, macht das aus, was die *Racenverschiedenheit* der Menschen genannt worden ist.

§

Diese Unterschiede sind Qualitäten, weil sie der natürlichen Seele, dem blossen *Seyn* des Geistes angehören; aber der Begriff des Geistes, Denken und Freyheit, ist höher als das blosse Seyn, und der Begriff überhaupt und näher die Vernünftigkeit ist eben diß nicht qualitativ **bestimmt** zu seyn. Die Unterschiede fallen in die *besondere* Natur des Menschen oder in seine Subjectivität, die sich als Mittel zur Vernünftigkeit verhält, wodurch und worin diese sich zum Daseyn bethätigt. Diese Unterschiede betreffen deswegen nicht die Vernünftigkeit selbst, sondern die Art und Weise der Objectivität derselben, und begründen nicht eine ursprüngliche Verschiedenheit in Ansehung der Freyheit und Berechtigung unter den sogenannten Racen. Aber die Unterschiedenheit ist darum, daß sie die Objectivirung der Vernünftigkeit betrifft, noch groß genug, denn die Vernünftigkeit ist wesentlich diß, sich zum Daseyn zu bethätigen; — eine bloß mögliche Vernünftigkeit wäre gar keine, und alle die ungeheuren Verschiedenheiten unter den Nationen und Individuen reduciren sich allein auf die Art und Weise des Bewußtseyns, d. i. der Objectivirung der Vernunft.

Die Frage, ob das Menschengeschlecht von Einem Paare abstamme, welche mit der Racenverschiedenheit zusammenhängt, hat für uns kein philosophisches, sondern, ausserdem wie sie sich auf die religiöse Geschichte bezieht, nur ein historisches und verständiges Interesse. Ohnehin würde es nur eine müssige Frage seyn, ob die mannichfaltig verschiedenen Menschenstämme von verschiedenen ursprünglichen Menschenpaaren ihre Herkunft haben, und ein näheres Interesse für den Verstand kann die Frage nur in Beziehung auf die gemeinsame oder unterschiedene Abkunft der verschiedenen Menschenracen haben. Die historische Untersuchung müßte die geschichtlichen Daten oder Spuren, soweit sich deren vorfinden, verfolgen, und die Verschiedenheit als eine geschehene oder nicht geschehene Veränderung auf ihre Weise anzugeben bemüht seyn. Aber solche Untersuchung liesse schon darum nicht erwarten, zu etwas Schließlichem zu kommen, weil das Resultat, daß soweit die Geschichte oder Sage zurükgehe, sich nur die Verschiedenheit schon als vorhanden finde, auf diesem Felde immer schlechthin die Möglichkeit übrigläßt, daß noch ältere Begebenheiten uns nur unbekannt seyn. Aber ohnehin

ing qualitatively different masses, and what has been called the *racial variety* of men consists of the physical differences of this foundation coming into existence as anthropological particularization in the general nature of man.

§

These differences are qualities, for they pertain to the natural soul, 5
to the mere *being* of spirit; the Notion of spirit, thought and freedom, is higher than mere being however, and it is precisely the absence of qualitative determination which characterizes the Notion in general, and more particularly rationality. The differences fall within the *particular* nature of man or within his subjectivity, which conducts itself 10
as the instrument of rationality, the means whereby and within which this rationality activates itself into determinate being. Consequently, these differences are only relevant to the way and manner in which rationality is objectified; they are irrelevant to rationality itself, and provide no basis for any radical variety of freedom and rights between 15
the so-called races. Since the difference touches upon the objectification +
of rationality however, it is still great enough, for rationality is essentially a matter of activating itself into determinate being; — rationality cannot be simply a possibility, and all the tremendous varieties of nations and individuals resolve themselves into nothing more than the 20
mode and manner of consciousness i.e. the objectification of reason.

Although the question of whether or not the human race is descended +
from a single couple has a bearing upon racial variety, it is of no philosophic interest to us, its only interest, apart from its relevance to religious history, being an historical and general one. It would in 25
any case be idle to enquire into whether or not the widely differing human tribes are ultimately descended from various human couples. Such an enquiry can only be of more specific interest to the under-standing, and on account of the light it throws upon the common or separate lineage of the various human races. The historical investiga- 30
tion would involve following up the recorded data or evidence in so far as this is extant, and deciding in its own way whether or not the transmutation has taken place. No final outcome could be expected from such an investigation however, since in this field the conclusion that there was nothing but variety present at the dawn of 35
history or legend is always open to the possibility of our simply being ignorant of earlier events. In any case, historical research

muß die geschichtliche Forschung auf solchem alten Boden des noch ganz trüben Bewußtseyns der Begebenheiten und Thaten bald abbrechen, und es ist nur der Verstand, der gegen das geschichtliche Datum der Abstammung von Einem Paare, seine Zweiffel erhoben, indem er die vorhandene Verschiedenheit geltend macht, und auf seine Weise dieselbe sich begreifflich zu machen bestrebt ist. Diß Verstehen steht auf dem Felde natürlicher Einwirkungen und äusserlicher Ursachen für die vorhandene Verschiedenheit, und befindet sich also auf einem andern Felde als das philosophische Denken, welches die Verschiedenheiten nur in der Bestimmtheit des Begriffs aufsucht, aber um das geschichtliche Entstehen und die natürlichen Ursachen derselben unbekümmert ist. Zugleich aber kennt der Begriff des Geistes sein Verhältniß zu den geschichtlichen und verständigen Forschungen. Die Vernünftigkeit des Geistes und ebendeßwegen seine an sich Qualitätslose Allgemeinheit steht für sich selbst über diesen unterschiedenen Qualitäten, weil sie Besonderheiten sind, dem natürlichen Daseyn angehören, und daher ein natürliches Entstehen haben. Sie befinden sich daher auf einem Felde, wo der natürliche Zusammenhang und die Wirksamkeit natürlicher Ursachen Statt hat. Nach der Seite des Daseyns und ihrer Entstehung in demselben ist deßwegen die Aufsuchung der natürlichen Ursachen und die verständige Betrachtung ihrer Wirksamkeit hier an ihrem Platz, und eine solche Betrachtung ist es allein, die hier gültig seyn kann; Vorstellungsweisen, worin die Idee und natürliche Existenz ineinander gebraut sind, wenn sie nicht mythologisch sind und gar die Prätension haben, etwas philosophisches zu seyn, sind nur phantastisch und unwürdig, daß der Gedanke einige Rüksicht auf sie nimmt; denn es ist die Unfähigkeit, den denkenden Begriff zu fassen und von ihm sich leiten zu lassen, welche dergleichen phantastische Extravaganzen hervorbringt.

§

Das *Allgemeine, von welchem* die sich unterscheidende, individuelle Seele sich unterscheidet, ist zunächst das unmittelbare Seyn ihres in sich noch eingehüllten Lebens, welches zu einer Form, einem blossen *Zustande* derselben, als *Schlaff* herabgesetzt ist. Aber diß Allgemeine ist auf der andern Seite, die Substanz der Seele selbst, und so unterschieden von ihrer leeren Hülse, der Allgemeinheit als bloßer Form von Unmittelbarkeit oder Seyn, ist sie das *innere* Allgemeine, die concrete Natur der Seele, und im Verhältniße zu dem Unmittelbaren, welches die *unmittelbare Einzelnheit* der Seele ist, — die *Gattung* im Verhältniße zum *natürlichen Individuum* als solchem.

based on such an ancient and still wholly benighted consciousness of events and actions cannot get far. It is only the understanding which, by emphasizing the present variety and attempting in its own way to come to grips with it, has cast doubt upon the historical datum of descent from a single couple. Since such understanding involves the   5
natural influences and external causes operative in respect of the variety present, the field with which it is occupied is not that of philosophic thought, which seeks the varieties only in the determinateness of the Notion and is not concerned with their historical emergence and natural causes. At the same time however, the Notion of spirit   10
acknowledges its relationship to historical and unphilosophical research. The rationality of spirit, which involves the implicit absence of quality in its universality, stands by itself over and above these various qualities, for since they are particularities pertaining to natural determinate being, they have a natural origin. The field in   15
which they occur is therefore that of natural connectedness and the efficacy of natural causes. Here, therefore, it is in the aspect of determinate being and of the emergence of these qualities within it, that the tracing of natural causes and the understanding's consideration of their efficacy have their place, and only such a consideration can   20
have validity. Modes of presentation in which there is a mixing of the Idea and natural existence, if they are not mythological, and even pretend to be in some way philosophical, are simply fantastic, and are unworthy of being considered by thought; for it is the inability to grasp the thinking Notion and to allow oneself to be guided by it   25
which gives rise to such fantastic extravaganzas.

§

The *universal*, *from which* the self-distinguishing individual soul   +
distinguishes itself, is initially the immediate being of the soul's still internally enveloped life, which as *sleep* is reduced to a form, a mere *condition* of this soul. On the other hand however, this universal is the   30
very substance of the soul, and distinguished as such from the soul's empty shell, from universality as a mere form of immediacy or being, it is the *inner* universal, the concrete nature of the soul, and is in relationship with the immediate being constituting the soul's *immediate singularity*, — it is the *genus* in relationship with the natural *individual* as such.   35

§

Diß Verhältniß begründet den *Lebensproceß* aller lebendigen wie der geistigen Natur, weil es dessen Gegensatz, die innere substantielle Allgemeinheit und die unmittelbare Einzelnheit enthält. Er ist die Thätigkeit, die erste, nur unmittelbare Einheit zu dem Gegensatze zu bringen, und sie zu einer aus demselben sich hervorbringenden Einheit zu erheben, die unmittelbare Einzelnheit dem Allgemeinen einzubilden und gemäß zu machen, und ebendamit das Allgemeine in dem Individuum zu realisiren. Er ist die *Entwicklung* des Lebendigen überhaupt, und im Geistigen, als Seele oder unmittelbar existirendem Individuum die *Bildung*.

§

Dieser Proceß als am natürlichen Individuum erscheint in der Zeit, und die früher nur qualitative Unterschiede (§) als eine Reihe unterschiedner Zustände, in denen sich der Proceß zur Totalität vollendet. Er ist die Reihe der Lebensalter, welche mit der unmittelbaren noch unterschiedslosen Einheit der Gattung und der Individualität als einem abstracten unmittelbaren *Entstehen* der unmittelbar seyenden *Einzelnheit*, der Geburt, beginnt, und ebenso mit der Einbildung der Einzelnheit in die Gattung, welche hiebey als an der seyenden, hiemit der Allgemeinheit nicht adäquaten noch adäquat werden könnenden, nur als *Macht* erscheinen kann, — hiemit der abstracten Negation der Einzelnheit, dem *Tode*, endigt.

§

Was die Gattung am Lebendigen als solchem, ist die objective Vernünftigkeit im Geistigen, und weil jene gleichfalls schon innere Allgemeinheit ist, so entsprechen sich hier die anthropologische Erscheinungen der Entwicklung im Physischen und Intelligenten inniger. Allein die geistige Natur zeigt sich zugleich unabhängiger, und es finden sich eine Menge Ausnahmen, daß Kinder sich geistig früher entwickeln, als ihr Körper zu einer entsprechenden Ausbildung gelangt ist. Doch behauptet sich dabey auch das Sprichwort, daß der Verstand nicht vor den Jahren kommt.

Es sind vornemlich entschiedene künstlerische Talente, und besonders das musicalische, die sich oft durch die Frühzeitigkeit ihrer Erscheinung ausgezeichnet haben. Auch die Intelligenz hat durch Interesse und leichteres Auffassen von mancherley Kenntnißen, und einem verständigen Räsonnement darin, besonders im mathematischen

§

Since this relationship contains the opposition in the life-process   +
between inner and substantial universality and immediate singularity,
it is the foundation of the *life-process* of all that is of a living and spiritual
nature. This process is the activity which brings the initial and merely
immediate unity into the opposition and raises it to a unity which   5
brings itself forth from it, forming the immediate singularity within
the universal and making it adequate to it, and thereby realizing the
universal within the individual. It is the *development* of living being in
general. In spiritual being, as soul or the immediately existent individual, it is *cultivation*.   10

§

In the natural individual this process appears in time, while the   +
differences, which were formerly merely qualitative, (§   ) appear as a
sequence of different states, within which the process rounds itself off
into a totality. It is the sequence of the stages of life, which begins with
birth, in which the immediate and still undifferentiated unity of genus   15
and individuality has being as an abstract and immediate *emergence* of
the immediate being of *singularity*, and ends therefore with the merging
of singularity in the genus, the genus here only being able to appear as a
power in that which has being, and which is therefore not only not
adequate to universality but also incapable of becoming so, — with   20
the abstract negation of singularity i.e. *death*.

§

What the genus is in living being as such, objective rationality is in   +
spiritual being, and since the former is also already inner universality,
there is an intimate correspondence here between the anthropological
appearances of development in physical and intelligent being. At the   25
same time, however, the spiritual aspect shows itself to be more independent, exceptional cases, in which the spiritual development of
children outstrips the corresponding development of their bodies being
not uncommon. Nevertheless, the proverb to the effect that understanding never comes before its time is also borne out.   30   +
In the main it is predominantly artistic talents, and particularly
musical ones, which have most frequently distinguished themselves
in respect of precociousness. On account of an interest in and an
aptitude for acquiring various kinds of knowledge, together with an
effectively facile manner of reasoning, especially in the field of   35

Fache, selbst auch in den sittlichen und religiösen Gegenständen eine solche frühzeitige Stärke gezeigt. Evelyn...
Jedoch sind es vornemlich artistische Talente, wo die frühe Erscheinung eine Vorzüglichkeit angekündigt hat. Frühe Entwicklung allgemeinerer Intelligenz dagegen hat nicht etwa die Folge gehabt, daß solche Individuen im Jünglings- und Mannesalter vor mittelmässigen Talenten vorausgewesen und eine besondere Auszeichnung bewiesen hätten. Fertigkeit in Kenntnissen und im Räsonnement ist noch verschieden von dem Verstand im Charakter, sowohl dem intelligenten als dem praktischen, und solcher Verstand erfodert, daß der ganze Mensch fertig sey.

§

Der Proceß der Entwicklung des Individuums (§   ) hat näher zu seinem Ziele, daß einerseits dasselbe zu dem Gegensatze seiner Selbstständigkeit gegen das Allgemeine, als die an und für sich seyende, fertige und bestehende Sache komme, und andererseits derselbe so in ihm versöhnt sey, daß es in ihr seine wesentliche Thätigkeit und seine eigene Befriedigung allein zu finden, das Bewußtseyn habe. Die Entwicklung unterscheidet sich daher in die *drey* Perioden, 1) die der Entwicklung der zuerst nur natürlichen Einheit des Individuums mit seinem Wesen bis zu jener geistigen Vereinung, 2) die der objectiven Geistigkeit selbst, und dann 3) die der Rükkehr derselben zu der Interesselosen, die Thätigkeit darum aufgebenden Einheit, — das *Kindes*alter, das *Mannes-* und das *Greisen*alter.

§

Die physische Entwicklung beginnt mit dem Heraustreten, das ein Sprung ist, aus dem Zustande einer vegetativen, elementarischen Ernährung und gegensatzlosen Lebens überhaupt in den Zustand der Absonderung, des Verhältnißes zu Licht, Lufft, und einer vereinzelten Gegenständlichkeit, und durchs Athmen zunächst constituirt es sich zu einem Selbstständigen, welches die elementarische Strömung unterbricht, an einem einzelnen Punkte seines Organismus Speise einzieht, und ebenso Lufft einathmet und ausstößt.
Gegen das bloß quantitative Zunehmen und nur formelle Wachsthum, wozu die Vollendung der Knochenbildung überhaupt und insbesondere das Hervortreten der Apophysen der Rückenwirbel, zur Befestigung und Haltung der Rückenwirbelsäule gerechnet werden kann, ist die nächste qualitative Stuffe, daß das Kind *Zähne* bekommt,

mathematics, but even in respect of ethical and religious matters, intelligence has also displayed precocious ability of this kind. Evelyn...   +
It is, however, mainly in artistic talents that precociousness has been the forerunner of excellence; during their youth and manhood,   5
individuals whose more general intelligence has developed early do not seem to have been anything more than mediocre talents, or to have given proof of any particular distinction. Readiness in respect of knowledge and facile reasoning is certainly different from the intelligent and practical understanding involved in character, and   10
such understanding requires the maturity of the whole man.

§

More precisely considered, the one aspect of the goal of the process   +
of the individual's development (§   ), is that the individual should attain to the opposition of its independence to the universal as the being-in-and-for-self of the mature and subsistent factor; the other is   15
that this process should be so reconciled within the individual that it is only within the development that the individual finds its essential activity and proper satisfaction i.e. possesses consciousness. It is on account of this that the development falls into *three* periods: 1) that of the development of the initially merely natural unity of the individual   20
with its essence up to this spiritual unification, 2) that of objective spirituality itself, and then 3) that of the return of this spirituality to the unity which relinquishes activity on account of its being devoid of interest, — *childhood, manhood* and *old age.*

§

The physical development begins with emergence, which constitutes   25   +
a sharp change from the state of vegetative, elementary nutrition, of life in which opposition is generally absent, to that of separation, of relationship with light, air, and a singularized objectivity. The child first makes itself independent by breathing, interrupting the elementary flow, taking in food at a specific point in its organism, and inhaling   30
and exhaling the air.
As distinct from simply quantitative increase and merely formal   +
growth, which may be taken to include the general completion of bone formation, and especially the consolidation of the shape of the spinal column through the emergence of the spinal apophyses, the first   35
qualitative stage is the child's *teething, standing,* and learning to *walk,* by

sich *aufrecht* stellt, und zu *gehen* vermag, so daß es itzt seine Richtung nach Aussen und seine Individualität gewinnt.

Der *Knabe* ist zum *Jüngling* gereifft, indem bey Eintritt der Pubertät das Leben der *Gattung* in ihm sich regt, und seine Befriedigung sucht. Der Übergang des Jünglings oder Mädchens zum *Mann* oder *Frau*, besteht nur darin, daß die Bedürfnisse von jener Stuffe befriedigt seyen, nicht in neuen Bedürfnissen, und ist darum durch kein physisches Entwicklungs-moment bezeichnet; ausser daß etwa die vollendete Entwicklung der subjectiven Individualität, sich sträubend gegen ihr Aufgehen in der Allgemeinheit und Objectivität, noch ein Ansichhalten und Verweilen in leerer Subjectivität, — eine Hypochondrie zu bekämpfen hat. Diese Hypochondrie fällt meist etwa um das 27ste Jahr des Lebensalters oder zwischen dasselbe und das sechsunddreissigste; — sie mag oft unscheinbarer seyn, aber es entgeht ihr nicht leicht ein Individuum; und wenn dieses Moment später eintritt, zeigt es sich unter bedenklichern Symptomen; aber da es zugleich wesentlich geistiger Natur ist, und vielmehr nur von dieser Seite her zur körperlichen Erscheinung wird, kann sich jene Stimmung unter die ganze Flachheit eines Lebens, das sich nicht zum Momentanen concentrirt hat, vertheilen und hindurchziehen.

Indem nun aber das subjective Interesse und Princip der Thätigkeit befriedigt und sich in die objective Welt und physisch zunächst in seinen Organismus eingelebt hat, so löscht sich der bisherige Gegensatz der Lebendigkeit aus; und endigt sich in die Verknöcherung und Unerregsamkeit, und diese zur Unmittelbarkeit gediehene Allgemeinheit endigt mit dem Verschwinden der daseyenden, und nur durch den Gegensatz zu Interesse, Thätigkeit und Lebendigkeit erregten Einzelnheit.

Um von dem natürlichen Verlauffe des Geistigen durch seine Lebensalter bestimmter und concreter zu sprechen, muß, wie zu der Schilderung der Racenverschiedenheit, die Kenntniß der concretern Geistigkeit, wie sie in der Wissenschaft auf dem Standpunkte der Anthropologie sich noch nicht gefaßt hat, anticipirt und mit zur Unterscheidung der Stuffen genommen werden.

c. *Die empfindende Seele*

§

Die Seele is erstens bestimmt (§   ) aber zweytens ist sie zur Individualität bestimmt, und die Bestimmtheiten verändern sich zunächst an sich, so daß die Seele die allgemeine Substanz dieser Veränderungen

means of which it becomes orientated outwards and acquires its individuality.

The *boy* matures into the *youth* with the onset of puberty, in that the +
life of the *genus* begins to work and seek satisfaction within him. Since
the transition of the youth or girl into the *man* or *woman* consists not of 5
new needs, but simply of the satisfying of the needs of this stage, it is not
distinguished by any moment of physical development; except, perhaps
that the completed development of subjective individuality, in that it
resists being assimilated into universality and objectivity, still has to
struggle with a hypochondria, a reserve, a lingering in empty sub- 10
jectivity. This hypochondria usually occurs at about the age of twenty
six, or between then and the age of thirty five; — it may not often be
apparent, but it is not easy for an individual to avoid it, and if this
moment occurs later, the symptoms of it are so much the more serious.
However, since at the same time it is a moment which is essentially 15
spiritual in nature, and since it tends to be this aspect alone that gives
rise to the corporeal appearance, this is a general mood which can
disperse into and pervade the whole ineffectiveness of a life which has
not concentrated upon what is momentary.

The opposition in which animation has been involved now resolves 20
itself however, in that the subjective interest and principle of activity
has found satisfaction, initially physically in its organism, in accustoming itself to the objective world; it ends in ossification and inexcitability,
and this universality turned immediacy terminates with the disappearance of the singularity, which is a determinate being, and which is only 25
excited into interest, activity and animation through opposition.

As with the delineation of racial variety, a more specific and concrete +
account of what is spiritual throughout the natural course of the stages
of its life, requires that knowledge of a more concrete spirituality than
that which has yet been grasped scientifically from the standpoint of 30
anthropology, should be anticipated and made use of in the distinguishing of the levels.

c. *The sentient soul*

§

The soul is initially determined (§   ), but in the second instance it +
is determined into individuality, and at first the determinatenesses alter
implicitly, so that the soul is the universal substance of these alterations 35

und die Totalität der Bestimmtheiten ist. Die Wahrheit dieses Verhältnisses ist daher, daß die Bestimmtheit nicht durch eine andere
verändert wird, sondern in der allgemeinen Seele unmittelbar eine
aufgehobene, und diese darin in sich reflectirt ist, und so in ihrer
Allgemeinheit die Bestimmtheit negirend, erst als *für sich seyende*
Individualität, nicht mehr nur als Individualität *ansich* oder Zustand
bestimmt ist. Oder die Seele ist und bleibt diß allgemeine durch-
gängige Wesen, in dem alle Besonderheit aufgelöst; in ihrer Individualität aber ist solche Besonderheit nun gesetzt, und *für die Seele.*

§

Die Seele *empfindet,* nicht indem sie nur als wach einer Welt von
Bestimmtheiten sich gegenüber, sondern indem sie *sich* selbst bestimmt
findet. Sie ist selbst das Gedoppelte, Unterschiedne, einmal die *bestimmte*
Seele, und das andremal die *allgemeine,* aber indem diese und die
unterschiedene Seele eins und dasselbe ist, so ist sie in dieser Bestimmtheit bey sich selbst. Aber nicht nur ist auf diese Weise die Seele nur an
sich Eine, sondern daß die Bestimmtheit, als unterschieden von der
Allgemeinheit der Seele, und als *ideelle in ihr selbst* ist, dadurch ist die
Seele in ihrer Bestimmtheit für sich.

Wenn das neutrale Wasser, indem es z. B. gefärbt, und so nur in
dieser Qualität oder Zustand ist, nicht nur *für uns* oder was dasselbe
ist, der Möglichkeit nach, von diesem seinem Zustande unterschieden,
sondern selbst von sich als so bestimmtem, zugleich unterschieden
wäre, so würde es empfindend seyn. Oder die Gattung Farbe existirt
nur als blaue oder als irgend eine bestimmte Farbe; sie bleibt die
Gattung Farbe, indem sie blau ist. Wenn aber die Farbe als Farbe,
d. i. nicht als Blau, sondern zugleich als Farbe gegen sich als blaue
Farbe bliebe, — der Unterschied ihrer Allgemeinheit und ihrer Besonderheit nicht bloß *für uns* sondern in ihr selbst existirte, so wäre sie
Empfindung des Blauen.

§

Die Bestimmtheit oder der Inhalt der Empfindung ist noch ein
Seyendes; die Seele *findet* sich so oder anders afficirt. Daß die Seele *sich*
so bestimmt findet, diß ist es, daß die Bestimmtheit zugleich als *ideell* in
der Seele gesetzt, nicht eine Qualität derselben ist, und indem die
Idealität dieser Bestimmtheit nicht eine andere Bestimmtheit, die an
deren Stelle träte und die erstere verdrängte, sondern daß die Seele
selbst die Idealität dieser Bestimmtheit, das in ihr, dem Endlichen, in
sich reflectirt, d. i. unendlich ist, ist diese Bestimmtheit auch nicht ein

and the totality of the determinatenesses. The truth of this relationship is therefore that the determinateness is not altered by means of another, but is an immediate sublatedness within the soul in general, which in this is intro-reflected. Consequently, in that in its universality the soul negates the determinateness, it is no longer determined merely as   5
an *implicit* individuality or condition, but for the first time as an individuality which *is for itself*. Differently expressed: although the soul is and remains this universal and permeating essence within which all particularity is dissolved, in its individuality such particularity is now posited, and has being *for the soul*.   10

§

The soul does not *sense* through simply being awake and finding a   +
world of determinatenesses over against itself, but in that it determinately *finds itself*. It is itself double, differentiated, being both the *determinate* soul and the soul *in general*, although since the general and the differentiated soul are one and the same, it is with itself in this   15
determinateness. This does not merely mean that the soul is only implicitly a unit however, for since the determinateness differs from the universality of the soul, and, being *of an ideal nature*, is *in itself*, the soul in its determinateness is a being-for-self.
If neutral water is coloured, for example, and so simply possesses this   20   +
quality, is simply in this condition, it would be sentient if it were not only *for us* or as a matter of possibility that it differed from the condition, but it were at the same time to distinguish itself from itself as being so determined. Differently expressed: the genus colour only exists as blue, or as a certain specific colour; in that it is blue, it   25
remains the *genus* colour. But if the colour as colour, i.e. not as blue, but at the same time as colour, persisted in opposition to itself as blue colour, — if the difference between its universality and its particularity were not simply *for us* but existed within itself, it would be sensation of blue.   30

§

The determinateness or content of sensation is still a being; the soul   +
finds itself affected in one way or another. This finding of *itself* to be so determined is not a quality of the soul, but that which at the same time posits the determinateness in the soul as being *of an ideal nature*. Nor is this determinateness a mere condition, for its ideality is not another   35
determinateness, replacing and driving out the first, but consists of the soul itself being the ideality of this determinateness, i.e. infinite, in that

blosser Zustand. — Die Seele ist somit *freye* Lebendigkeit in der Empfindung, und zugleich als *seyend* bestimmtes, als abhängiges. Der Inhalt
der Empfindung ist ein Gegebenes, und die Empfindung selbst ist der
Widerspruch, der Reflexion der Seele in sich selbst, und der Äusserlich-
keit derselben; — ein Widerspruch, der in der Empfindung noch nicht
aufgelöst, sondern seine Auflösung in einer höhern Weise der Seele hat.
Die Endlichkeit einer Existenz es sey einer natürlichen oder geistigen
besteht in einem Widerspruche, der sie in sich selbst ist, und es ist
wesentlich diß überhaupt aber vornemlich den bestimmten Widerspruch einzusehen, der die Natur einer bestimmten Existenz ausmacht. Die Empfindung ist diese erste Gestalt, in welcher die Seele
als concret, als Individualität, oder somit eigentlich erst Seele ist.
Aber die Empfindung ist eben darum zugleich diese ganz untergeordnete Weise der Seele, weil sie dieser unittelbare Widerspruch
ist, das ganz Freye und zugleich als *seyend* bestimmt zu seyn, so daß
dieser Inhalt der Empfindung noch ganz unversöhnt, noch auf keine
Weise geistiger Inhalt ist. Der Widerspruch der Empfindung allein
ist es, welcher den Geist aus derselben hinaus, oder vielmehr dazu
treibt, sie aufzuheben, wie alles Höhere nur dadurch entsteht, daß das
Niedrigere sich als Widerspruch in sich zu dem Höhern aufhebt.
Diejenigen, welche die Empfindung oder das Gefühl für die wahre
Weise des Geistigen, und damit für die Weise, in welcher die Wahrheit für den Geist ist, halten, haben über das, was die Natur der
Empfindung ist, so wie überhaupt über das, was Geist und Wahrheit
ist, noch wenig nachgedacht.

§

Weil das was einen Inhalt zum Inhalt der Empfindung macht, als ein
Seyendes ist, ist es in vollkommener *qualitativer Beschränktheit*, eine
unmittelbare Einzelnheit. Ein solches Beschränktes ist aber nur so, daß
seinem Andern ebenso gut das Seyn zukommt; es ist ein Daseyn, das nur
den Werth der Möglichkeit hat, — *ein Zufälliges*. Diß macht die
Zufälligkeit der Empfindung überhaupt aus, und die Empfindung
heißt darum auch etwas bloß Subjectives, weil die Seele als empfindend
überhaupt in beschränkter Qualität sich befindet, und darum sich
nach unmittelbarer Einzelnheit verhält.
Die Subjectivität der Empfindung bedeutet die Beschränktheit und
Zufälligkeit derselben im Gegensatz gegen das *Objective*, den Inhalt,
insofern er *an und für sich* selbst ist; seine Wahrheit besteht darin,
daß er als blosser Inhalt in sich selbst, dessen Realität mit seinem Begriffe zusammenstimmt. Solches Wahre kann und ist denn auch...
Das Wahre, Gewußte oder Gewollte, muß wesentlich ebenso ein sub-

it is intro-reflected in the finite being of this determinateness. — In sensation therefore, the soul is *free* animation, and at the same time a determinate, a dependent *being*. The content of sensation is something given, and sensation itself is the contradiction between the intro-reflection of the soul and its externality; — the resolution of which 5 takes place not in sensation, but in a higher mode of the soul.

A natural or a spiritual existence is finite in that it is in itself a contradiction, and it is essentially the awareness of this in general, but especially of the determinate contradiction, which constitutes the nature of a determinate existence. Sensation is this primary shape, 10 within which the soul is first properly the soul in that it is a concrete individuality. At the same time however, since sensation is this immediate contradiction of being that which is wholly free and yet determined as *being*, so that this its content is still wholly unreconciled and in no respect a matter of spirit, it constitutes this wholly 15 subordinate mode of the soul. It is solely sensation's contradiction which drives spirit out of it, or rather to sublate it, for all that is higher only occurs in that what is lower sublates itself into what is higher through internal contradiction. Those who maintain that sensation or feeling is the true mode of what is spiritual, and therefore the 20 mode in which truth has being for spirit, have given as little consideration to the nature of sensation as they have to that of truth and of spirit in general.

§

Since that which makes a content the content of sensation is as a *being*, it is an immediate singularity in its complete *qualitative limitation*. 25 However, such a limited being only has being in that being also pertains to its other; it is a determinate being, which merely has the status of a possibility, — *a contingency*. It is this that constitutes the general contingency of sensation, and sensation may also be said to be something merely subjective, for in that the soul senses, it generally finds itself to 30 be qualitatively limited, and therefore conducts itself in accordance with immediate singularity.

The subjectivity of sensation has the significance of being its limitation and contingency in the face of *objective being* or the content, in so far as this content is *in and for itself*; the truth of this content 35 consists of its according within itself merely as content, the reality of which accords with its Notion. Such a truth can... and is then also... In essence, what is true, known or willed must belong to the sub-

jectives, der Intelligenz oder dem Willen angehöriges, seyn, als es seinem Inhalte nach objectiv ist. Aber eine solche Subjectivität, wie sie vernünftige Einsicht und vernünftiger Wille ist, ist eine ganz andere Subjectivität, als die der blossen Empfindung; diese letztere ist eben die nur ganz abstracte Subjectivität, welche der Seele in ihrer noch ungeistigen nur *unmittelbaren Einzelnheit* zukommt, und einen wahren ebensowohl als einen falschen, einen guten sowohl als einen schlechten Inhalt haben kann.

§

Die Empfindung ist zunächst überhaupt unendlich *mannichfaltiger* Art, weil der Inhalt derselben eine *seyende Bestimmtheit* ist, dieser aber zunächst den formlosen Unterschied, die vielfache Mannichfaltigkeit an ihr hat. Indem die Empfindungen nach diesem ihrem mannichfaltigen Inhalte betrachtet werden, so wird von demselben, die Form, nach der er Empfindung ist, weggelassen, und es wird also von den Bestimmtheiten in ihrer sonstigen objectiven Form die Rede.

§

Insofern dabey die Empfindung nach dem Gegenstande, innerem oder äussern, von welchem sie erregt werde, bestimmt wird, so liegt in dieser Betrachtung ein Unterschied von dem Empfindenden und den Empfundenen, dem fühlenden Subjecte und gefühlten Objecte, so wie ein Verhältniß, so daß das Object einen *Eindruck* auf das Subject mache, dieses von dem Gegenstande *afficirt* werde, der Gegenstand Ursache oder Erregendes u. s. f. sey. Alle diese Unterschiede aber gehören noch nicht dem Standpunkte der Empfindung selbst an, sondern einer spätern Reflexion der Seele, insofern sie sich weiterhin zu *Ich* und dann zum *Geiste* bestimmt hat, — oder wenn *wir* sogleich bey der Empfindung sie nach solchen Unterschieden betrachten, *unserer äussern* Reflexion. Wenn ich sage: ich fühle *Etwas* Hartes oder Warmes, oder sehe *Etwas* Rothes, oder ich habe ein Gefühl von *Recht* und *Unrecht*, so gehört diese Unterscheidung meinem Bewußtseyn oder Reflexion an, welche die Unterscheidung von subjectiver Empfindung und deren Gegenstand macht, — eine Unterscheidung, welche der Empfindung als solcher noch nicht angehört. — Ich empfinde Freude, Schmerz, Zorn u. s. f. ist insofern ein pleonastischer Ausdruck, als Freude, Schmerz, Zorn u. s. f. selbst Empfindungen sind, und dieser Ausdruk spricht nur zuerst mein Empfinden überhaupt und dann die besondere Empfindung aus, die ich habe.

jectivity of the intelligence or the will to the same extent as it is in its content objective. A subjectivity such as rational insight and rational will is, however, wholly distinct from that of mere sensation; it is precisely this latter subjectivity which is merely the wholly abstract one pertaining to the soul in its unspiritual and simply *immediate*   5
singularity, and which can have a true as well as an erroneous, a good as well as a bad content.

§

Initially, sensation is generally of an infinitely *manifold* kind, for its content is the *being* of a *determinateness*; even initially however, this determinateness includes formless difference, variegated multiplicity.   10
In that sensations are regarded in respect of this their manifold content, the form in accordance with which this content is sensation is left out, and it is therefore the determinatenesses in their remaining objective form which are being spoken about.

§

In so far as this involves sensation's being determined in accordance   15
with the internal or external general object by which it is excited, such a consideration involves a difference between what senses and what is sensed, the feeling subject and the felt object, as well as a relationship in which the object makes an *impression* upon the subject, which is *affected* by the general object, the general object being the cause or   20
exciting principle etc. All these differences do not yet pertain to the standpoint of sensation itself however, but to a subsequent reflection of the soul in so far as it has determined itself further as *ego* and then *spirit*, — or to *our external* reflection if *we* consider the soul directly in accordance with such differences, as sensation. If I say that I feel   25
something hard or warm, or see *something* red, or have a feeling of what is *right* and *wrong*, this distinguishing pertains to my consciousness or reflection, which distinguishes between subjective sensation and its general object, — a distinction which is as yet absent in sensation as such. — In so far as joy, pain, rage etc. are themselves sensations, it is   30
pleonastic to speak of one's sensations of joy, pain, rage etc., for one is merely speaking of one's sensing in general, and then of one's particular sensation.   +

§

Indem das, was ich empfinde, als *Seyendes* in mir ist, welchen Inhalt auch dasselbe weiter an sich habe, so bin ich als empfindend, nur als *Seele* bestimmt. *An sich* ist die Seele *Ich*, *Geist*; aber die Unterschiede als Seele, Ich und Geist betreffen eben die unterschiedene Bestimmtheit, in welcher diß Ansich existirt. Die Seele aber überhaupt, oder der noch als *seyend* bestimmte Geist, ist noch der Geist in Leiblichkeit, und die Empfindung ist daher unmittelbar zugleich ein *Leibliches*. Die Empfindung gehört der noch *unmittelbaren Einzelnheit* des Geistes an, und diß ist die nähere Bestimmung der Subjectivität (§), die der Empfindung zukommt.

§

Die Bestimmtheit der Empfindung ist noch als eine *unmittelbare* Affection in der Seele; der Geist selbst damit noch als unmittelbarer Geist bestimmt. In dem Empfinden als solchen ist daher die Seele noch nicht *frey*. Selbst im Gefühle der Freyheit ist die Seele nach der Seite unfrey, nach welcher sie die Freyheit fühlt; diese Seite der Unmittelbarkeit ist es deßwegen, an welcher alle Zufälligkeiten und Particularitäten des Subjects in die Freyheit sich einmischen. — Ferner aber ist die Unmittelbarkeit des Geistes als empfindenden, in ihrem bestimmten Sinne zu nehmen, sie ist die *Leiblichkeit*; die Empfindung muß daher wesentlich als *leibliches* gefaßt werden. Welchen Inhalt die Empfindung sonst auch habe, zum Beyspiel auch wenn sie religiöse Empfindung ist, ist sie unmittelbar zugleich in einer Leiblichkeit.

§

Die Empfindung, weil sie leiblich ist, ist insofern *animalisch*. Aber ein Anderes ist die Animalität des Thieres, welches nicht Mensch ist, und ein anderes die Animalität des Menschen. Die anthropologische Betrachtung kann deßwegen nicht bey der Animalität des Empfindens stehen bleiben, sondern hat dasselbe als Empfinden der Seele zu fassen, und deßwegen als *zweyseitig* zu erkennen. Nemlich es ist vorhin (§) zwischen der *bestimmtseyenden* Seele und zwischen der Seele als *allgemeiner*, für welche jene ist, unterschieden worden. In der Seele tritt diese Unterscheidung erst in dem Empfinden ein, und sie ist es, welche zugleich schon in dieser Sphäre die Seele des Thieres von der geistigen unterscheidet.

§

In that what I sense is in me as a *being*, and this being is also implicit within this content, as a sentient being I am determined merely as *soul*. The soul is *implicitly ego, spirit*; but as soul, ego and spirit, the precise relatedness of the differences is to the differentiated determinateness in which this implicitness exists. In corporeity however, the soul in general, 5
or spirit which is still determined as *being*, is still spirit, and sensation in its immediacy is therefore at the same time *corporeal*. Sensation belongs to the still *immediate singularity* of spirit, which is the more precise determination of the subjectivity (§   ) proper to sensation.

§

Since the determinateness of sensation is still determined as an 10
immediate affection within the soul, spirit itself is still determined as being immediate. In sensing as such therefore, the soul is not yet *free*. Since even in the feeling of freedom, the soul is not free in that aspect of it which feels, it is in this aspect of immediacy that all the contingencies and particularities of the free subject intermingle. — In that the 15
immediacy of spirit is sentient moreover, it is also to be considered in its specific sense as *corporeity*, and sensation has therefore to be grasped as being essentially *corporeal*. Although sensation may have further content, that of being religious for example, in its immediacy it is at the same time located in a corporeity. 20

§

Sensation is *animal* to the extent that it is corporeal. The animality of the non-human creature is not the same as that of man however, so that anthropological consideration has to pass beyond the animality of sensing to grasp the sensing of the soul, and so recognize that sensing has *two aspects*. A distinction was previously drawn (§   ) between the 25
determinate and the *universal being* of the soul, the former of which has being for the latter. The difference first occurs in the soul in sensing, and is that which, already in this sphere, distinguishes the animal soul from that of spirit.

§

Die Empfindung überhaupt ist zwar die Rüknahme und Aufheben der unmittelbaren Wirklichkeit der organischen *Einzelnheit* in der *Allgemeinheit* oder Gattung, so daß die Einzelnheit nunmehr als concretes Moment der Allgemeinheit ist (Encyklop. der philos. Wissensch. § 273. u. 276). Aber im Thiere ist und bleibt diese *Einheit* des Individuums und der Gattung selbst in ihrer Unmittelbarkeit, und die *Gattung* ist nicht *für sich* in ihrer einzelnen Bestimmung, oder die bestimmte Seele ist nicht *für* die *allgemeine* Seele. Die geistige Seele aber ist eben diß, *als allgemeine für sich zu seyn*.

Diß Fürsichseyn der allgemeinen Seele aber ist *zunächst abstract*; — hier nemlich, wo sich noch keine Bestimmtheit in diesem allgemeinen Medium gesetzt hat. Dieser ideelle Raum ist daher noch unbestimmt und leer, — er ist die tabula rasa, welche erst erfüllt werden soll und die als die abstracte Idealität zugleich absolut weich genannt werden kann. Aber freylich wird diß Erfüllen nicht durch sogenannte Eindrücke von aussen geschehen, etwa in der Weise wie durch ein Petschaft Bilder auf Wachs abgedruckt werden. Was in dem Geiste zur Existenz kommen kann, kann nur so in ihn kommen, daß er dasselbe selbstbestimmend in sich setzt. Die Empfindung ist daher als Empfinden der geistigen Seele das Zweytheilige, das einemal als *Affection* zu seyn, welche der empfindenden, individuellen Seele überhaupt angehört, das andremal aber...

γ) Die reale Individualität der empfindenden Seele

§

Die Empfindungen, sowohl äussere als innere, sind bestimmte, und zunächst als der formellen Individualität, dem Empfindenden überhaupt angehörig schließen sie sich gegenseitig aus, verdrängen einander und sind so in der Zeit spurlos verschwindende äussere Begebenheiten an dem Subject. Die Seele aber ist nicht seyende, unmittelbare, sondern *allgemeine* Substanz, somit ist sie in sich das *Bestehen* des Mannichfaltigen, und nicht ein bloßes Durchlauffen von seyenden Empfindungen, sondern das Aufbewahren von ideell gesetzten. Denn die blosse, abstracte *Negation* des Seyenden wird in der Seele zu einem Aufgehobenen als *aufbewahrten*; — ein Übergang, der im Begriffe und zeitlos ist, und bey welchem daher auch die Bestimmung des Seyenden als eines *Itzt*, und desselben, insofern es ideell ist, als eines *Vergangenen* und *Gewesenen* nicht das Wesentliche, sondern vielmehr das erst in der

§

Sensation in general certainly consists of the immediate actuality of organic *singularity* being so taken back and sublated within the *universality* or genus, that the singularity now has being as a concrete moment of the universality ('Encyclop. of the philos. Sciences' § 273 and 276).   +
In the animal however, this *unity* of the individual and the genus itself   5
is in its immediacy, and remains so, — the *genus* is not *for itself* in its single determination, the determinate soul is not *for* the *universal* soul. The spiritual soul, is, however, precisely its *being for itself as a universal*.

In the *first instance*, however, no determinateness having as yet posited itself here within this universal medium, this being-for-self of   10
the universal soul is *abstract*. The ideal nature of this space is therefore still indeterminate and empty, — it is the tabula rasa which first has to be filled, and which as abstract ideality may at the same time be said to be absolutely receptive. It is certainly the case however, that this filling does not take place through so-called impressions from without   15
in the way that a signet impresses images upon wax. That which can   +
come into existence in spirit can only do so in that spirit self-determiningly posits it within itself. As the sensing of the spiritual soul, sensation therefore has the two aspects of being firstly the *affection* which pertains in general to the individual sentient soul, while secondly it is...   20

γ) The real individuality of the sentient soul

§

Both external and internal sensations are determinate, and initially   +
it is in that they pertain to the formal individuality, the generally sentient being, that they are mutually exclusive, displace one another, and in respect of the subject are therefore external occurrences disappearing in time without trace. However, since the substance of the soul is not a   25
being or an immediacy but *universal*, it is in itself the *subsisting* of what is manifold, not a mere traversing of sensations which are, but a preserving of sensations which are posited as being of an ideal nature. This is because in the soul, the naked and abstract *negation* of that which has being is *preserved* and so sublated. Since this is a transition within the   30
Notion and is timeless, it also involves the determination of what *is* having being as a *now*, and in so far as this being is of an ideal nature, as being not what is essential, but a *past* and a *has-been*, as being that which

weitern Form des äusserlichen Sinnlichen Hinzukommende ist. Die Seele ist als diese insichseyende Allgemeinheit des Bestimmten der *unendliche Raum*, in welchem der Inhalt unmittelbar als aufbewahrter ist; der Durchgang einer Affection zur eigentlichen Erinnerung, welcher durch das Bewußtseyn und Anschauung eines äusserlichen Gegenstandes *vermittelt* ist, hat hier noch nicht seine Stelle, sondern gehört einer entwickeltern Stuffe des Geistes an. — Dieser so in der Seele ein Bestehen erhaltende Inhalt der Affectionen gehört nun zu dem *eigenen Bestimmtseyn* der Seele, wie die Bestimmungen, welche die Grundlage der Triebe, Neigungen u. s. f. überhaupt der innern Empfindungen der Seele ausmachen; und daß dieser Stoff als Inhalt empfunden werde oder aus dem *Ansichseyn* in das *Fürsichseyn* der Seele heraustrete, ist ein Übergang und vermittelnde Thätigkeit, welche erst später als *reproducirende* Thätigkeit des Geistes überhaupt, zu betrachten ist.

§

Ferner nun sind die Empfindungen, wie sie als Arten bestimmt worden sind, beschränkte, qualitativ unterschiedene, auseinander fallende Bestimmungen. Die Seele aber ist individuell überhaupt und das Mannichfaltige, das an sich zum Kreis der Totalität gehört, in ihr zur Einheit verbunden. Der Inhalt ist an ihm selbst nur das Concrete jener vereinzelten Bestimmungen und diese *ansichseyende* Verknüpfung macht die *Objectivität* desselben aus. Die Seele ist jedoch noch nicht als *Subject* und nicht als Geist bestimmt, darum ist der Inhalt *für dieselbe* noch nicht in einer eigenthümlichen Objectivität, d. i. entwickelten und in ihre Bestimmungen zugleich ausgelegten Einheit (wovon nachher noch näher); die Seele hat aber überhaupt die Bestimmungen der Empfindung als einen zum Concreten vereinigten Inhalt in ihr, und was sie aus sich reproducirt, sind solche Ganze von Inhalt.

§

Dieser concrete Inhalt hat hier noch keine der weitern nähern Bestimmungen, welche er daher erhalten wird, daß er durch das Bewußtseyn und den Geist hindurchgegangen und durch sie gebildet worden wäre. Er ist aber auch nicht nur irgend ein Inhalt, sondern *an sich* der allgemeine Inhalt, aber zugleich für die einzelne Seele individualisirt; die ganze — zunächst noch *zukünftige* — Welt des Individuums liegt in seiner Seele. Aber diß, was noch in ihr eingehüllter Stoff ist, wird ihm erst durch das Bewußtseyn und die Thätigkeit des Geistes als seine Welt vorgeführt werden.

accrues first in the further form of external sensuous being. As this universality of determinateness which is in itself, the soul is *infinite space*, in which content has being immediately, as that which is preserved; this is not yet the place for an affection's passing into recollection proper, for this is *mediated* by consciousness and the intuition of an 5 external general object, and belongs to a more developed stage of spirit. — This content of affections, which acquires a subsistence within the soul, now pertains to the *proper determinateness* of the soul, as do the determinations which constitute the basis of impulses, inclinations etc., i.e. of the broadly internal sensations of the soul. The sensing of this 10 material as content, or its emergence from *implicitness* into *being-for-self*, is a transition and a mediating activity which does not have to be considered until later, is a *reproducing* activity of spirit in general.

§

Now sensations which have been determined as species are also determinations which are limited, which are qualitatively distinct, and 15 which fall apart. The soul in general is individual however, and the manifoldness which implicitly pertains to the circle of totality is bound up within it into a unity. The content by itself is merely the concrete being of these singularized determinations, and this *implicit* linking together constitutes the *objectivity* of this content. But since the soul is 20 not yet determined as *subject* and as spirit, the content does not yet have being *for it* in a proper objectivity i.e. in a developed unity which is at the same time displayed in its determinations (more of this later); for in general, the soul has the determinations of sensation within itself as a content united into concrete being, and what it reproduces from 25 out of itself consists of the wholes of this content.

§

At this juncture, this concrete content still has none of the further and more precise determinations which it would possess had it been permeated and formed by consciousness and spirit. Nevertheless, it is not simply any content, but *implicitly* the universal one, which is however 30 at the same time individualized for the single soul; the whole of the individual's world, which is at first still in the *future*, lies in his soul. It is however through consciousness and the activity of spirit that this material still enveloped within the soul is first brought before the individual as his world. 35

§

Insofern das Individuum noch als empfindendes existirt, ist noch nicht an den Unterschied eines objectiven Daseyns und äusserer oder innerer, überhaupt gegebener Dinge, gegen die Subjectivität der Seele zu denken. Die Empfindungen sind seyende Affectionen, ob ihre Bestimmtheit späterhin als durch ein Object veranlaßt, als Eindruck von Aussen, oder ob als durch vorhandene innere Affectionen bewirkt angesehen werde.

§

Indem vorhin das Empfinden überhaupt betrachtet worden ist, so ist nunmehr das Empfindende als Individuum bestimmt, welches zu betrachten ist, und es ist zunächst die Bestimmung anzugeben, welche in das Empfinden kommt, dadurch daß es ein individuelles ist; wie der Stoff des Empfindens durch die Individualität der Seele bestimmt wird ist so eben angegeben worden. Das Empfinden aber als der individuellen Seele ist, daß sie als empfindend *für sich selbst* ist, — d. i. daß sie sich empfindet, und in dieser Unterscheidung zwar, aber darin in unmittelbarer Beziehung auf sich, und bey sich ist.

§

Die *sich* empfindende Seele aber ist bestimmt und beschränkt überhaupt, weil sie nur erst auf unmittelbare Weise ist, näher aber ist sie ihrer *selbst nicht mächtig*. Daß sie frey und ihrer mächtig wäre, dazu gehört, daß ihr Inhalt und ihre natürlichen Bestimmungen sich in ihr als ideell bestimmt hätten, und sie abstracte, bestimmungslose Beziehung auf sich selbst, — *als Ich wäre*. Hiemit wäre verbunden, daß sie ihre Bestimmungen von sich abgetrennt, sie ausser ihr selbst gesetzt hätte, und daß sie als *andere* denn sie ist, als für sich seyende Objecte ihr gegenüberstünden. So wäre sie *Bewußtseyn*, das abstracte Ich, für welches der Inhalt als für sich seyender Gegenstand, als eine vorhandene Welt, ist. Daß die Seele ohne Freyheit, und daß ihr Inhalt ohne seine von ihr unterschiedene Objectivität an ihm selbst ist, ist eins und dasselbe. Erst als *Bewußtseyn*, nur als diese Negativität ihrer Bestimmungen, das abstracte nur bey sich seyende *Ich*, ist die *Macht* über dieselben, welche sie von sich ausgeschlossen hat. Aber nicht frey und mächtig *ihrer* selbst ist die Seele, insofern ihr noch unabgeschiedene, unmittelbar ihr immanente Bestimmungen zukommen, insofern sie somit überhaupt noch auf unmittelbare, natürliche Weise existirt.

§

In so far as the individual still exists as a sentient being, there can as yet be no thought of the subjectivity of the soul being confronted with the difference of an objective determinate being and of external or internal things which are generally given. Sensations are the being of affections, irrespective of whether their determinateness is subsequently   5
regarded as being occasioned by an object, as an impression from without, or as produced by the presence of internal affections.

§

Since sentience in general has already been considered, what now comes under consideration is the individuality of the sentient being, and the first determination to be adduced is that which occurs within   10
sentience on account of its being individual. The way in which the material of sentience is determined through the individuality of the soul has just been indicated. In that the individual soul is sentient however, it senses as a *being-for-self*, — i.e. it senses itself, and while certainly differentiated within this difference, is immediately self-   15
related and with itself.

§

The *self*-sensing soul is generally determined and limited however, since in the first instance it only has being in an immediate manner, and, moreover, is *not in control* of itself. In order to be free and in control of its own, it would have to have its content and its natural determina-   20
tions determined within it as of an ideal nature, and *be* abstract, indeterminate self-relation i.e. *ego*. This would involve its separating its determinations from itself, its having posited them outside itself, as well as its standing over against itself as the other of what it is, as objects which are for themselves. It would then be *consciousness*, the abstract   25
ego, for which the content constitutes a general object which has being for itself as a world which is present. The soul's being devoid of freedom is the same as its content by itself being devoid of its objectivity in respect of the soul. *Power* over that which the soul has excluded from itself occurs first as *consciousness*, as simply this negativity of the soul's   30
determinations, the abstract ego which is merely with *itself*. However, the soul is not free and in control of *itself*, in so far as in that it still accommodates determinations which are unexcluded and immediately immanent within it, it still exists in a generally immediate and natural manner.   35

§

Es ist gerade um der noch unmittelbaren Einheit der erst empfinden-
den Individualität mit sich, daß die Seele in dieser Form als *subjective*
Seele zu bestimmen ist, — zum Unterschiede von der Objectivität des
Bewußtseyns und dann des Verstandes.

Es ist eine alte Vorstellung, daß der primitive Zustand der Menschen
als ein Zustand der Unschuld, oder als ein goldenes Zeitalter von
einfacher Lebensweise, einfachen, genügsamen, von Leidenschaften
freyen Sitten aufgefaßt wird. Dieser Vorstellung ist in neuerer Zeit
zuerst als einer geschichtlichen Theorie, die nachher von da auch
in die Philosophie überging, die Bedeutung gegeben worden, daß
dieser Zustand ein geistiger Zustand sowohl der Reinheit des Willens
als eines ungetrübten Durchschauens der innern Lebendigkeit der
Natur, und eines klaren Anschauens der göttlichen Wahrheit
gewesen sey. So daß der spätere Aufgang des Bewußtseyns, alle
Kenntniße von Gott und von Pflichten seiner Verehrung, wie von
den Gesetzen der Natur, einerseits nur eine Trübung und Verderben
jenes göttlichen Lebens und Schauens gewesen, andererseits alles,
was unter solchem Vorkommen, noch von höherem Inhalt und
Wissen sich zeige, nur nachgelassene Trümmer und Spuren aus
jener ersten Reinheit und Klarheit seyen. Es soll diese Vollkommenheit wesentlich nicht als eine selbstbewußte Sittlichkeit des vernün-
ftigen Willens, noch als eine gedachte und wissenschaftliche Einsicht
in die Gesetze der Natur und des Geistes, noch als ein begreiffendes
Erkennen des göttlichen Wesens bestimmt werden, sondern im
Gegentheil ist dasjenige, wodurch solcher Zustand ein Leben in der
Wahrheit sey, eben die noch ungetrennte Einheit des intelligenten
und natürlichen Lebens, des Denkens und Empfindens. — So leicht,
faßlich und selbst anmuthig sich solche Ansicht für die Vorstellung
macht, so zeigt sie sich doch bey näherer Betrachtung, nicht nur
oberflächlich zu seyn, sondern selbst auf der gänzlichen Verkennung
der Natur des Geistes, auf der Verkennung des Begriffes überhaupt
zu beruhen. Denn der Begriff, und dann der als Begriff existirende
Begriff, der Geist, *ist* nur, insofern die durch Aufheben der Unmittelbarkeit für sich seyende Idee ist. Die unmittelbare Idee überhaupt
ist nur die Natur, und der unmittelbare Geist, nur der schlaffende,
nicht der selbstbewußte, noch weniger der wirklich denkende,
wissende und erkennende Geist. Die Natur aber in ihrer Wahrheit
ist sie die an sichseyende *Idee*, das Leben des *Allgemeinen* in sich. Aber
eben das Allgemeine ist nicht das Unmittelbare des Daseyns; die
Natur, wie sie in ihrer Unmittelbarkeit ist, bietet sie das Schauspiel
der sinnlich bunten Welt dar. Sinnliches Daseyn heißt nichts anderes,

§

It is precisely on account of the sentient individuality's unity with +
itself still being immediate, that the soul in this form has to be determined as being *subjective*, — as distinct from the objectivity of consciousness and subsequently of understanding.

There is a long tradition of men's primitive state being presented as 5 +
one of innocence, or conceived of as a golden age, an uncomplicated
way of life, a state of simple and sober manners, undisturbed by the
passions. In more recent times this presentation has been taken over
from historical theory into philosophy, the state being given a
spiritual interpretation, and regarded as having been one of purity of 10
will, unclouded insight into the inner life of nature, and a clear
intuition of divine truth. There are two aspects here therefore, — the
subsequent dawning of consciousness, all cognition of God and of
the duties involved in worshipping Him, like all cognition of the laws
of nature has simply been a sullied perversion of that divine life and 15
vision. On the other hand, all which still gives evidence of a higher
content and knowledge in circumstances such as these, consists of
nothing but the surviving remnants and traces of this initial purity
and brightness. — Rather than this perfection being determined as
essentially the self-conscious ethics of the rational will, as a thought- 20
out and scientific insight into the laws of nature and spirit, or as a
comprehending cognition of the Divine Essence, that through which
such a condition constitutes a living within truth is supposed to be
merely the still undivided unity of intelligent and natural life, of
thinking and sensing. — Although such a view is easily grasped, and 25
therefore recommends itself to the presentative faculty, if it is more
closely considered it not only shows itself to be superficial, but to be
founded on a complete misunderstanding of the nature of spirit, on a
miscomprehension of the Notion. The Notion, and the Notion existing
as Notion, or spirit, only has *being* in that the Idea which has being- 30
for-self through the sublation of immediacy has being. The immediate Idea in general is merely nature, and immediate spirit is merely
dormant spirit, not spirit which is self-conscious, and certainly not
spirit which actually thinks, knows and cognizes. Although nature
in its truth is the implicit being of the *Idea* however, the *universal's* 35
living within itself, it is precisely the universal which is not the
immediacy of the determinate being, for nature as it is in its immediacy presents the pageant of the sensuously variegated world.
Sensuous determinate being is nothing other than the... in the

als die in das Aussersichseyn des Begriffs..., der in die Verworren-
heit und Vergänglichkeit der Erscheinung verlorne Begriff. Wenn
aber der Geist sich *anschauend* verhält, so verhält er ebendamit sich
nur auf eine unmittelbare, d. i. sinnliche, sich selbst und seiner
Freyheit äusserliche Weise, und nur *zu* jener äusserlichen Weise und
unvernünftigen, unwahren Gestalt der Natur. Nur erst für denkenden
Geist ist die Wahrheit, die Idee als Idee; der denkende Geist aber ist
nicht der empfindende und anschauende. Es hilft nichts zu sagen,
jenes primitive Anschauen der Natur, sey nicht ein sinnliches,
äusserliches Anschauen, sondern ein Schauen durch die Äusserlich-
keit der Natur, eine Gegenwart ihres Centrums, ein intellectuelles
Anschauen, indem eben in dieser Ursprünglichkeit das Denken sich
noch nicht von dem Anschauen losgerissen, und zum reflectirenden
Erkennen sich bestimmt habe. Allein eben diese *nur unmittelbare*
Einheit des Denkens und Anschauens, ist es, worin nur das Anschauen gesetzt ist; es ist ein leeres Wort, davon zu sprechen, daß
es nicht bloß Anschauen, sondern vielmehr das Denken darin
enthalten sey. *An sich* ist freylich die Natur sowohl als der Geist
Denken; aber das Denken ist eben diß, nicht bloß ansich, nicht in der
Gegensatz-losen Einheit und Unmittelbarkeit zu seyn, und wenn es
nicht bloß als innere Natur, sondern existirendes Denken seyn soll,
so ist es nicht in seiner nur an sich seyenden Einheit mit dem Anschauen geblieben. Dieses Denken, um Wissen von dem Wahren zu
seyn, überhaupt daß das Wahre für dasselbe sey, muß freylich nicht
auf dem Standpunkt der nur trennenden Reflexion stehen bleiben,
sondern, als Idee, zur objectiven Einheit sich hindurchgearbeitet
haben. Das Denken ist nur Wissen und Erkennen, insofern es sich
befreyt hat, und zwar befreyt wesentlich von der Weise der blossen
Unmittelbarkeit der Seele; diese Unmittelbarkeit werde nur als
Anschauen, oder als Einheit des Anschauens und Denkens genommen.

self-externality of the Notion, the Notion lost in the confusion and transitoriness of appearances. It is precisely when spirit conducts itself *intuitively* however, that it does so in a way which is merely immediate i.e. sensuous, external to itself and its freedom, merely relative *to* the external mode and irrational, untrue shape of nature. Truth first has being, the Idea is as Idea, only for thinking spirit; but thinking spirit is not sensing and intuiting spirit. There is no point in asserting that this primitive intuiting of nature is not sensuous and external but a penetration of the externality of nature, an invocation of its centre, or intellectual intuition, for it is precisely in this origination that thought has not yet torn itself free from intuitiveness and determined itself as reflecting cognition. Since it is precisely and solely in this *simply immediate* unity of thought and intuition that only intuition is posited, it is meaningless to assert that the unity contains thought as well as intuition. *Implicitly*, both nature and spirit are certainly thought; but it is precisely its not being implicit, its not being in oppositionless unity and immediacy, which characterizes thought, and if it is supposed to be not merely as inner nature is, but existent thought, it has not remained in its merely implicit unity with intuition. In order to be knowledge of what is true, in order that what is true may generally have being for it, this thought must certainly not remain at the standpoint of simply divisive reflection, but must have worked itself through to objective unity as Idea. Thought is only knowledge and cognition in so far as it has freed itself, and indeed freed itself essentially from the mode of the mere immediacy of the soul; this immediacy being taken merely as intuition, or as the unity of intuition and thought.

NOTES

cli, 15

Volumes 8, 9, 10, 11–12 and 13–14 respectively of the edition of Hegel's 'Works' published at Berlin between 1832 and 1845. Boumann edited volume 7 pt 2, which was the last to appear.

cli, 19

Hegel lectured twice on his system as a whole in Heidelberg (1817/18), and twice in Berlin, during the Winter Terms of 1818/19 and 1826/27. See F. Nicolin 'Hegel als Professor in Heidelberg' ('Hegel-Studien' vol. 2 pp. 71–98, 1963); 'Berliner Schriften' (ed. J. Hoffmeister, Hamburg, 1956) pp. 743–9.

cliii, 10

These note-books seem to have been lost, see F. Nicolin and H. Schneider 'Hegels Vorlesungsnotizen zum subjektiven Geist' ('Hegel-Studien' Vol. 10 pp. 11–77, 1975) p. 15, although the surviving note-book material dates from approximately the same period. Cf. 'Unveröffentliche Diktate' (ed. H. Schneider, 'Hegel-Studien' vol. 7, 1972).

cliii, 19

Hegel's Berlin lectures on 'Subjective Spirit' were delivered during the Summer Terms of 1820, 1822 and 1825, and during the Winter Terms of 1827/28 and 1829/30. K. G. von Griesheim's (1798–1854) record of the 1825 series is the only one of Boumann's sources currently extant. The recorder of the 1828 lectures would appear to have been the classicist F. W. A. Mullach (b. 1807): see, W. Kroner 'Gelehrtes Berlin im Jahre 1845' (Berlin, 1846) pp. 251–2.

clvii, 2

The first two parts of the 'Encyclopaedia' appeared as volume 6 (Logic) and volume 7 pt. 1 (Philosophy of Nature) of the 'Works'.

clvii, 5

C. L. Michelet (1801–93) 'Anthropologie und Psychologie oder die Philosophie des subjectiven Geistes' (Berlin, 1840); J. K. F. Rosenkranz (1805–79) 'Psychologie, oder die Wissenschaft vom subjektiven Geist' (Königsberg, 1837; 2nd ed. 1843); Carl Daub (1765–1836) 'Vorlesungen über die philosophische Anthropologie' (ed. Marheineke and Dittenberger, Berlin, 1838). J. G. Mußmann's (1798–1833) 'Lehrbuch der Seelenwissenschaft oder rationalen und empirischen Psychologie' (Berlin, 1827), though approved of by Hegel himself ('Berliner Schriften' p. 646), was not well thought of by these later followers. Boumann might, however have added J. E. Erdmann's (1805–92) 'Grundriss der Psychologie' (Leipzig, 1840), to this list of 'highly distinguished works' by Hegelians dealing with the general subject-matter of 'Subjective Spirit'.

By the 1840's therefore, quite a school had developed, and it had given rise to a lively exchange of views. The theses criticized and defended were, however, almost entirely irrelevant to an understanding of Hegel's own work: see, F. Exner 'Die Psychologie der Hegelschen Schule' (Leipzig, 1842); K. Rosenkranz 'Widerlegung der von Herrn Dr. Exner gegebenen vermeintlichen Widerlegung der Hegel'schen Psychologie' (Königsberg, 1843); C. H. Weiße 'Die Hegel'sche Psychologie und die Exner'sche Kritik' ('Zeitschrift für Philosophie und spekulative Theologie' ed. I. H. Fichte vol. XIII pp. 258–97, Tübingen, 1844).

clvii, 6

Ludwig Boumann (1801–1871) was of Dutch extraction, his grandfather having been called to Berlin by Frederick the Great, and employed in an architectural capacity. His father was ennobled, but Boumann himself relinquished the title, deeming it to be incompatible with the sedentary and retired life of a scholar and man of letters. He was educated at the Greyfriars Grammar School in Berlin, and studied languages and philosophy at Berlin University. In May 1827 he submitted 'De physiologia Platonis' for his doctorate, and was backed by Hegel, but the classicists objected so violently that he subsequently decided to withdraw the thesis. He submitted 'Expositio Spinozismi' in the August of 1827, again backed by Hegel, and despite complaints about the obscurity of the work, was eventually awarded his doctorate in August 1828. See 'Berliner Schriften' pp. 659–64.

In appearance, he bore a remarkable resemblance to Goethe. He was lively and humorous in private conversation, but lived a solitary and retired life in Berlin, and never married. A certain donnish ineptitude evidently prevented him from carving out a career for himself. He worked thoroughly, but slowly, and wrote very little, —mainly articles on aesthetics and literature, but including the translation of a life of Marlborough.

He was evidently given the job of editing this part of the Encyclopaedia simply because no one else wanted to take it on. Since he does not seem to have been qualified for it in any particular way, it is very doubtful if it was simply 'lack of time' which forced him to 'drop the idea of providing a scientific consideration of the work.'

3,3

Cf. Lord Lytton (1803–73), 'The Caxtons' 1850 XVI X: "'Know thyself', saith the old philosophy. 'Improve thyself' saith the new." Diogenes Laertius, Life of Thales, I.40; J. Fontenrose 'Python. A Study of the Delphic Myth and its origins' (Univ. California Press, 1959).

Since Hegel goes on to criticize the simply piecemeal accumulation of facts and observations in order to achieve this end, it seems reasonable to suppose that he has in mind 'Γνωθι σαυτον, oder Magazin zur Erfahrungsseelenkunde' (6 vols. Berlin, 1783–88), edited by C. P. Moritz (1757–93) and C. F. Pockels (1757–1814). This periodical was planned, 'einige Materialien zu einem Gebäude zusammen zu tragen' (I p. 2), but never got beyond the preliminary procedures of such a worthy enterprise.

Many of the contemporary works on anthropology took Thales' command as their motto, see J. C. Goldbeck (1775–1831) 'Grundlinien der Organischen Natur' (Altona, 1808) p. iii; W. Liebsch (d. 1805) 'Grundriß der Anthropologie' (2 pts. Göttingen, 1806/08); F. Gruithuisen 'Anthropologie' (Munich, 1810); J. Salat (1766–1851) 'Lehrbuch der höheren Seelenkunde' (Munich, 1820); H. B. von Weber 'Handbuch der psychischen Anthropologie' (Tübingen, 1829) etc.

3, 26

Just as the Idea (§§ 213–44) presupposes and is the truth or sublation of all the preceding *categories* of the Logic, so Spirit (§§ 377–577) presupposes and is the truth or sublation of all the preceding levels of Logic (§§ 19–244) and Nature (§§ 245–376). To work out or grasp the truth of this, is to grasp the actualization of the Idea in Spirit.

5, 13

On the Oriental religions, see Phil. Rel. I.317–II.122; on the Greek religion II.224–88.

5, 22

Phil. Rel. II.327–III.151. The Incarnation being here interpreted as a prefiguration of the actualization of the Notion, the communion of believers as an anticipation of human freedom (note 3, 26; §§ 481–2).

5, 25

For Hegel, religion grasps by means of sensuous images such as Incarnation, Crucifixion, Resurrection, Ascension etc., what philosophy makes thoroughly explicit and universal as human enlightenment, absolute finitude, the negation of negation, the truth of spirit etc. See, for example, Phil. Rel. III.91.

5, 35

In the dialectical progression, spirit's *initial* attainment of the complete exclusion of nature is in the ego (§§ 413–17).

5, 36

Hegel may well be referring to 'aesthetic romances', in which case, he could have had Hölderlin's 'Hyperion' (1797/99) in mind. The translation is based on the assumption that he was referring to the theory behind the 'idyllic' literature of the time: note III.335, 21.

7, 15

Enc. §§ 377–577.

7, 22

The 'beginning' and 'end', 'whence' and 'whither' here, are simply figurative terms relating to the systematic progression in degree of complexity traced in the Philosophy of Spirit.

7, 26

Enc. §§ 350–412.

7, 38

Enc. §§ 160–244.

9, 3

It was necessary to emphasize this point as against some of the over-cautious and dogmatic theorizing then being put forward: see H. B. Weber 'Anthropologische Versuche' (Heidelberg, 1810) pp. 11–12, "It is therefore better to have nothing but fragments, separate biographies and characterizations, simply the materials provided by reliable observations and experience, than it is to systematize too quickly." Cf. C. F. Pockels (1757–1814), 'Beiträge zur Beförderung der Menschenkenntnis' (2 pts. Berlin, 1788/89).

9, 10

Cf. Christian Weise (1642–1708) 'Der kluge Hofmeister' (1675); Lord Chesterfield (1694–1773) 'Letters to His Son' (London, 1774; Germ. tr. 6 pts. Leipzig, 1774/77); A. F. F. Knigge (1752–96) 'Ueber den Umgang mit Menschen' (2 pts. Hanover, 1788, 11th ed. 1830).

9, 26

World Hist. 20.

9, 30

Enc. § 34. Hegel probably has in mind Kant's critique of the rational doctrine of the soul in the 'Critique of Pure Reason' A 341–405. Cf. 97, 30. C. A. Crusius (1715–75), 'Entwurf der notwendigen Vernunftwahrheiten' Leipzig, 1745; 3rd ed. 1766) p. 852 defined pneumatology as the, "science of the necessary essence of a mind and of the differences and properties which may be understood *a priori* on the basis of it." For Crusius, therefore, as for many of his school, rational psychology was very much a matter of logic, and had little use for empirical information. See Max Dessoir 'Geschichte der Neuern Deutschen Psychologie' vol. I p. 102 et seq. (Berlin, 1902). Cf. note III.107, 8.

11, 7

See note 3, 3 and I. D. Mauchart (1764–1826) 'Allgemeines Repertorium für empirische Psychologie' (6 pts. Nuremberg and Tübingen, 1792–1801). For a good survey of the genre, see Dessoir op. cit. pp. 283–300.

11, 39

Although Griesheim gives *none* of the Greek quoted here by Kehler, it looks as though Hegel must have read out this passage during the lecture. For the original, see 'De Anima' (ed. R. D. Hicks, Amsterdam, 1965) II. ch. 4 (415 a 23–6), ch. 5 (416 b 33–4), III ch. 4 (429 a 10–21). Kehler inserts the first sentence in the margin in note form, punctuates the passage, in general, as if it were German, and in one or two instances evidently corrects what he has written *while writing*. Since his version has been brought into line with Hicks', it may be of interest to give the original:

1) νοητικον   2) αισθητικον   3) θρεπτικον deren ἀντικειμενα sind:
νοητον, αἰσθητον τροφη 3) ἡ γὰρ θρεπτικὴ ψυχη, καὶ τοῖς ἄλλοις ὑπάρχει, καὶ πρωτη καὶ κοινοτάτη δύναμίς ἐστι ψυχης; καθ' ἥν ὑπάρχει τὸ ζῆν ἄπασι, ἧς ἐστιν ἔργα, γενῆσαι καὶ τροφη χρησασθαι. 2) ἡ αισθησις εν τω κινεισθαι τε καὶ πασχειν συμβαινει. 1) τὸ μόριον τῆς ψυχης, ᾧ γιγώσκει τε ἡ ψυχη τε καὶ φρονεῖ, ἀπαθὲς ἄρα δεῖ εἶναι; ἀναγκη ἄρα, ἐπεὶ πάντα νοεῖ ἀμιγῆ εἶναι, ὥσπερ φησὶ 'Αναξαγόρας, ἵνα κρατῇ, τοῦτο δέ ἐστι, ἵνα γνωρίζῃ· παρεμφαινόμενον γὰρ κωλύει καὶ ἀντιφράττει.

The superiority of Aristotle's method is, therefore, that by recognizing spirit's progression in degree of complexity (note 7, 22), it allows rationality and system to enter into easy commerce with empiricism. Its tripartite structure must also have pleased Hegel. For a contemporary interpretation of the 'vegetable soul', see F. Arnold (1803–90) 'Der Kopfteil des vegetativen Nervensystems beim Menschen' (Heidelberg, 1831); cf. E. H. Ackerknecht 'The history of the discovery of the vegetative (autonomic) nervous system' ('Medical History' vol. 18 pp. 1–8, 1974).

13, 6

Cf. Enc. §§ 47, 389, 445. Hegel quite evidently has in mind the way in which the soul was treated during the period between Descartes, who regarded it as a spiritual substance, created by God, and constituting nothing more than a *unio compositionis* together with the body, and Kant's critique of rational psychology.

13, 17

Eliciting its Notion involves recognition of its presupposing the body (§§ 350–76).

13, 31
Note III.107, 8.

14, 32
Insert 'werden' after 'gesetzt'.

15, 4
Cf. 97, 12.

15, 8
F. E. Beneke (1798–1854), 'Erfahrungsseelenlehre' (Berlin, 1820), 'Über die Vermögen der menschlichen Seele' (Göttingen, 1827), rejected the traditional concept of faculties, and postulated certain predispositions or impulses which are harmonized in the self. For a detailed account of Hegel's reaction to his ideas, see 'Berliner Schriften' pp. 612–26; cf. note III.107, 8. Nevertheless, Hegel was not entirely averse to associating the soul with harmony: Phil. Nat. II.69, 19; 75, 30; III.127, 30; cf. Subj. Sp. II. 265, 24; 321, 16.

J. C. Hoffbauer (1766–1827), in his 'Naturlehre der Seele' (Halle, 1796), gave classic expression to the then already dated theory of perfection. He maintained that two sorts of inclination arise out of intellectual activity, — objective pleasure, derived from the purposeful accomplishment of something, and subjective pleasure, derived from the understanding's dwelling upon such an accomplishment, further, that the pleasure increases together with the purposefulness and *perfection* of the accomplishment. K. H. Heydenreich (1764–1801) effectively criticized the theory in some articles published in K. A. Cäsar's 'Denkwürdigkeiten' (6 vols. Leipzig, 1784/88) vol. 5.

15, 17
On self-awareness, see §§ 407–8.

15, 21
See Hegel's criticism of Ernst Stiedenroth's (1794–1858) 'Psychologie' (2 pts. Berlin, 1824/25), in 'Berliner Schriften' pp. 569–70. Cf. note III.107, 8.

15, 35
§ 406.

17, 8
Christian Wolff (1679–1754), the methodizer of Leibniz. In his 'Psychologia rationalis' (Frankfurt, 1734), the pre-established harmony of body and soul is presupposed, the soul taken to be an immaterial substance able to represent the universe and express free will, and an attempt made to deduce the *laws* of psychology. In his 'Psychologia empirica' (Frankfurt, 1732) he lays emphasis upon the *physiological* basis of psychology, and explores the *cognitive* and *volitional* activity of the soul in the light of his 'rational' theory.

For Hegel's general assessment of his thought, see Hist. Phil. III.348–56; for an account of his very considerable influence on eighteenth century German psychology, see Dessoir op. cit. pp. 64–108.

Hegel may have in mind the 'Kurzer Begriff der Gelehrsamkeit' (1759) by J. G. Sulzer (1720–79), which deals with pneumatology and psychology in a Wolffian manner: see 'Dokumente' (1936) p. 110; A. Palme 'Sulzers Psychologie' (Diss. Berlin, 1905).

17, 10
§ 415.

17, 22
It should be noted that the *content* necessarily develops *itself*. This is evidently a reference to the 'Encyclopaedia', not to the 'Phenomenology' of 1807: note III.3, 1.

17, 30
Since eliciting the Notion of any particular sphere necessarily involves reference to every other, each sphere is an aspect of the Idea. In that the Notion of Spirit involves recognising its distinctness from and relationship to Logic and Nature for example, it has to be regarded as an integral part of the universal Notion. The *contents* of Logic, Nature and Spirit are distinct, but they have one Notional or logical inter-relationship. Cf. § 243.

19, 2
Phil. Nat. III.68, 16.

19, 27
Phil. Nat. III.98, 36.

19, 37
The context here is *Spirit* (§§ 574–7), not Logic (§§ 160–244).

21, 8
The Notion which *sublates*, that is to say, *includes* all other Notions within itself. *Truth* has this rather technical meaning in Hegel.

21, 18
Note III. 139, 35.

21, 32
Enc. §§ 153–4.

23, 5
Note II.243, 31.

23, 14
§§ 413–23.

24, 11
  Read 'nöthig, um sie'.

25, 5
  §400.

25, 18
  See § 398 and § 413; § 408 and § 422.

These opening §§ (377–80) first appeared in the second edition of the 'Encyclopaedia' (1827), and are all Hegel published of the much more elaborate 'Introduction' he had planned (see pp. 90–103). It is perhaps worth noting, that although their position within the 'Encyclopaedia' might lead us to regard them as an introduction to the Philosophy of Spirit as a *whole*, they are in fact concerned predominantly with *Subjective* Spirit.

25, 19
  § 299 (1817).

25, 21
  Note 21, 8.

27, 6
  Enc. §§ 226–232.

27, 27
  Griesheim's version must be the more accurate. Spirit is *the* product of Logic and Nature in that these less complex spheres are sublated within it.

29, 13
  Phil. Nat. § 247.

29, 27
  Hist. Phil. I.315. Hegel's main source there is Aristotle's 'De Anima' I.ii. He quotes what Aristotle has preserved of Empedocles' saying:
    "By Earth we see Earth, by Water Water,
      By Air the divine Air, by Fire destroying Fire,
      Love by Love, and Strife by bitter Strife,"
and then comments as follows: "Through our participation in them they become for us. There we have the idea that spirit, the soul, is itself the unity, the very totality of elements, in which the principle of earth relates to earth, water to water, love to love, etc. In seeing fire, the fire is in us for whom objective fire is, and so on."

29, 36
  Phil. Nat. §§ 263–4. There is no mention there of atoms. Cf. § 98. Boumann may have used early notes here (cf. 105, 30), since there is plenty of evidence that the mature Hegel had no very high opinion of atomistic metaphysics in any context: Phil. Nat. II.213; Hist. Phil. I.300–10; II.288–90.

31, 11
Phil. Nat. §§ 274–85 and §§ 269–71.

31, 22
Phil. Nat. §§ 343–6.

31, 32
Phil. Nat. §§ 350–2.

33, 4
Phil. Nat. §§ 359–362.

33, 26
Phil. Nat. §§ 363–6.

33, 36
Phil. Nat. §§ 368–9.

35, 18
Phil. Nat. §§ 375–6.

35, 27
This outline of the Philosophy of Nature establishes the main reason for distinguishing between nature and spirit. The essence of it is the progression, in degree of inner co-ordination. There is no hint of Cartesian dualism here, simply a comprehensive and informed attempt to avoid any confusion of levels, any 'category-mistakes'. See G. Ryle 'The Concept of Mind' (1949) ch. I.

Cf. F. A. Carus (1770–1807) 'Psychologie' (2 vols. Leipzig, 1808) vol. I p. 78: "The general doctrine of the soul is the natural doctrine of man. We find that by its very nature it has within itself its own teleology; not a chimerical teleology, justifiably rejected by philosophy, but a *real* teleology, for here —within the highest cycle of the earthly spheres of formation, crowned as this always is by man, — nature itself is *teleological*."

37, 17
§§ 413–39.

37, 23
It is more complex than nature, and has the potentiality of assimilating it i.e. of working out the philosophy of Nature.

37, 26
Phil. Rel. I. 80–1.

37, 34
Phil. Nat. §§ 254–9.

39, 7
It is, therefore, only with the 'Phenomenology' (§§ 413–39) that the Philosophy of Spirit proper may be said to begin. Cf. note III.79 ,1. The 'Anthropology' is still very much part of nature.

39, 11
Phil. Nat. II. 13, 34; Phil. Hist. 173–81; Phil. Rel. II.70–82.

39, 28
§§ 438–9.

39, 36
Enc. §§ 564–71.

41, 19
This is evidently a summary of the Philosophy of *Subjective* Spirit: Anthropology, Phenomenology, Psychology.

41, 25
Enc. §§ 553–77: Art, Religion and Philosophy.

43, 19
See the extensive exposition of this in Phil. Rel. II.327–58: III.1–151.

45, 1
Enc. §§ 548–52.

45, 13
Phil. Hist. 271–4; 306–13.

47, 8
The Logic (§§ 19–244) presupposes Spirit (§§ 377–577) within the whole *cycle* of the Encyclopaedia. Since the Philosophy of Spirit is simply a more precise and elaborate exposition of the 'Phenomenology of Spirit' of 1807, there is no reason why this early work should not be regarded as a useful introduction to Hegel's system for those who have not yet attempted to master the subject-matter of the Encyclopaedia. Cf. note III. 3, 1.

47, 15
There could hardly be a clearer confirmation than this of Otto Pöggeler's thesis: "In opposition to the usual interpretation of the Hegelian text, I should like to propose the following: that the actual science of Spirit is not the Logic, but the Philosophy of Spirit." See 'Zur Deutung der Phänomenologie des Geistes' ('Hegel-Studien' vol. I pp. 282–3, 1961).

47, 31

A *systematic* progression in degree of complexity or inwardness, not simply a process taking place in time, although there may, of course, be a certain correspondence between such a progression and such a process.

49, 6
Phil. Nat. III.147, 8.

49, 9
§§ 465–68.

49, 17
§ 300 (1817).

49, 23
Cf. the turgid peroration which concludes 'Glauben und Wissen' (1802): "But as the abyss of nothingness within which all being sinks away, the pure Notion or infinitude must define *infinite pain*... purely as a moment... of the supreme Idea." Directly or indirectly, the concept almost certainly derives from Plato's 'Philebus', in which pain, like all else capable of degrees of intensity, is classified as being unlimited and therefore inferior to knowledge. See 32a: "Whenever in the class of living beings, which, as I said before, arises out of the natural union of the infinite and the finite, that union is destroyed, the destruction is pain, and the passage and return of all things to their own nature is pleasure."

49, 29
Realized, within Subjective Spirit, in different ways at different levels: see § 362 (III.353, 39), §§ 481–2.

49, 33
John VIII. 32: "And ye shall know the truth, and the truth shall make you free." Cf. note III.357, 11.

51, 19
See the treatment of the Syrian religion of pain in Phil. Rel. II.82–85.

51, 26
Phil. Right §§ 139–40.

51, 34
Enc. §§ 48, 60.

53, 15
Note 49, 29.

53, 19
§ 301 (1817).

55, 34
Enc. §§ 131–54.

57, 11
Hist. Phil. III.20: "The Arians, since they did not recognize God in Christ, did away with the idea of the Trinity, and consequently with the principle of all speculative philosophy."

The Eastern Church was united against the heresy at the first Council of Constantinople (381), but Ulphilas (d. 383) had so propagated it among the Goths, that after the Teutonic invasions of the Western Empire, theologians continued to concern themselves with it for several centuries.

57, 30
§ 302 (1817).

59, 27
Enc. §§ 19–244.

59, 32
Enc. §§ 245–376.

59, 38
§§ 388–412.

61, 32
§§ 413–39.

63, 9
§§ 440–82; see especially the last two §§.

63, 17
§§ 548–52.

63, 19
Phil. Rel. I.317–49; II.1–224. The Judaic religion is succeeded in the dialectical progression by the 'religion of beauty' i.e. that of the Greeks.

63, 20
See the critique of Spinoza's conception of God in Hist. Phil. III.264–69.

63, 35
Phil. Rel. II.239–43: "This unity, as being absolute necessity, has universal determinateness within it. It is the fulness of all determinations; but it is not developed in itself, the fact rather being that the content is divided in a

particular way among the many gods who issue forth from this unity. It is itself empty and without content, despises all fellowship and outward embodiment, and rules in dread fashion over everything as blind, irrational, unintelligible power."

64, 39
Original: 'seiner Idialität'.

65, 2
Enc. §§ 564–71; Phil. Rel. II.327–58; III.1–151.

65, 5
Enc. §§ 572–7.

65, 19
Despite the appearance of the first edition of the Encyclopaedia in June 1817, Hegel found it necessary to *dictate* supplementary passages in order to bring out the *general drift* of his lectures. This passage supplemented § 304, and was subsequently (1827) reformulated as the § 385 published above.

67, 4
Enc. §§ 160–244. The Subjective Notion (§§ 163–93), as the first major sphere of this section of the Logic, corresponds to Being (§§ 84–111), the first major sphere of the Logic as a whole. Taking Spirit as a whole (§§ 377–577), Subjective Spirit (§§ 387–482) is the first major sphere, and therefore corresponds to Being and the Subjective Notion. Such *analogical* references to the *structure* of the Logic are common throughout the whole Philosophy of Subjective Spirit.

67, 16
II.107, 32.

67, 27
Throughout the Anthropology (§§ 388–412), the soul is mediated by natural factors such as geography, climate, biology, sensation, feeling, habit etc.

67, 30
§§ 413–39.

69, 1
What we, in pursuing enquiry into the Philosophy of Spirit, grasp of its subject-matter and structure: hence the command of the Delphic Apollo (§ 377).

69, 10
See, especially, §§ 391–402.

69, 27
Phil. Right §§ 34–71.

69, 32
Phil. Right §§ 257–360.

71, 2
Enc. §§ 553–77, and the corresponding lectures.

71, 3
§ 305 (1817).

73, 1
Enc. §§ 92–5, where reference is made to Plato's 'Philebus' (note 49, 23).

73, 11
This succinct and illuminating definition of dialectic was first inserted in 1827. In the major sphere now under consideration, the soul, mediated as it is by natural factors, 'passes away' by means of the pure, inner self-certainty of the ego, and is sublated, together with the ego, in psychology. Presented in such an abstractly schematic way, the dialectic is quite obviously not a good topic for constructive debate and enquiry. It comes into its own as a living philosophical issue once the judgements involved in the arrangement of its subject-matter are brought into a consideration of its *general* significance.

73, 21
Note III.241, 25.

75, 15
Phil. Nat. §§ 269–71.

75, 30
Cf. 109–111. There has been much confusion on this point. In making this simple and perfectly valid observation, Hegel is not denying the *possibility* of an unknown, but merely of our being aware of it. And in any case, to know a limit could be to know no more than what is not beyond it. He is not saying, as Feuerbach and others seem to have assumed, that we can actually do away with limits simply by being aware of them. Stanley Rosen, 'G.W.F. Hegel. An Introduction to the Science of Wisdom' (Yale University Press, 1974), one of the more adventurous of the present-day interpreters of Hegel's works, has even maintained, "that to know anything is to know everything" (p. 62), and that Hegel's wisdom is, "the final justification of God" (p. 238) etc. Cf. note III.277, 11.

77, 25

It is for the *theoretical structure* of the Philosophy of Spirit that Hegel is claiming infinitude, not the *content*. This structure contains the subject-matter, which is always to some extent idiosyncratic and historically contingent i.e. finite. Although it depends upon this finitude for its *full* significance, it is also the structure of the Logic and the Philosophy of Nature, and cannot therefore be regarded as *completely* dependent upon the subject-matter of any one of these three spheres. Cf. 83, 6–15.

79, 1

§ 307 (1817). Cf. Hegel's lecture-notes ('Hegel-Studien' vol. 10 pp. 18–19, 1975).

81, 21

See, however, § 396 (II.111–15).

87, 11

Note III.3, 1.

89, 2

Note III.229, 27.

89, 4

Phil. Nat. §§ 262–8; Phil. Right §§ 34–40.

89, 11

§§ 445–68; 469–80.

89, 32

§§ 481–2.

91, 1

As early as August 1811 ('Briefe' I.389) we find Hegel referring to a forthcoming work on Psychology. In § 4 of the Phil. Right (1820), he refers to the treatment of this science in the Encyclopaedia, and adds: "I hope by and by to be able to elaborate upon it still further. There is all the more need for me by so doing to make my contribution to what I hope is the deeper knowledge of the nature of spirit, in that... scarcely any philosophical science is so neglected and so ill off as the *doctrine of spirit* usually called psychology". The fragment published here, which seems to date in the main from the summer of 1822, is evidently the outcome of this long-standing plan to publish a separate work on 'Subjective Spirit', comparable in detail and scope to the Philosophy of Right. Had it been realized, the overall context of Hegel's political philosophy would have been much more widely understood, and the general nineteenth and twentieth century conception of his thought would certainly have been rather different.

The manuscript is now in the possession of the Staatsbibliothek Preussischer Kulturbesitz, Archivstrasse 12–14, 1 Berlin 33 (Hegel-Nachlaß vol. 2, acc. ms. 1889, 252), and consists of sixty folio sheets. The fragment is not, however, a continuous text, but consists of seven separate sections. F. Nicolin has published a critical edition of it: 'Ein Hegelsches Fragment zur Philosophie des Geistes' ('Hegel-Studien' vol. 1 pp. 9–48, 1961).

91, 13

Cf. § 377.

93, 36

§§ 376–7 deal with the transition from Nature to Spirit, §§ 572–7 with the philosophical restatement of religion.

93, 46

Cf. Phil. Nat. I.206, 23: "God is subjectivity, activity, infinite actuosity, within which the other is only momentary." The meaning of this word, which is extremely rare in Hegel, depends upon Aristotle's distinction between potentiality and actuality. Potentiality, as aptitude to change, is the mark of imperfection; *God* alone, in that He is all He can be, is free from potentiality, and is therefore pure actuality, infinite actuosity.

In using 'actuosity' with reference to *spirit*, Hegel may have in mind 'De Anima' III.vii 431a: "Knowledge, when actively operative, is identical with its object. In the individual, potential knowledge has priority in time, but generally it is not prior even in time; for everything comes out of that which actually is. And clearly the sensible object makes the sense-faculty actually operative from being only potential."

97, 5

Note 9, 26.

97, 12

Cf. 15, 4. In a note in the margin at this point, Hegel makes mention of a work by Johann Heinrich Campe (1746–1817) which we know he read and took notes upon on October 10th 1786 ('Dokumente' 1936, pp. 101–4): 'Kleine Seelenlehre für Kinder' (1779; 3rd. ed. Brunswick, 1801). It is a charming and well-written little book, but as its author admits in the third edition, it was based to too great an extent upon the Wolffian system (note 17, 8) to be at all satisfactory after the publication of Kant's critical work (pp. 13–14). Nevertheless, the distinction between a mirror which simply 'presents' things, and a soul which is *conscious* of what it presents (pp. 10–11), may well have started the line of thinking which eventually gave rise to the subsumption of consciousness under psychology in Hegel's mature system.

97, 21

Note 15, 8.

97, 30

Cf. § 378.

99, 2
Cf. Note II.11, 13 and Campe op cit. p. 10:
"*Johannes*. Die Seele ist ein Ding, das sich etwas vorstellen kann.
Vater. Ein solches Ding pflegen wir eine Vorstellungskraft zu nennen."

99, 13
Cf. § 379.

99, 32
Hegel presents it (§ 406) as the immediate *presupposition* of derangement (§ 408).

101, 12
Notes II. 279, 20; 289, 22; 299, 34; 315, 24.

101, 39
A. C. A. Eschenmayer (1768–1854): for further details, see Phil. Nat. I.285. The 'Psychology' (2nd ed. 1822) touches upon many of the topics dealt with in Hegel's lectures on 'Subjective Spirit', — the ages of life, the senses, dreaming and waking, animal magnetism etc., as well as more strictly psychological topics such as memory, recollection, attention, language, reason, phantasy, will etc., — but it is evidently written under the spell of Schelling (p. 110), and as Hegel notes, is completely devoid of any truly scientific lay-out or procedure.

The second part opens with an interesting line of argument (p. 258): "Cogito, ergo sum, says Descartes, and we may well add the two further propositions: Sentio, ergo sum, Valo, ergo sum. Here, the utterance: *I am* lies at the base of three others — and the three functions appear merely as modifications of the spiritual being. This raises the question of what spiritual being is."

The § to which Hegel takes particular exception reads as follows (p. 281): "The second part deals with 1) the relations of the faculties to one another, the value and significance of their combination, and their disturbances; 2) the searching out of the sources of all the functions distinguished within empirical psychology, and the establishing of the principles by means of which the connectedness of the whole is to be grasped. Since the soul is a primary force, which in combination with a corporeal principle distributes itself in various primary and secondary directions, it is the task of pure psychology to recombine into the form of the whole all these various radiations of the psyche, and to demonstrate how the phenomenon of man has its connectedness in his noumenon. Cognition of this kind is no longer factual, but speculative, for it is based 1) on reflecting by means of notions, judgements and syllogisms, and 2) on ideal intuition, and these are no longer accessible to experience and observation from the empirical standpoint."

103, 6

Henrik Steffens (1773–1845): for further details, see Phil. Nat. II.253. He takes Terence's noble declaration, 'Homo sum, humani nil a me alienum puto' as the motto of 'Anthropology' (II.396), and then anticipates nineteenth and twentieth century developments in this field by treating it as a medley, a gallimaufry, an omnium gatherum of the most diverse and ill-assorted disciplines: "*Anthropology*, if one considers the meaning of the word, is so boundless in scope, that it might well be used in order to refer to human cognition of the highest kind. Anthropology would then be philosophy in the widest sense of the word." (I.1).

Nevertheless, the general sequence of the work is broadly speculative, and is not completely at odds with the lay-out of Hegel's Encyclopaedia. The first volume (pp. 476) corresponds to Hegel's treatment of the Terrestrial Organism (§§ 339–42), though there are diversions into discussing the nature of innocence (p. 286), Hume (pp. 365–73) and wilfulness (pp. 399–400). The first part of the second volume (pp. 1–352) corresponds broadly to Enc. §§ 350–76, though there is much confusion of widely separated levels of enquiry. It is presented as a 'Physiological Anthropology', but concludes with a discussion of speech, music, the plastic arts etc. (pp. 306–65). The second part of volume two (pp. 353–456) is headed 'Psychological Anthropology', and includes surveys of human geography, racial variety, the temperaments, the ages of life etc. It therefore corresponds very roughly to the first major sphere (§§ 391–402) of Hegel's Anthropology.

The work was reviewed by J. F. Herbart (1776–1841), whose general assessment of it was not very different from Hegel's. He saw that it was well-intentioned, but regarded it as contributing to the general confusion in philosophy ('Werke' XII pp. 436–61).

103, 13

Hist. Phil. III. 252–90: "If, in conclusion, we sum up this criticism that we have offered, we would say that on the one hand with Spinoza negation or privation is distinct from substance; for he merely assumes individual determinations, and does not deduce them from substance. On the other hand the negation is present only as Nothing, for in the absolute there is no mode; the negative is not there, but only its dissolution, its return: we do not find its movement, its Becoming and Being." (p. 289). Cf. Subj. Sp. II. 7, 11.

103, 15

Note II. 7, 18.

103, 21

Note 11, 39. Cf. Hist. Phil. II.117–231: "In Aristotle the Idea is at least implicitly concrete, as the consciousness of the unity of subjective and

objective, and therefore it is not one-sided. Should the Idea be truly concrete, the particular must be developed from it. The other relation would
be the mere bringing of the particular under the universal, so that both
should be mutually distinguished; in such a case the universal is only a
formal principle, and such a philosophy is therefore one-sided." (p. 230).

103, 22
    Cf. § 378. § 380 seems to have no forerunner in this fragment.

103, 35
    Note 77, 25.

105, 1
    Cf. § 381.

105, 10
    Cf. 29, 20 et seq.

105, 17
    Cf. 29, 27 et seq.

105, 30
    Note 29, 36.

107, 5
    29, 31.

107, 8
    Cf. § 386, and Enc. §§ 92–5. This § begins the *second* main section of the
fragment.

107, 17
    Cf. 73, 22 et seq.

107, 30
    Cf. 31, 7.

109, 1
    Cf. 75, 15.

109, 6
    Cf. 75, 17 et seq.

111, 18
    Cf. § 387.

111, 28
Cf. § 393. This § begins the *third* main section of the fragment.

113, 16
Cf. II.45, 37.

113, 22
Cf. II.45, 25 et seq.

115, 27
Cf. § 398. This § begins the *fourth* main section of the fragment.

117, 1
Cf. § 395.

117, 11
Cf. § 396.

117, 22
Cf. II.99, 25 et seq.

117, 30
Cf. II.101, 16. The English equivalent is, 'Reason grows with the years': Claude Berry 'Cornwall' (London, 1949) p. 169, but it does not seem to be well-known. Cf the French proverb, 'Avec l'âge on devient sage'.

119, 3
Richard Evelyn (1653–1658), see note II.101, 22. Hegel's source was evidently 'The Quarterly Review' of April, 1818 (vol. xix p. 30).

119, 12
Cf. II.101, 23 et seq.

119, 25
Cf. 105, 10 et seq.

119, 32
Cf. II.109, 7 et seq.

121, 3
Cf. II.115, 38 et seq.

121, 27
Cf. II.125, 14.

121, 33
Cf. § 399. This § begins the *fifth* main section of the fragment. The main difference between the lay-out here and that of the later Encyclopaedia lies

in the dialectical placing of *waking and sleeping*. In this fragment, as in the 1817 Encyclopaedia (§§ 315–16), they constitute the transition from natural changes to the stages of life, whereas in the later formulation they succeed the stages of life, and constitute the transition to sensation (§ 398).

123, 11
 Cf. II.149, 8.

123, 20
 Cf. II.151, 11.

123, 31
 The following §§ on sensation (pp. 123–135) correspond to what was published as § 318 in 1817. They therefore constitute a quite extraordinary extension of the treatment of the subject, which is even more remarkable in that by the summer of 1825 it had undergone yet another full-scale reformulation (Kehler ms. pp. 97–109; Griesheim ms. pp. 132–52). The whole section was re-written in 1827 (§§ 399–402), next to nothing of the 1817 text being incorporated. Further minor changes were made in 1830.

 The main difference between these §§ and the later lectures, is that whereas here Hegel is simply intent upon formulating his attitude in *general terms*, from 1825 on there is a marked increase in his concern with what is *specific* and *concrete*.

127, 33
 Cf. II.163, 27. This is one of the few passages in which there is *any* similarity between this treatment of sensation and that in the later editions of the Encyclopaedia.

131, 4
 1817 edition. The § correspond to §§ 350 and 353 of the 1830 edition (Phil. Nat. III.102 and 109).

131, 16
 This is a reference to Aristotle, 'De Anima' II xii. cf. Hist. Phil. II.189: "In speaking of sense-perception, Aristotle... makes use of his celebrated simile, which has so often occasioned misapprehension, because it has been understood quite incorrectly... For the form is the object as universal; and theoretically we are in the position, not of the individual and sensuous, but of the universal."

 Cf. Plato's 'Theaetetus' 191c. The reference to Homer (194c) is spurious, Plato is simply using an archaism in order to make a pun concerning wax and mind.

131, 21
 This begins the *sixth* main section of the fragment.

137, 1
 Cf. § 405.

137, 5
 Cf. II.231, 9 et seq.

139, 5
 See §§ 446–50.

139, 27
 See §§ 465–8.

INDEX TO THE TEXT

INDEX TO THE INTRODUCTION AND NOTES

The page numbers in italics refer to the Introduction